Advanced and Intelligent Manufacturing in China

Series Editor

Jie Chen, Tongji University, Shanghai, Shanghai, China

This is a set of high-level and original academic monographs. This series focuses on the two fields of intelligent manufacturing and equipment, control and information technology, covering a range of core technologies such as Internet of Things, 3D printing, robotics, intelligent equipment, and epitomizing the achievements of technological development in China's manufacturing sector. With Prof. Jie Chen, a member of the Chinese Academy of Engineering and a control engineering expert in China, as the Editorial in Chief, this series is organized and written by more than 30 young experts and scholars from more than 10 universities and institutes. It typically embodies the technological development achievements of China's manufacturing industry. It will promote the research and development and innovation of advanced intelligent manufacturing technologies, and promote the technological transformation and upgrading of the equipment manufacturing industry.

Maoai Chen · Wenjian Ren · Yuanning Jiang

Technologies of Robotic Welding

Chemical Industry Press Co., Ltd.

Springer

Maoai Chen
Institute for Materials Joining
Shandong University
Jinan, Shandong, China

Wenjian Ren
Shandong Aotai Electric Co., Ltd.
Jinan, Shandong, China

Yuanning Jiang
Shandong University
Jinan, Shandong, China

ISSN 2731-5983 ISSN 2731-5991 (electronic)
Advanced and Intelligent Manufacturing in China
ISBN 978-981-97-3148-0 ISBN 978-981-97-3149-7 (eBook)
https://doi.org/10.1007/978-981-97-3149-7

Jointly published with Chemical Industry Press
The print edition is not for sale in China (Mainland). Customers from China (Mainland) please order the print book from: Chemical Industry Press.

B&R Book Program

This Springer imprint is published by the registered company Springer Nature Singapore Pte Ltd.
The registered company address is: 152 Beach Road, #21-01/04 Gateway East, Singapore 189721, Singapore

Preface

In recent years, with the development of industrial technology, especially sensing technology, welding robot technology is becoming more and more sophisticated and cost effective, and thus its application in various industrial fields is being increased dramatically. At present, welding robots have been widely used in automobile manufacturing, construction machinery, electronic communication, aerospace, national defense, energy equipment, rail transit, marine heavy industry, and other fields, with a rapid development momentum. Welding robot technology has become one of the most popular technologies in the welding field. It integrates materials, control, machinery, computer, and other interdisciplinary knowledge. Welding robots are gradually evolving from teaching-and-playback type to intelligent one. At present, the increasing shortage of labor and the increasing requirements of workers for working environment conditions have rapidly increased the necessity of welding robots to replace humans. And with the release of the "Made in China 2025" plan, the state's support for the localization of industrial robots has gradually increased. Domestic robot manufacturing technology will become increasingly mature, and the cost of welding robots will further decrease. In the future, welding robots will fully replace welding workers. In this situation, it is urgent for welding technicians and operators to learn the relevant knowledge and technology of welding robots.

The purpose of this book is to systematically introduce welding robot technology. On the basis of briefly explaining the basic theoretical knowledge of robots, it introduces in detail the body structure of industrial robots, robot sensing technology, welding robot system configuration and requirements, welding robot application operation technology and maintenance technology, as well as common robot welding processes. A large number of typical application cases of welding robot system in manufacturing industry are given in this book. We strive to avoid esoteric theoretical derivation and explanation, and introduce the basic theoretical knowledge needed to understand the welding robot in a simple way, focusing on practicality, novelty and advancement. This book can be used as a reference for engineers and operators engaged in welding work, as well as for undergraduates majoring in welding engineering in colleges, universities and higher vocational colleges.

The participants in the preparation of this book are Chen Maoai, Ren Wenjian, Jiang Yuanning, Jia Chuanbao, Chen Shaoqing, Li Chaofan, Chen Dongsheng, Zhang Dong, and Gao Haiguang.

Jinan, China
May 2023

Maoai Chen

Contents

Chapter 1
Robot Overview

The development and application level of robots is an important indicator for measuring the manufacturing level of a country. In China, robots are generally considered to be the crown jewel of manufacturing. At present, robots have been widely applied in all aspects of society, economy and life and are showing a trend of rapid development. Welding robots are one of the most widely used robots today, and their application will become increasingly widespread as labor costs continue to rise and the demand of workers on working environment continue to increase. This chapter mainly introduces the basic concept and development history of robots and looks into the development direction of welding robots.

1.1 What is a Robot?

The robot is a kind of machine which is capable to perform specific tasks in place of people. Although robots are widely used in all aspects of social life and are familiar to most people, there is yet currently no universally recognized definition of robot. This is not only because different countries and organizations define robots from different perspectives, but also because robots themselves are constantly evolving and developing.

1.1.1 Definition and Classification of Robots

(1) Basic concepts of robots

Robot is an automation equipment integrating machinery, automatic control, computer, artificial intelligence and bionics and other multidisciplinary technologies. The term robot was first put forward by the Czech playwright Carel Keppek in

M. Chen et al., *Technologies of Robotic Welding*, Advanced and Intelligent Manufacturing in China, https://doi.org/10.1007/978-981-97-3149-7_1

his science fiction drama "Rossnm's Universal Robots". In Czech, Robot means "servant of human beings", while in this drama, Robot is a humanoid machine made by Rosam to work for human beings. The following definitions of robots are commonly used today:

(a) Chinese national standard Classification of Robots (GB/T 39045-2020) defines a robot as an actuator with two or two programmable axes and a certain degree of autonomy that can move within its environment to perform tasks.
(b) The Robotics Society of America defines a robot as a reprogrammable and multifunctional manipulator used for handling materials, parts or tools; Or a specialized device with changeable or programmable actions to perform multiple tasks.
(c) National Standards Bureau of America defines a robot as a mechanical device that is programmable and capable of automatically performing desired operations or actions under programmed control.
(d) JIRobot defines a robot as a general purpose machine equipped with memory device and end effector, which can move automatically and replace human labor through the movement.
(e) International Organization for Standardization (ISO) defines robot as an automatic, reprogrammable (programmable control of three or more axes) multipurpose manipulator, which may be fixed or mobile in industrial applications.
(f) Wechsler's Dictionary defines a robot as a computer-controlled machine that resembles a human or animal in being capable of moving independently or performing complex actions, or a computer-controlled device capable of performing various complicated, often repetitive tasks automatically.

Robots have been defined from different perspectives by different organizations, but all of them emphasize the programmable and automatic control features of robots and their ability to perform task in place of humans or animals.

Robots similar in appearance to humans are called human-like robots. Robots does not necessarily resemble human or animal appearance, but take on a variety of forms, i.e., industrial robots often look like human arms. Figure 1.1 shows typical legends for humanoid robots and humanoid-arm robots. Whatever the shape, the basic characteristics of robots are that they can imitate the action of humans (or animals), perform repetitive work in place of human, have the ability to sense and identify, and even to be intelligent. Robots can be used in industrial and agricultural production, education, service, medical and military industries.

(2) Classification of robots

There are many methods for the classification of robots. The most basic and common method is to classify robots according to their fields of application. Robots can be classified into industrial robot, service robot and special robot according to their application fields.

Industrial robot refers to a mechanic with multiple joints or a mechanical device with multiple degrees of freedom applied in industry. It is mainly used for handling,

(a) Humanoid robot

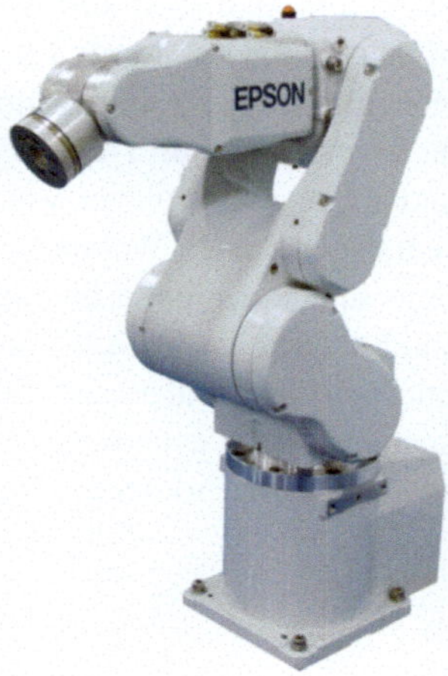

(b) Humanoid arm-like robot

Fig. 1.1 Illustrations of typical robots

welding, cutting, spraying, loading and processing. The welding robot accounts for the largest proportion of industrial robot.

Service robot refers to robot used for personal or public service, such as domestic robot, entertainment robot, and explanation & guiding robot.

Special robots refer to robots used for certain special operations, such as underwater robot, blasting robot, search and rescue robot, military robot, digging robot, and surgical robot, etc.

1.1.2 A Brief History of Robots

(1) Origin of robots

Robots in the broad sense originated in China. As early as Western Zhou Period, Chinese skillful craftsmen have invented a dancing robot. Luban in Spring and Autumn period made flyable wooden birds. In the Han Dynasty, there is a legend that Chen Ping, a great general, used robot dancers to defeat the Xiongnu army. The Prime Minister of Shu, Zhuge Liang, designed and built the wooden oxen and horses for military logistics. During the Three Kingdoms period, Ma Jun of the State of Wei invented a mileage-recording drum cart, which had a small wooden man on board to measure the mileage and beat the drums and bells, beating the drums once for every 10 miles, and beating the bells once for every 100 miles. In the Tang Dynasty, Luozhou County Magistrate Yin Wenliang created a service robot that could sing, play the Sheng and respond with wit; Hangzhou craftsman Yang Wulan created a monk-like robot who held a bronze bowl and smiled at pedestrians, bowed to the donors, and automatically collected money after the bowl was full; Liuzhou craftsman Wang Zhi made an otter-like fishing robot, which can jump into the water to catch fish, and expose its head after catching a fish. Figure 1.2 shows several ancient Chinese robots.

An automatic machine player was invented in Nakami, Japan, in 1662. More than 200 years ago, French Wokonsson made a humanoid robot capable of playing 12 flute songs in succession, a humanoid orchestral performer capable of playing 20 pieces and a duck that can digest. This duck is the first machine capable of simulating the physiological process of an animal. It is also capable of flapping wings, drinking, spattering, and even making expressions. These ancient robots were only mechanically controlled to accomplish a specific task, without electrical control,

(a) Wooden bird (b) Wooden horse

(c) Mileage-recording drum cart (d) Alms-begging robot

Fig. 1.2 Chinese ancient robots

let alone program control. In 1774, Jacotte Droz invented a writing robot-a "Writer". The Writer was the first programmable robot.

(2) Development of modern robots

American George Divo applied for a robot patent in 1954, which specified the three essential feature of the robot: servo control of the joints, need for teaching and saving and play-back of the teaching actions. Up to now, most of the industiral robots still operates with these thress features. In 1958, Georges Dewo and Ingelberg jointly established Unimation, a company that in the following year made the first modern robot, as shown in Fig. 1.3. The robot is a five-axis hydraulic drive robot whose teaching and operation are controlled by a computer, and the path is stored by a magnetic drum. Unimate, the first robot for industrial production, was born in 1961. This robot was installed on the production line of General Electric Company for the production of die castings. In 1969, the first welding robot, a spot welding robot, was put into service in the vehicle body production line of General Motors USA. Since then, vehicle body welding has been fully automated. In 1973, the German company Kuka manufactured the first electric-driven 6-axis robot. In 1979, Unimation launched the six-axis electric robot PUMA robot with visual, tactile and force sensors, see Fig. 1.4. Since then, with the development of computer, modern control, sensing and artificial intelligence technologies, robot technology has also experienced rapid development. So far, robots have developed into the third generation.

The first generation of robots is characterized by teach and playback. This type of robot needs to be taught by an operator via a teach pendant. Operators guide the robot to move by using the control button on the teach pendant and set parameters at some key nodes. The robot will automatically record the path and parameters of key nodes in the form of program, and repeat the path and parameters during operation. This generation of robots has little ability to perceive and react to the surrounding environment. However, due to high reliability, precision and cost-effectiveness, they are still the mainstream robots in industrial applications up to now.

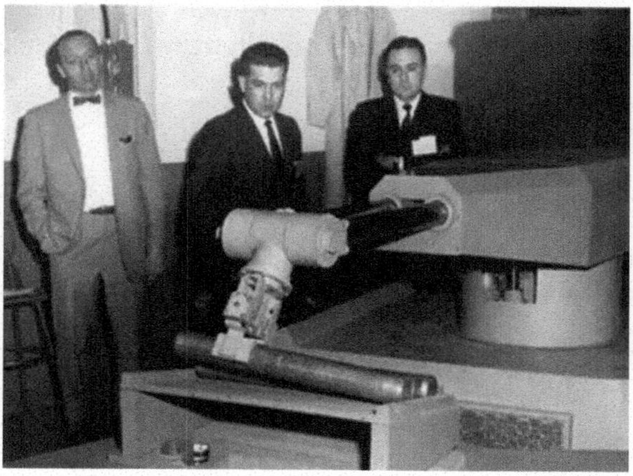

Fig. 1.3 The first modern robot in the world

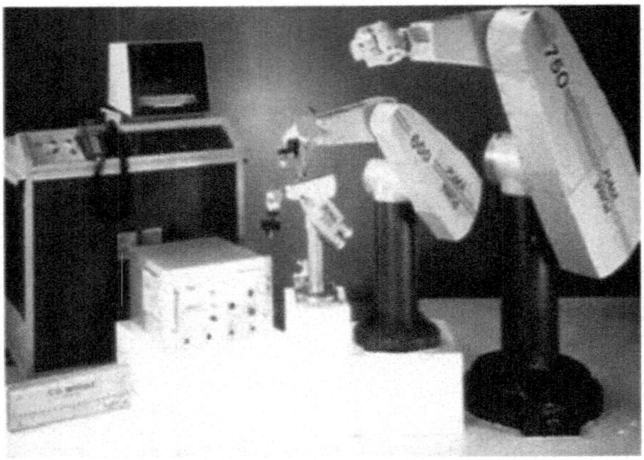

Fig. 1.4 PUMA robot

The second generation robot is perceptive robot, which detects the external environment through various sensors. The controller can respond to the change of environment conditions detected and adapt to the change of environment well. The welding robot adopting weld seam tracking and sensing technology is the second generation of robot. During the welding process, the seam tracking sensor detects the change of position of welds. The robot controller adjusts the movement of the welding torch according to the detected information, so that the welding torch is always aligned with the seam to ensure the welding quality.

The third generation robot is "intelligent robot", which adopts artificial intelligence technology. These robots not only have sensing capability, but also have the ability to perceive, think, reason, judge and make decisions. And they are able to perform complex tasks under changing conditions or with insufficient information. These robots are generally used in special industries, such as medical robots, ground sweeping robots, service robots and tour-guiding robots, but are rarely used in industrial production.

1.2 Structure and Classification of Industrial Robots

1.2.1 Structure of Industrial Robots

Industrial robots generally consist of three basic parts: robot proper, control system and mission control terminal. During performing the task, the robot needs to work in a certain environment. Therefore, from the control theory, the industrial robot system shall also include external sensors to detect changes in environment in addition to the above three parts, as shown in Fig. 1.5.

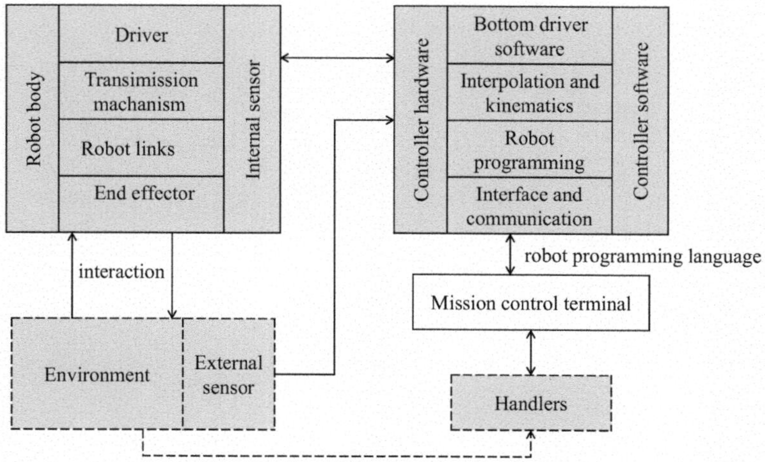

Fig. 1.5 Block diagram of industrial robot system

(1) Robot Body

The robot body is also called a manipulator, but it cannot be called robot by itself if there are no other parts. Its role and task is to achieve the exact position, orientation, and path required by the end effector. According to the type of motion synthesis, the robot body can be of Cartesian coordinate type, polar coordinate type, cylindrical coordinate type and articulated type, among which, most of the robots are of articulated type.

The body of a articulated robot usually consists of body (base), arm (upper arm and forearm), hand (terminal actuator), shoulder joint, elbow joint and wrist joint, etc., as shown in Fig. 1.6. The body, arms and hands are connected by the waist, shoulder, elbow and wrist joint. A drive system is installed at the joint to drive the joint to rotate. The position and orientation of the terminal manipulator are synthesized by motion of each joint during work. The robot body generally has 3–6 degrees of freedom, including 1–3 at the wrist, which is used to synthesize the position & orientation of the end effector.

The drive system usually consists of a power unit and a transmission mechanism, and is used to drive the actuator to perform and complete the corresponding action. There are three types of power devices used: Electric, hydraulic and pneumatic. Whether a servomotor is used, or a hydraulic cylinder or an air cylinder is used as the power unit, it is generally required to be connected to an actuator through a transmission mechanism. There are several types of transmission mechanisms, including gear drive, harmonic gear drive, chain drive, screw drive and belt drive.

The robot itself is also equipped with internal sensors, which are used to detect the status of the robot (such as position and speed) and provide the data to the control system.

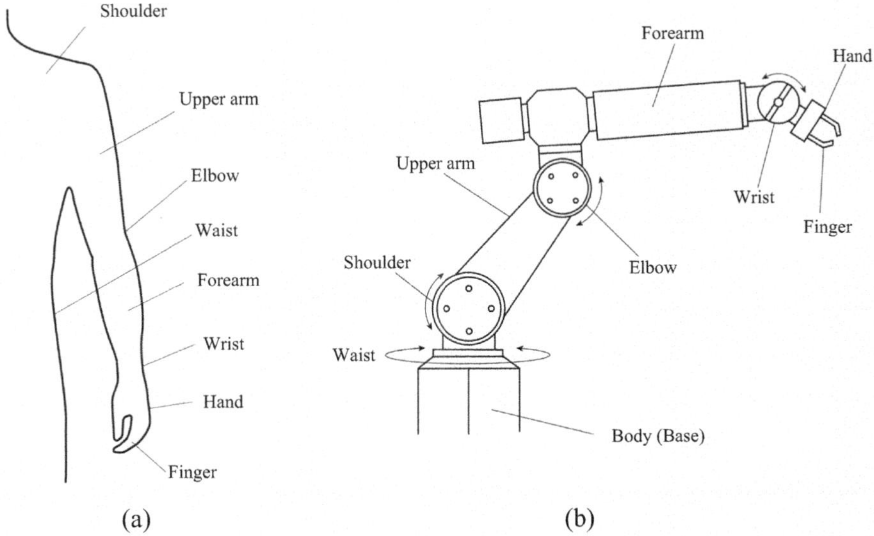

Fig. 1.6 Comparison of **a** Human arm and **b** Industrial robot body

(2) Environmental or external sensing systems

External sensors are used to monitor the operating environment of robots. The commonly used sensors include visual sensor, proximity sensor and force sensor. During working, a robot needs to make appropriate response actions based on the current state of the work object and environment, for example, making avoidance actions when encountering obstacles. Robots can make use of kinematic model and dynamic model to make recalculation based on the obstacle information detected by external sensors, form control signals, and send them to various joints for servo control so that a motion plan can be formed to avoid obstacles.

(3) Control system

The control system is the command center of robot and consists of hardware and software. It is actually a dedicated industrial computer. The hardware system consists of central processing control unit, memory unit, servo control unit, sensing control unit and communication bus. The software consists of controller system software, robot dedicated language, robot kinematics software, robot dynamics software, robot control software, self-diagnosis software and self-protection software. The control system is responsible for receiving the operator's work instructions and the environmental information fed back by internal and external sensors, making judgments and decisions on the instructions and feedback signals according to a predetermined strategy, and then outputting corresponding control signals to each motion actuator. Then, each motion actuator is controlled to perform a defined motion to complete a specific job.

(4) Mission control terminal

The mission control terminal is the interaction interface between the robot and humans, which is an intuitive program-controlled interface for control and applications. Its functions are to give task instructions to robots, perform task simulation, display and analyze status data etc. The most typical example of the mission control terminal is the teach pendant, which is used to control the path as well as all the positions and orientations of each important node of the end effector during the teaching process, and it is used to input process parameters (e.g. welding parameters). Also, it can be used to store all information in the control system memory after completion of the teaching.

1.2.2 Classification of Industrial Robots

There are various classification methods for industrial robots, including drive mode, motion path control mode, control method, coordinate form and intelligent degree.

(1) Classification by drive mode

Industrial robots can be classified into electric drive, hydraulic drive and pneumatic drive in terms of drive mode.

(a) Electric driven robots are driven by servo motors or step motors. Advantages include high control accuracy, quick response, large driving force, flexible and convenient detection and control. Currently, most of the welding robots in use are electric driven robots.
(b) Hydraulic driven robots are driven by servo controlled hydraulic cylinders. Although having the advantage of high power, the hydraulic drive has the drawbacks of low speed, easy leakage of hydraulic oil, high noise, unwieldy control unit and high cost. Therefore, they are only used as some heavy-duty robots, such as heavy-duty handling robots and spot welding robots.
(c) Pneumatic-driven robots are driven by air compressors and cylinders. The pneumatic drive has the advantages of low cost, simple structure, quick speed and easy maintenance, but it has the drawbacks of low control accuracy and high noise. It is currently rarely used in industry, and mainly used for gripping and assembling small workpieces.

(2) Classification by motion path control

According to the motion path control mode, industrial robots can be classified into three types, namely, point-to-point control (PTP), continuous path control (CP) and controllable path robots.

(a) Point-to-point control (PTP) robots only provide programmable control of the starting point and end point of the end effector during one motion, with the most direct and economical motion path adopted. This kind of robot has the advantages of simple structure and low cost. Spot welding robots and handling robots are usually PTP type.
(b) Continuous path control (CP) robot is also called controllable path robot. It can control the end effector to stay for a certain period of time at a certain number of points on a certain path during one motion, and smoothen the path between these points, allowing the end effector to travel smoothly along the specified path. These points need to be programmed in advance.
(c) Controllable path robot is also called computed path robot. It can accurately calculate the motion path, according to the mission requirements, with high precision. When using this kind of robot, all that need to be done is set the starting point and end point coordinates. The robot can calculate the optimal path by itself.

Arc welding robots are usually continuous path control (CP) robots, and resistance spot welding robots are usually point control robots.

(3) Classification by control method

According to control methods, industrial robots can be classified into program-controlled robot, teaching-playback Robot, numerical control robot, adaptive control robot and intelligent robot.

(a) Program-controlled robot is also called sequence control robot, which performs a series of specific actions according to the preset program. Usually, the logic control device, programmable controller or single-board processor are used as control units to set work sequence and realize position control by means of mechanical devices such as limit switch, cam, limit stop, matrix latch board, and step line selector, etc. This kind of robot has simple structure and low cost, and is suitable for simple and repetitive operations in mass production.
(b) Teaching-playback Robot, also called playback robot, is used to program the work tasks through manual teaching process, in which the end effector is controlled to travel along the predetermined path by using a teach pendant, and process parameters are set at several key nodes to simulate the completion of specified tasks. The information sent by the position sensor is recorded in the memory and saved as a program. During the operation of robot, the path and process parameters that have been taught are played back through this program.
(c) Numerical control robot, also known as controllable path robot, is the earliest to be used in industry. This type of robot also requires teaching, but the teaching process is not manual. The path between key points are determined via programming, where the operator only needs to specify these key points and the curve type between them.
(d) Adaptive robots can automatically sense changes in surrounding working conditions and make adjustments accordingly to adapt to such changes and better perform tasks.

(e) Intelligent robot is not only able to sense the change of surrounding conditions and make adjustments, but also can make in-depth analysis and make decisions under the condition of insufficient information or rapidly changing environment to better perform work tasks. This kind of robot is closer to human, but it is still difficult to make it to think like a human.

(4) Classification by coordinate system type

Industrial robots can be classified into Cartesian coordinate robot, cylindrical coordinate robot, spherical coordinate robot, and articulated robot according to the coordinate characteristics of manipulator, as shown in Fig. 1.7.

Cartesian coordinate robot is also called machine tool robot. Its hand can move along the three coordinate axes of Cartesian coordinates, as shown in Fig. 1.7a. It has the advantages of high structural rigidity, relatively independent motion of joints, simple structure, easy control, simple coordinate calculation and high precision. The disadvantage is that it cannot move at high speed, has poor operation flexibility and limited movement space.

Spherical robots offer rotation, pitching and extending/retracting motion. as shown in Fig. 1.7b. This type of robot has the advantages of high flexibility and large operating space.

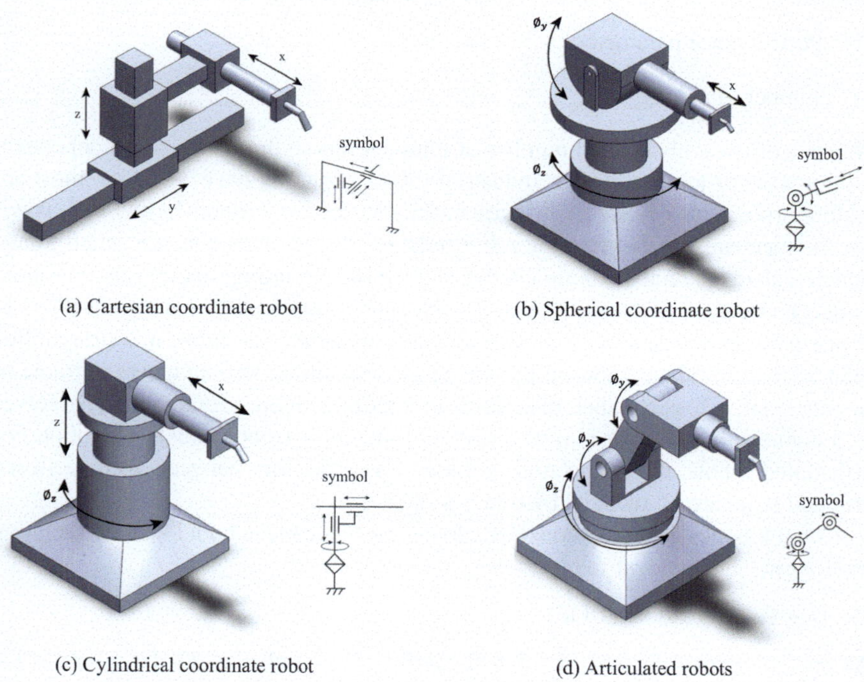

Fig. 1.7 Industrial robots with different coordinates

Robots with cylindrical coordinates offer lifting, rotation and extending/retracting motions as shown in Fig. 1.7c. This type of robot has the advantages of compact structure and easy control.

The arm of multi-articulated robot has several rotary joints, as shown in Fig. 1.7d. This kind of robot has large operating space, and can easily realize various positions and orientations within its specific movement space to perform various operations. Nonetheless, its coordinate calculation and control are very complicated, and it is hard for its control accuracy to reach that of Cartesian coordinate robot.

In addition to the above methods, industrial robots can also be classified according to their application fields. For example, industrial robots can be classified into welding robots, handling robots, assembly robots and painting robots.

1.2.3 Performance Parameters of Industrial Robots

As mentioned above, there are many types of industrial robots. Different industrial robots have different functions and performance characteristics. This section introduces the performance parameters of industrial robots by taking articulated robot as an example, mainly including number of axes, degree of freedom, operating space, resolution, positioning accuracy, repetition accuracy, load, and duty cycle, etc.

(1) Performance parameters

1) Number of axles

Number of axes, also called number of joints, refers to the number of independent motion axes or joints of robot. Industrial robots generally have 3–6 axes, most arc welding robots have 6 axes, while resistance spot welding robots have 5 or 6 axes.

The mechanisms that link the robot rods and their movements are called joints, which realize the relative motion between the rods. According to the type of motion, joints can be classified into prismatic, swivel and rotary joints, as shown in Fig. 1.8. A prismatic joint refers to a joint that allows two members to move in a straight line relative to each other. A swivel joint is a joint that allows two members to rotate or revolve relative to each other. In addition to rotating motion, a rotary joint also serves as a connection point for two or more rigid members (robot arms) of a robot. For articulated robots, the connections between base and arm, between arms, between arm and wrist are all realized by rotary joints.

Robot kinetics generally consider a joint that can achieve both linear and rotating motion two independent joints.

2) Degree of Freedom (DOF)

As an important indicator of robot flexibility, the number of DOFs refers to the number of axes or joints, and it is the number of independent coordinates required to describe the motion of an object. The greater the DOF, the more smart and adaptable the robot. However, the increased degree of freedom comes at the expense of

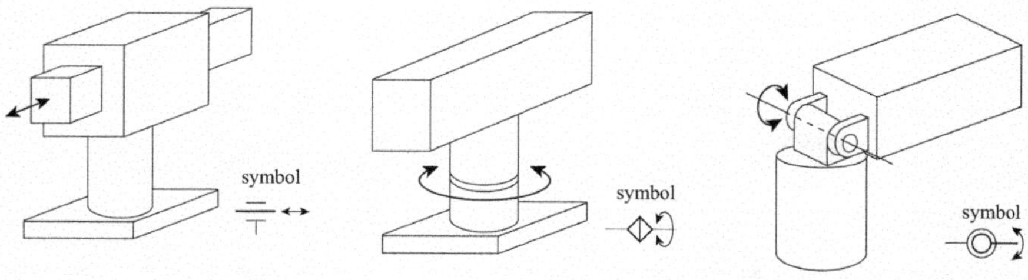

(a) Prismatic joint (b) Swivel joint (c) rotary joints

Fig. 1.8 Schematic diagrams and symbols of prismatic, swivel and rotary joints

complicated mechanisms and high costs. Industrial robots generally have 3–6 DOFs, as shown in Fig. 1.9. Arc welding and cutting robots generally require 6 DOFs, and spot welding robots require 5 DOFs.

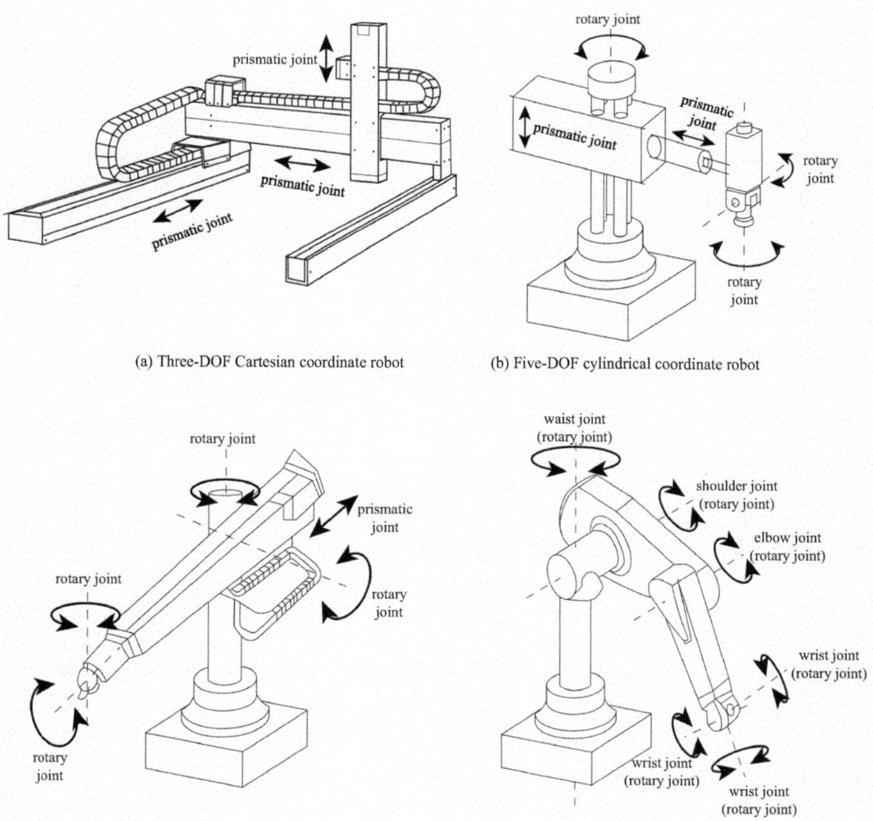

(a) Three-DOF Cartesian coordinate robot (b) Five-DOF cylindrical coordinate robot

(c) Five-DOF spherical coordinate robot (d) Six-DOF joint robot

Fig. 1.9 Robots with different DOFs

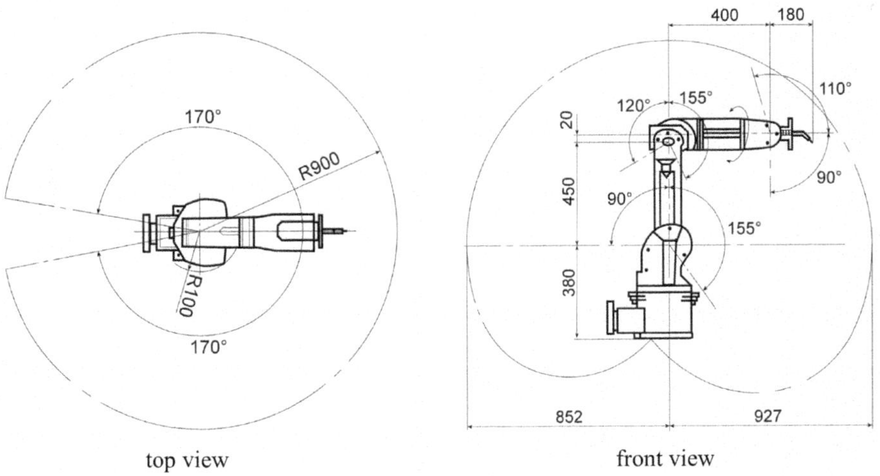

Fig. 1.10 Schematic representation of robot operating space

3) Work space

Work space refers to the the spatial range in which the wrist-axis intersection point of an industrial robot can move during the execution of task, which is usually represented by the graphics, as shown in Fig. 1.10. For simplification, it can also be indicated by maximum vertical motion range and maximum horizontal motion range. The maximum vertical motion range is the range between the lowest point (usually below the base of the robot) and the highest point the the robot wrist can reach. The maximum horizontal motion range refers to the distance between the farthest point that the robot wrist can reach and the center line of the robot base.

The operating space of a robot can be expanded by mounting the robot on a track or other walking devices.

4) Rated load

The maximum load (including end effector) that an industrial robot can withstand at its mechanical interface when it meets the specified operating accuracy is usually expressed by mass, moment or moment of inertia. The rated load of robots is specified mainly to ensure that the forces and moments on individual motion axes of robots will not affect their motion accuracy. In addition to the loads on robots, the inertia forces and moments caused by the change of motion speed shall also be considered. Under a certain load, the higher the motion speed, the higher the force and moment on each motion shaft. Therefore, the maximum load that the mechanical interface can withstood during high speed operation is usually taken as the rated load.

The load capacity of different robots varies greatly, with the maximum load capacity currently reaching up to 1000 kg.

5) Resolution

The resolution of a robot refers to the minimum moving distance or minimum rotation angle of each motion axis of the robot, which determines the minimum displacement of the actuator of the robot, i.e. the minimum step distance of motion. The resolution is a parameter determined during the design of robot system based on the test parameter.

The resolution can be classified into programming resolution and control resolution. Programming resolution is the minimum moving distance or angle set in the robot control software, also known as reference resolution. The control resolution depends not only on the programming resolution, but also on the accuracy of the position feedback detection unit.

The resolution determines the difference between the expected motion position and the actual position achieved by the operation instruction and it influences the positioning accuracy to some extent.

6) Positioning accuracy

The difference between the actual position or path of the mechanical interface center of a robot and the expected position or path is called positioning accuracy. It is used to indicate the ability of a robot end effector or its mechanical interface center to achieve the specified position. At present, the positioning accuracy of welding robots is generally 0.01 mm.

The major factors affecting positioning accuracy include mechanical error, control algorithm error and resolution error, etc. The mechanical error is the main factor, which involves the tranmission error, the joint clearance and the flexibility of linkage mechanism.

The transimission error depends on the gear tooth error and thread pitch error. The joint clearance depends on the bearing clearance and the harmonic tooth clearance of the joint transmission components, etc. The flexibility of linkage mechanism mainly depends on the machining and manufacturing accuracy and material quality of mechanical structure of robot, and is also affected by the magnitude and change of load.

7) Repeat accuracy

Repeat accuracy refers to the consistency of position or path of an industrial robot measured through repeating execution of an operation instruction n times under the same condition.

Repeat accuracy is further classified into repeat positioning accuracy and repeat path accuracy. It shall be noted that repeat accuracy and positioning accuracy are completely different concepts. For most industrial robots, there may be large error between the actual orientation or motion path and the expected position or path set by the instruction, i.e., the positioning accuracy may be high, but repeat orientation or path error between successive identical motions is usually small. The requirements of practical production can be met as long as the repeat accuracy is high enough.

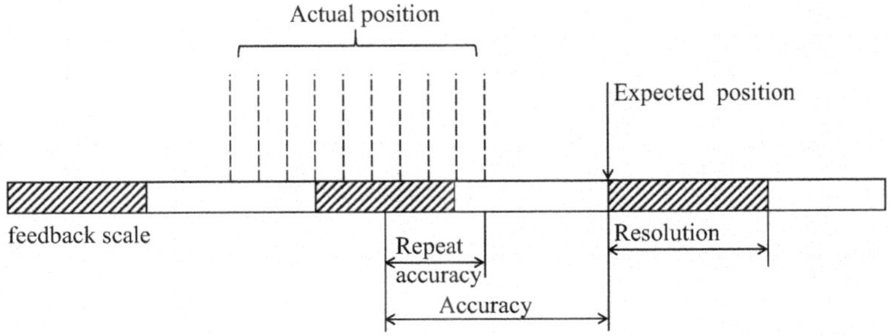

Fig. 1.11 Positioning accuracy, repeat positioning accuracy, and resolution

The positioning accuracy is greatly affected by gravity deformation, while the repeat positioning accuracy is immune from gravity deformation, since the error caused by gravity deformation exists and remains the same for every positioning. Since it is difficult to measure positioning accuracy, only repeat positioning accuracy is provided in the technical parameters of robots. Figure 1.11 shows the relationship between positioning accuracy, repeat positioning accuracy, and resolution.

The repeat positioning accuracy of industrial robots is generally ±0.01 to ±0. 5 mm.

8) Duty cycle

Duty cycle, also called duty cycle time, refers to the time taken to complete a task or operation. This is a very important indicator. The shorter the work cycle, the higher the work efficiency and competitiveness. The duty cycle consists of three stages, i.e. acceleration start-up, constant speed operation and deceleration stop. The acceleration and deceleration process usually takes a long time to ensure the positioning accuracy. Work efficiency can be improved by increasing driving power, reducing the mass of joints and rods, and adopting more effective control methods. Increase of drive power: Use the drive motors with high torque, good torque characteristics, low mass and low inertia to increase the driving power. Reducing the mass of joints and rods: while reducing the mass of joints and rods, their strength and stiffness should be ensured, i.e. rods and joints require a high specific strength and stiffness, for which rods are usually tapered. Advanced robots also use fiber-reinforced composites to effectively improve the specific strength and stiffness of the rods. More effective control methods can also be adopted, which should have fast computational speed and sophisticated path calculation methods to save path and mission planning time. At present, the duty cycle of robots can be made very short, for example, ABB's small robot "IRBI20" only takes 0.58 seconds to pick up 1 kilogram of materials.

9) Other parameters

(a) Maximum operating speed: The maximum operating speed refers to the maximum allowable moving speed or rotating speed of the main joint or mechanical interface center of a robot under rated load.

(b) Maximum operating acceleration refers to the maximum acceleration of the main joint or mechanical interface center of a robot.
(c) Types of motion control: PTP control and CP control.
(d) Drive type: the driver types mainly include electric drive, hydraulic drive, and pneumatic drive.
(e) Flexivity: The flexivity refers to the change in angle or position of a certain axis caused by the deformtion under the effect of external force or moment.
(f) Service life, reliability and maintainability: The average service life of industrial robots is generally more than 10 years, with some up to 15 years. As the robots are designed to use as few replaceable parts as possible, only a small number of spare parts need to be stored for replacement, thus the average repair time is generally not exceed 8 h.

(2) Factors influencing the motion accuracy of robots

Positioning accuracy and repeat accuracy are important performance indicators of robots and have great influence on the operation quality of robots and the quality of products manufactured by robots. The positioning accuracy and repetition progress not only depend on the design and manufacturing quality of robots, but also are affected by the operating conditions and environment. Therefore, it is very important to understand the factors influencing the motion accuracy of a robot.

Factors affecting the positioning accuracy and repeat positioning accuracy of robots mainly include:

(a) Robot mechanical structure design and manufacturing quality, control methods, and control system errors. When designing robots, the allowable errors and fit tolerance of various components should be strictly controlled based on the factors affecting motion accuracy, and suitable materials should be selected according to work requirements. The design dimensions and performance requirements shall be strictly guaranteed during manufacturing. In addition, the motion accuracy of robots should be improved by optimizing control methods and reducing control system errors.
(b) Gravity applied at the mechanical interface of the robot during operation (i.e. the robot's load) and the vertical deformation of the arm caused by the robot's load. Gravity mainly affects the positioning accuracy of robots. Generally, if the load of robot does not exceed the specified value, the gravity has little influence on positioning accuracy. However, the positioning accuracy will drop significantly if the load exceeds the specified limit. The larger the load or arm length of robot, the more serious the positioning accuracy drop. Under the condition of constant load, the robot's load has little influence on the repeat positioning accuracy. As long as the robot's load remains the same, the deformation of the arm is the same. Therefore, even if deformation is large (i.e., the positioning accuracy is poor), the repeat accuracy of the robot is high since the same deformation occurs during every repeat positioning.
(c) Looseness of drive gear and drive belt caused during applicaiton. Such deformations can lead to drive errors, i.e. speed ratio errors, which can further lead

to position errors. Gear meshing will inevitably result in clearance, which is related to machining accuracy and fit tolerances, as well as service time. Poor machining accuracy or long service time can lead to increased clearances and position errors. Generally, gear clearance should be controlled below 0.1 mm.

(d) Radial deformation caused by inertial force and distortion of long-sized rotating elements. When the robot arm rotates around its axis, inertial force will be generated in the radial direction of the robot arm, which will cause radial deformation and bending deformation of other members. In most cases, these deformations are negligible due to the small movement speed of the robot arm. The effect is greater at high speeds.

In addition, the expansion or contraction of the robot arm members, bearing clearance, and error in control method or control system caused by thermal effects may also cause movement error.

1.3 Development Trend of Welding Robot and Its Application

1.3.1 Welding Robot

(1) Structure of welding robot

Robots used for welding are called welding robots. Like other industrial robots, welding robots also consist of robot body, control system and task control terminal (usually an teach pendant). Welding robots need to be accompanied with relevant welding equipment (including welding power supply, welding torch, wire feeding mechanism and shielding gas delivery system), welding fixtures, welding sensors and system safety protection devices to complete welding.

Welding robots are basically articulated robots with 5 or 6 degrees of freedom (axes), as shown in Fig. 1.12. Resistance spot-welding robots are usually 5-DOF robots and electric arc welding robots are usually 6-DOF robots. The three degrees of freedom of the waist, shoulder and elbow joint are used to bring the torch to the desired position, while the two or three degrees of freedom of the wrist joint are used to determine the orientation of the torch.

Compared with other industrial robots, welding robots have a harsh working environment with adverse effects such as arc, spatter, smoke and dust, high frequency interference and high temperature during operation. In addition, the uncertainty of workpiece assembly error and thermal deformation during welding also complicate the working environment of robots. These also place high demands on the sensing system of welding robots.

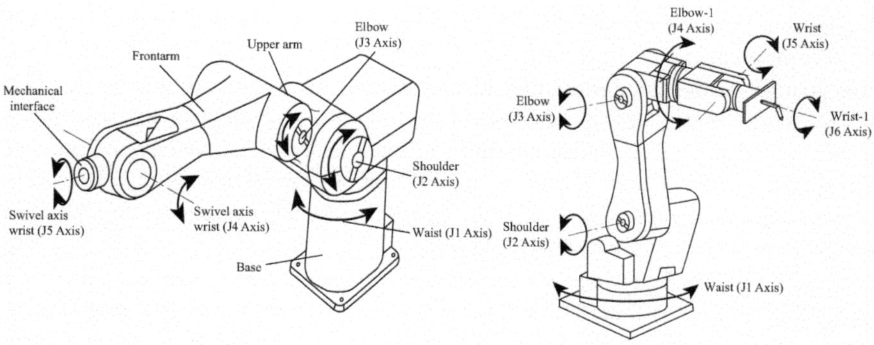

(a) 5-DOF robot (b) 6-DOF robot

Fig. 1.12 DOF of welding robots

(2) Classification of welding robots

Welding robots can be classified according to welding process in addition to industrial robots. welding robots can be classified into arc welding robot, resistance spot-welding robot, friction stir welding robot and laser beam welding robot according to the welding process.

1.3.2 Application Status of Welding Robot

Among all robots, industrial robots account for 67%, service robots 21% and special robots 12%. More than 40% of industrial robots are welding robots, which have been widely used in various fields of manufacturing industry.

(1) Development history of welding robot

In 1969, the first electric resistance welding robot was put into service on the vehicle body production line of General Motors, USA. In 1974, Kawasaki developed the first arc welding robot in the world, which is used for welding the frames of motorcycles. In the middle and late 1980s, welding robot technology in advanced industrialized countries has been highly mature and widely applied in automobile and motorcycle industries. At the end of 1970s, a Cartesian coordinate manipulator was also successfully developed in China for welding of car chassis. In 1984, China FAW took the lead in introducing the German KUKA welding robot for welding the body of the Red Flag car. And in 1988, this company developed a robot production line for vehicle body welding. Since the early 1990s, with the birth of the joint venture automobile factory, the application of welding robot has entered the period of rapid development.

In addition, as early as 1980s, domestic universities and research institutes started the development and research of welding robot technology. The national 8th and 9th Five-year Plan also listed robot technology and application research as a key R&D project. Through decades of continuous efforts, significant progress has been made in basic technology, control technology and manufacturing of the key components of welding robots in China. By the end of 2021, about 3 million welding robots were in use worldwide, among which, about 400,000 welding robots were in China. In recent years, the application of welding robots in China has exhibited a rapid growth trend with an annual average growth rate of more than 40%. Welding robots are currently most used in automobile, motorcycle, agricultural machinery, engineering machinery, locomotive and other industrial fields. Of the welding robots in China, about 90% are foreign brands, mainly Japanese brands (about 75%) such as OTC, Fanuc, Panasonic, Yaskawa, Fujikoshi, Kawasaki as well as European brands (about 20%) such as KUKA, ABB, CLOOS, IGM, and COMAU, and only about 10% of domestic welding robots are made in China. The main brands include Xinsong, Shidai, Huaheng, Jieboa, Xinshunda, Evert and Eston, etc. With the emphasis placed by the state on robot manufacturing technology and the increasing research and development investment from domestic robot manufacturers, it will not take long for home-made welding robots to replace imported robots.

Welding robot technology has gone through three stages. The first generation is a teach and play-back robot. The advantages of this type of robot are that it is easy to operate. Before welding, a teach process through a teach pendant is required in which the welding path and parameters are stored in the controller. During welding, the stored program is implemented to playback the taught and saved path and parameters. Its disadvantage is that it is not capable of external information perception and feedback, and cannot correct the path or welding parameters according to the change of working conditions. At present, such welding robots are still the most widely used in industrial production. The second generation is welding robots with sensing capabilities. Such robots are usually equipped with external sensors, which can detect and feedback changes in environmental conditions to a certain extent. During welding, the path or welding parameters can be corrected according to changes in the environmental condition. Good welding quality can be guaranteed even in situation where the pre-welding machining and assembly quality are not high. The third generation is intelligent robots. This type of robot not only have sensing capability, but also have the ability to independently judge, take actions, memorize, reason, and make decisions. They can automatically adjust the welding process parameters according to the change of external environment to fulfill more complex actions. At present, these robots are still under research and development and have not been widely applied in industry.

(2) Application significance of welding robot

Widespread application of welding robots has the following important significance:

(a) Improvement of welding quality and quality consistency. The welding of robots is less dependent on the technical level of operators and has a high repeatability, so the welding quality and quality consistency are greatly improved.

(b) High labour productivity. A welding robot station may be equipped with several assembly stations so that the robot can perform welding on a continuous basis. With the application of high-speed and high-efficiency welding techniques, robotic welding production efficiency will increase significantly.
(c) Good working conditions. During robotic welding, the operators only need to handle and unload the workpieces, and they can therefore keep away from welding arc, smoke and spatter, thus enjoy the better environment and reduced labor intensity.
(d) The production cycle is easy to control. The production schedule of robots will not be affected by external factors, and the production cycle is easy to be determined and fixed. Therefore, it is easier to implement the production plan.
(e) Shortened product modification cycle and reduced equipment investment. As compared with dedicated welding machine or dedicated production line, welding robot can adapt to the production of different workpieces by modifying its program.
(f) Welding robot-based welding production line facilitates the centralized collection of information on materials, semi-finished products, finished products and various process parameters during the manufacturing process, realizes information-based intelligent manufacturing, and benefits quality control, quality analysis and cost control.

1.3.3 Development Trend of Welding Robots

On one hand, with the continuous development of information technology, computer technology, control technology and welding technology, welding robot technology is becoming increasingly mature, with higher performance and lower cost. On the other hand, there is currently an increasing shortage of labor and climbing labor cost, and the ever-increasing competition in the market requires continuous improvement of welding quality. Therefore, it is urgent to use welding robots to perform welding instead of workers. In recent years, China has launched a series of industrial policies to promote the development and application of robots, such as Guidance on Promoting the Development of Industrial Robots, Robot Industry Development Plan (2016–2020) and Made in China 2025. With the support of the government, the manufacturing level and quality of robots in China will be rapidly improved and the market of industrial robots will continue to grow. According to the estimation of the Ministry of Industry and Information Technology of the People's Republic of China, the average annual sales growth rate of industrial robots will be more than 30% between 2015 and 2025, and the total market share of industrial robots and their peripheral components will reach about RMB 350 billion. Welding robots account for 40% of industrial robots. Calculated based on this proportion, the market share of welding robots is close to RMB 150 billion, and there is huge development space for welding robots.

As intelligent technologies such as smart perception, multi-modal human-machine interaction and cloud computing continue to mature, industrial robots will evolve rapidly to intelligent robots, and futuristic technologies such as in-depth learning of robots and multi-machine collaboration will be rapidly promoted in robots, making possible the wider application of robot systems. From the perspective of the demand for welding development in manufacturing industry, the development trend of welding robot system mainly includes:

(a) High efficiency robotic welding technology and process of medium-and thick-plates.
(b) Robotic welding of small-batched or single-piece large components (e.g. Marine engineering and shipbuilding industries);
(c) Further improvement of the process performance, adaptability, digitalization and intelligence of welding power supply;
(d) More intelligent welding robot systems;
(e) Various intelligent sensing technologies are used more widely in robots;
(f) More robust, adaptive software support systems;
(g) Integration of welding robot with upstream and downstream processing processes and bus control;
(h) The integration of information-based and intelligent welding with the Internet will finally realize the establishment of unmanned intelligent factories.

Chapter 2
Fundamentals of Robotic Kinematics

Robot is a complex system, including robot body, transmission mechanism, controller, end effector, and supporting process devices, etc. It involves many disciplines, such as mathematics, mechanics, electricity, automatic control, sensor technology, computer science, action planning and application engineering, etc. Therefore, robotic science is cross-disciplinary. The technology and science related to the design, manufacture and application of robot systems is known as robotics, also known as robotic engineering. The basis of robotics is the principles of motion control of robots and the relationship between robots and the objects they manipulate.

2.1 Description of Position and Orientation

2.1.1 Position and Orientation Description

A robot system executes and completes its work tasks through the complex space motion of the end effector, which is composed by the motion of various joints of the robot. The end effector shall also coordinate the movement with other machining or assembly devices connected with it. Therefore, robotic kinematics shall not only describe the position, displacement, speed and acceleration of the single rigid body of the manipulator, but also design the motion relationship between the rigid bodies of the manipulator and between the manipulator and other rigid bodies in the periphery.

In robotic kinematics, in order to easily describe and control the position and motion of each joint and member, each member needs a specific coordinate system. The position of each mass point on each member is first described by the vector in the member coordinate system. The relationship between individual members and between members and other rigid bodies is realized through mapping, and the commonly used one is homogeneous mapping.

M. Chen et al., *Technologies of Robotic Welding*, Advanced and Intelligent Manufacturing in China, https://doi.org/10.1007/978-981-97-3149-7_2

Fig. 2.1 Position vector P in coordinate system $\{A\}$

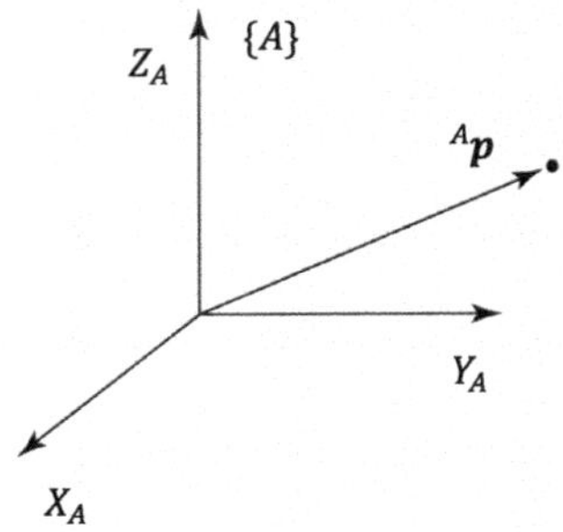

The available coordinate systems of robots include Cartesian coordinate system, cylindrical coordinate system and spherical coordinate system. Cartesian coordinate system is commonly used for industrial robots. The description method of position and orientation is described below by taking Cartesian coordinate system as an example.

(1) Description of point location

A Cartesian coordinate system is usually represented by three mutually perpendicular unit vectors. After a Cartesian coordinate system $\{A\}$ has been established, the position of any point in the space can be represented by a 3×1 position vector ${}^{A}p$, as shown in Fig. 2.1.

$$
{}^{A}p = \begin{bmatrix} p_x \\ p_y \\ p_z \end{bmatrix} \tag{2.1}
$$

where, ^{A} on the right indicates that vector p is a vector, in the coordinate system $\{A\}$; and P_x, P_y and P_z respectively represent the vector components of the vector p in the three axes of coordinate system.

(B) Orientation description

In order to determine the state of a rigid body such as a rod, an end effector or a part of a robot manipulator, it is not sufficient to describe the position of a point, but also its orientation. For example, in Fig. 2.2, the position of the lower-left finger of the end effector can be characterized by a 3×1 position${}^{A}p$, but the positions of other points on the end effector vary depending on its orientation, indicating that the position vector ${}^{A}p$ is not the indicator to fully determine the state of the end effector. In order to fully determine its state, a known coordinate system $\{B\}$, which is rigidly connected to the rigid member, is required, as shown in Fig. 2.2. The origin of the coordinate system $\{B\}$ is usually set at a characteristic point of the rigid body (end effector), such as center of mass, symmetrical center, or one end point. $\boldsymbol{i_B}, \boldsymbol{j_B}$ and $\boldsymbol{k_B}$ are unit vectors on the three axes of the coordinate system $\{B\}$. These three unit vectors are respectively designated as A_{i_B}, A_{j_B} and A_{k_B}, when written in term of the coordinate system $\{A\}$. The three unit vectors are represented by 3×3 matrices and expressed as:

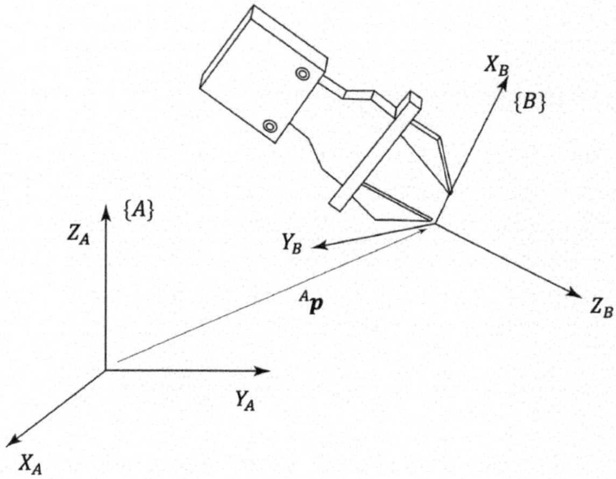

Fig. 2.2 Determining the position and orientation of a rigid body

$$ {}_B^A R = \left[{}^A i_B \ {}^A j_B \ A k_B \right] = \begin{bmatrix} r_{11} & r_{12} & r_{13} \\ r_{21} & r_{22} & r_{23} \\ r_{31} & r_{32} & r_{33} \end{bmatrix} \tag{2.2} $$

where, ${}_B^A R$ is known as the rotation matrix, with a superscript A on the left indicating that the $\{A\}$ coordinate system is the reference system and a subscript B on the left indicating that a vector in the $\{B\}$ coordinate system is being described; r_{ij} can be calculated using the unit vectors of the above two coordinate systems. $i_{\mathrm{A}}, j_{\mathrm{A}}$ and k_{A} are unit vectors on the three axes of the coordinate system $\{A\}$, then

$$ {}_B^A R = \left[{}^A i_B \ {}^A j_B \ {}^A k_B \right] = \begin{bmatrix} i_B \cdot i_A & j_B \cdot i_A & k_B \cdot i_A \\ i_B \cdot j_A & j_B \cdot j_A & k_B \cdot j_A \\ i_B \cdot k_A & j_B \cdot k_A & k_B \cdot k_A \end{bmatrix} \tag{2.3} $$

The dot product of the two unit vectors is equal to the cosine of the angle between the two vectors, so that the elements of the above matrix are also referred to as directional cosines. This matrix only represents the orientation of the end effector.

(C) Integral representation of position and orientation

If the position and orientation of a rigid body are determined, its state will be fully determined. In robotics, usually a characteristic point of a rigid body is selected, the position of which in the coordinate system $\{A\}$ is known, and the coordinate system $\{B\}$ is constructed with this characteristic point as the origin. The position and orientation of the rigid body are respectively described by the position vector ${}^A p_B$ and rotation matrix ${}_B^A R$ of the characteristic point. Then, the position and orientation of rigid body are described by the matrix formed by the position vector and rotation matrix.

$$ \{B\} = \left[{}^A p_B \ {}_B^A R \right] \tag{2.4} $$

2.1.2 Mapping

(1) Translational mapping

If the coordinate system $\{B\}$ and $\{A\}$ have the same orientation and only the coordinate origin is different, the position of $\{B\}$ with respect to $\{A\}$ can be described by a displacement vector ${}^{A}p_{BO}$, as shown in Fig. 2.3. And ${}^{A}p_{BO}$ is referred to as the translation vector of $\{B\}$ relative to $\{A\}$. For any point p known in the coordinate system $\{B\}$, its position vector is ${}^{B}p$, and its position vector at $\{A\}$ can be calculated using the following formula:

$$ {}^{A}p = {}^{B}p + {}^{A}p_{BO} \tag{2.5} $$

This formula is called the coordinate translation equation.

(B) Rotational mapping

If the coordinate systems $\{B\}$ and $\{A\}$ have the same origin, with only different orientations, as shown in Fig. 2.4, then the rotation matrix ${}^{A}_{B}R$ may be used to describe the orientation of $\{B\}$ relative to $\{A\}$. For any point p known in the coordinate system $\{B\}$, its position vector at $\{A\}$ can be calculated using the following formula:

$$ {}^{A}p = {}^{A}_{B}R\,{}^{B}p \tag{2.6} $$

This equation is called the coordinate rotation equation.

(C) General coordinate system transformation

The most common case is that the coordinate systems $\{B\}$ and $\{A\}$ are different in both orientation and origin, as shown in Fig. 2.5. In this case, a composite transformation is required. The position of the origin of the $\{B\}$ coordinate system relative to the $\{A\}$ is described by a displacement vector ${}^{A}p_{BO}$, and the orientation of the $\{B\}$ coordinate system relative to the $\{A\}$ is described by a rotation matrix ${}^{A}_{B}R$. By

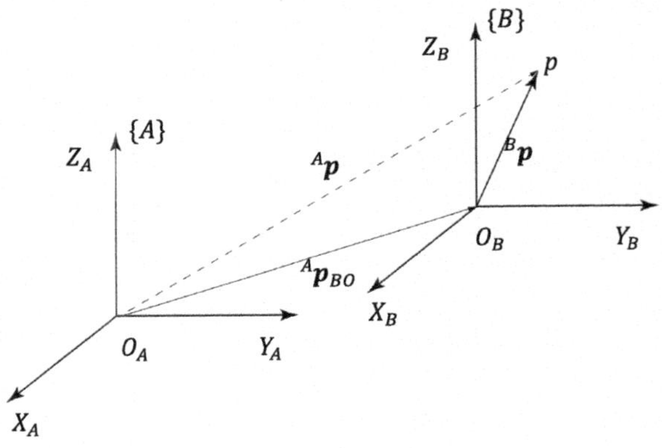

Fig. 2.3 Translational mapping

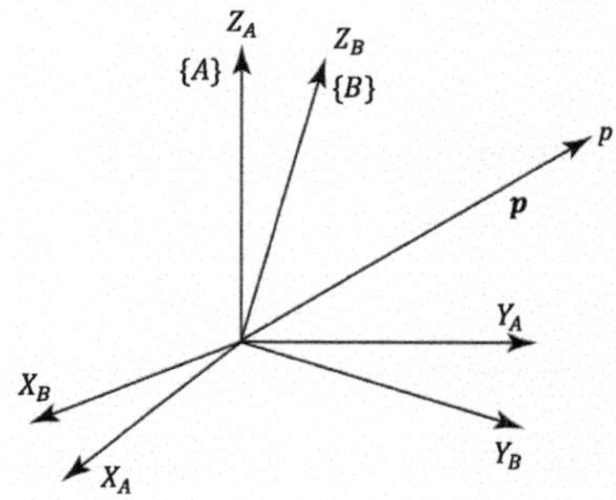

Fig. 2.4 Rotational mapping

introducing an intermediate coordinate system, the position vector ^{B}p of any known point p in the coordinate system $\{B\}$ can be converted into a position vector relative to $\{A\}$. A coordinate system $\{C\}$ is set, with the same origin as $\{B\}$ and the same orientation as $\{A\}$. ^{B}p is firstly transformed from $\{B\}$ to $\{C\}$ by the coordinate rotation equation, and then transformed from $\{C\}$ to $\{A\}$ by the coordinate translation equation, as follows:

$$^{A}p = {}^{A}_{B}R\,{}^{B}p + {}^{A}p_{BO} \tag{2.7}$$

(D) Homogeneous transformation

In robotic kinetics, there are several methods for analyzing the motion relationship between rigid members. Homogeneous transformation is the most commonly used one, which is characterized by intuitive, convenient and quick calculation. When a vector is transformed from one coordinate system to another by homogeneous transformation, translations and rotations can be performed simultaneously. To perform a homogeneous transformation, the vector must be described in a homogeneous coordinate system.

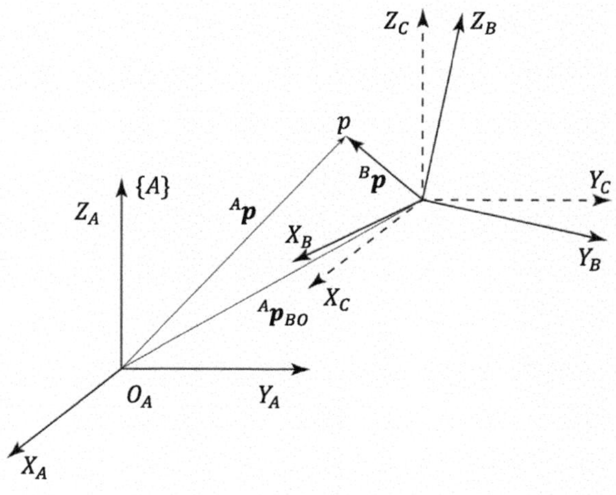

Fig. 2.5 General coordinate system transformation

The so-called homogeneous coordinate is used to express the n-dimension vector of the n-dimension rectangular coordinate system by (n + 1)-dimension vector. For space vector p

$$P = \begin{bmatrix} p_x \\ p_y \\ p_z \end{bmatrix} \tag{2.8}$$

A scale factor w is introduced, and then p is rewritten as

$$p = \begin{bmatrix} x \\ y \\ z \\ w \end{bmatrix} \tag{2.9}$$

where, $x = wp_x \quad y = wp_y \quad z = wp_z$.

This representation is called homogeneous coordinates. As the fourth element, w can be varied arbitrarily, and the vector magnitude will change with w. When $w = 1$, the vector magnitude remains unchanged when $w = 0$, the vector is infinite, $w > 1$, the vector is amplified, and when $w < 1$, the vector is scaled down. Here, the length of the vector is not important; what is important is the direction of the vector.

A) Translational homogeneous transformation

As shown in Fig. 2.3, the coordinate system $\{B\}$ has the same orientation as $\{A\}$, only the origin of the coordinates is different, and a displacement vector ${}^Ap_{BO}$ is used to describe the position of $\{B\}$ relative to $\{A\}$, and it is referred to as the translation vector of $\{B\}$ relative to $\{A\}$. In the coordinate system $\{B\}$, for any point p, whose position vector is ${}^Bp = [x \ y \ z \ \omega]^T$, if it is already known that ${}^Ap_{BO} = ai + bj + ck$, then the position vector in $\{A\}$ can be calculated using the following formula:

$$\begin{aligned} {}^Ap = {}^Bp + {}^Ap_{BO} &= \begin{bmatrix} a + \frac{x}{\omega} \\ b + \frac{y}{\omega} \\ c + \frac{z}{\omega} \\ 1 \end{bmatrix} = \begin{bmatrix} x + a\omega \\ y + b\omega \\ z + c\omega \\ \omega \end{bmatrix} = \begin{bmatrix} 1 & 0 & 0 & a \\ 0 & 1 & 0 & b \\ 0 & 0 & 1 & c \\ 0 & 0 & 0 & 1 \end{bmatrix} \begin{bmatrix} x \\ y \\ z \\ \omega \end{bmatrix} \\ &= \begin{bmatrix} 1 & 0 & 0 & a \\ 0 & 1 & 0 & b \\ 0 & 0 & 1 & c \\ 0 & 0 & 0 & 1 \end{bmatrix} \end{aligned} \tag{2.10}$$

That is, any vector known in $\{B\}$ can be converted into a vector in $\{A\}$ by $\begin{bmatrix} 1 & 0 & 0 & a \\ 0 & 1 & 0 & b \\ 0 & 0 & 1 & c \\ 0 & 0 & 0 & 1 \end{bmatrix}$, this matrix is called a homogeneous translational transformation matrix and is written as

$$H = Trans(a, b, c) = \begin{bmatrix} 1 & 0 & 0 & a \\ 0 & 1 & 0 & b \\ 0 & 0 & 1 & c \\ 0 & 0 & 0 & 1 \end{bmatrix} \tag{2.11}$$

B) Rotational homogeneous transformation

As shown in Fig. 2.4, for any point p known in the coordinate system $\{B\}$ with position vector ${}^{B}p = [x_B, y_B, z_B, 1]^T$, if it is assumed that its position vector in the coordinate system $\{A\}$ is denoted as ${}^{A}p = [x_A, y_A, z_A, 1]^T$, then x_A, y_A, z_A can be computed by the dot product formula of vectors.

$$x_A = i_A \cdot {}^{B}p = i_A \cdot i_B x_B + i_A \cdot j_B y_B + i_A \cdot k_B z_B \tag{2.12}$$

$$y_A = j_A \cdot {}^{B}p = j_A \cdot i_B x_B + j_A \cdot j_B y_B + j_A \cdot k_B z_B \tag{2.13}$$

$$z_A = k_A \cdot {}^{B}p = k_A \cdot i_B x_B + k_A \cdot j_B y_B + k_A \cdot k_B z_B \tag{2.14}$$

where, i_B, j_B and k_B are unit vectors on the three axes of the coordinate system $\{B\}$, and i_A, j_A and k_A are unit vectors on the three axes of the coordinate system $\{A\}$.

The expression in matrix form is as follows:

$$\begin{aligned} {}^{A}p = \begin{bmatrix} x_A \\ y_A \\ z_A \\ 1 \end{bmatrix} &= \begin{bmatrix} i_A \cdot i_B x_B + i_A \cdot j_B y_B + i_A \cdot k_B z_B \\ j_A \cdot i_B x_B + j_A \cdot j_B y_B + j_A \cdot k_B z_B \\ k_A \cdot i_B x_B + k_A \cdot j_B y_B + k_A \cdot k_B z_B \\ 1 \end{bmatrix} \\ &= \begin{bmatrix} i_B \cdot i_A & j_B \cdot i_A & k_B \cdot i_A & 0 \\ i_B \cdot j_A & j_B \cdot j_A & k_B \cdot j_A & 0 \\ i_B \cdot k_A & j_B \cdot k_A & k_B \cdot k_A & 0 \\ 0 & 0 & 0 & 1 \end{bmatrix} \begin{bmatrix} x_B \\ y_B \\ z_B \\ 1 \end{bmatrix} = \begin{bmatrix} i_B \cdot i_A & j_B \cdot i_A & k_B \cdot i_A & 0 \\ i_B \cdot j_A & j_B \cdot j_A & k_B \cdot j_A & 0 \\ i_B \cdot k_A & j_B \cdot k_A & k_B \cdot k_A & 0 \\ 0 & 0 & 0 & 1 \end{bmatrix} {}^{B}p \end{aligned} \tag{2.15}$$

Therefore, the rotational transformation matrix can be expressed as:

$$ {}_B^A R = \begin{bmatrix} i_B \cdot i_A & j_B \cdot i_A & k_B \cdot i_A & 0 \\ i_B \cdot j_A & j_B \cdot j_A & k_B \cdot j_A & 0 \\ i_B \cdot k_A & j_B \cdot k_A & k_B \cdot k_A & 0 \\ 0 & 0 & 0 & 1 \end{bmatrix} \tag{2.16} $$

The first column in ${}_B^A R$ is the projected component of the X-axis unit vector of coordinate system $\{B\}$ in the three axes of coordinate system $\{A\}$, and the second and third columns are the projected components of the Y-axis and Z-axis unit vector of coordinate system $\{B\}$ in the three axes of coordinate system $\{A\}$, respectively. The three columns respectively represent the direction of the three axes $\{B\}$ in the coordinate system $\{A\}$.

The rotation of coordinate system $\{B\}$ around of a single axis of coordinate system $\{A\}$ is called basic rotation, and the transformation matrix of this rotation is called the basic rotation matrix. If the rotation angle is α, then the three basic rotation matrices can be written as

$$ \mathrm{Rot}(x, \theta) = \begin{bmatrix} 1 & 0 & 0 & 0 \\ 0 & \cos\alpha & -\sin\alpha & 0 \\ 0 & \sin\alpha & \cos\alpha & 0 \\ 0 & 0 & 0 & 1 \end{bmatrix} \tag{2.17} $$

$$ \mathrm{Rot}(y, \theta) = \begin{bmatrix} \cos\alpha & 0 & \sin\alpha & 0 \\ 0 & 1 & 0 & 0 \\ -\sin\alpha & 0 & \cos\alpha & 0 \\ 0 & 0 & 0 & 1 \end{bmatrix} \tag{2.18} $$

$$ \mathrm{Rot}(z, \theta) = \begin{bmatrix} \cos\alpha & -\sin\alpha & 0 & 0 \\ \sin\alpha & \cos\alpha & 0 & 0 \\ 0 & 0 & 1 & 0 \\ 0 & 0 & 0 & 1 \end{bmatrix} \tag{2.19} $$

C) Composite homogeneous transformation

During robot operation, the motion of a rigid member in the reference coordinate system is usually a complex motion consisting of basic motions such as translation and rotation. The position before and after the motion can be described by composite homogeneous transformation. The coordinate system connected to rigid members is a moving coordinate system, and usually a fixed coordinate system is used as the reference coordinate system.

Composite homogeneous transformation is the product of basic homogeneous transformation matrices. It shall be noted that the position of basic matrices shall be arranged according to the order of transformation in calculation. As the transformation matrix is always left multiplied by the vector in the moving coordinate system at each transformation, the basic transformation matrix of the first transformation is arranged on the right, i.e., the composite transformation from moving coordinate system to reference coordinate system is carried out from right to left. The order must not be reversed because the matrix multiplication does not satisfy the commutative law.

For example, the reference coordinate system $\{A\}$ is $OXYZ$, and the moving coordinate system $\{B\}$ is obtained by the following operations: First, rotate α around the X axis, then rotate Φ around Y axis, and finally move the origin of the reference coordinate system by the position vector $[a, b, c]^{\mathrm{T}}$. Obtain the composite homogeneous transformation matrix ${}_{B}^{A}T$.

Solution:

The first basic transformation matrix is

$$T_1 = \mathrm{Rot}(x, \theta) = \begin{bmatrix} 1 & 0 & 0 & 0 \\ 0 & \cos\alpha & -\sin\alpha & 0 \\ 0 & \sin\alpha & \cos\alpha & 0 \\ 0 & 0 & 0 & 1 \end{bmatrix} \tag{2.20}$$

The second basic transformation matrix is

$$T_2 = \mathrm{Rot}(y, \phi) = \begin{bmatrix} \cos\phi & 0 & \sin\phi & 0 \\ 0 & 1 & 0 & 0 \\ -\sin\phi & 0 & \cos\phi & 0 \\ 0 & 0 & 0 & 1 \end{bmatrix} \tag{2.21}$$

The third basic transformation matrix is

$$T_3 = Trans(a, b, c) = \begin{bmatrix} 1 & 0 & 0 & a \\ 0 & 1 & 0 & b \\ 0 & 0 & 1 & c \\ 0 & 0 & 0 & 1 \end{bmatrix} \tag{2.22}$$

Any vector ${}^{B}p$ becomes $T_1^B p$ after the first transformation, $T_2 T_1^B p$ after the second transformation, and $T_3 T_2 T_1^B p$ after the third transformation is, then:

$${}_{B}^{A}T = T_3 T_2 T_1 = \begin{bmatrix} 1 & 0 & 0 & a \\ 0 & 1 & 0 & b \\ 0 & 0 & 1 & c \\ 0 & 0 & 0 & 1 \end{bmatrix} \begin{bmatrix} \cos\phi & 0 & \sin\phi & 0 \\ 0 & 1 & 0 & 0 \\ -\sin\phi & 0 & \cos\phi & 0 \\ 0 & 0 & 0 & 1 \end{bmatrix} \begin{bmatrix} 1 & 0 & 0 & 0 \\ 0 & \cos\alpha & -\sin\alpha & 0 \\ 0 & \sin\alpha & \cos\alpha & 0 \\ 0 & 0 & 0 & 1 \end{bmatrix} \tag{2.23}$$

Any vector in the moving coordinate system $\{B\}$ can be converted to a vector in the reference system $\{A\}$ by means of ${}^{A}_{B}T$.

It shall be noted here that the transformation operations described above are relative to the reference system, that is, all rotation and translation are measured relative to the reference system. If these operations are performed relative to the current coordinate system (moving coordinate system), the vector relative to the reference coordinate system shall be right-multiplied rather than left-multiplied by the fundamental transformation matrix. Therefore, when calculating the total transformation matrix, the basic transformation matrices of the preceding transformations are arranged on the left, and the left-to-right order is followed.

For a given coordinate system $\{A\}$, $\{B\}$ and $\{C\}$, if it is known that the composite transformation of $\{B\}$ relative to $\{A\}$ is ${}^{A}_{B}T$, and the composite transformation of $\{C\}$ relative to $\{B\}$ is ${}^{B}_{C}T$, then the composite transformation of $\{C\}$ relative to $\{A\}$ may be calculated as follows:

$$ {}^{A}_{C}T = {}^{A}_{B}T{}^{B}_{C}T \tag{2.24} $$

(E) Homogeneous inverse transformation

The vector transformation matrix from the moving coordinate system $\{B\}$ to the reference coordinate system $\{A\}$ is ${}^{A}_{B}T$. The vector transformation from the reference coordinate system $\{A\}$ to the moving coordinate system $\{B\}$ is called homogeneous inverse transformation and its transformation matrix is ${}^{B}_{A}T$. We get

$$ {}^{B}_{A}T = {}^{B}_{A}T^{-1} \tag{2.25} $$

Therefore, if the homogeneous transformation from the motion system $\{B\}$ to the reference system $\{A\}$ is:

$$ {}^{A}_{B}T = \begin{bmatrix} m_x & n_x & o_x & p_x \\ m_y & n_y & o_y & p_y \\ m_z & n_z & o_z & p_x \\ 0 & 0 & 0 & 1 \end{bmatrix} \tag{2.26} $$

Then

$$ {}^{B}_{A}T = \begin{bmatrix} m_x & m_y & m_z & -\boldsymbol{p}\cdot m \\ n_x & n_y & n_z & -\boldsymbol{p}\cdot n \\ o_x & o_z & o_z & -\boldsymbol{p}\cdot o \\ 0 & 0 & 0 & 1 \end{bmatrix} \tag{2.27} $$

where *m*, *n*, *o* and *p* are four column vectors.

$$ \boldsymbol{m} = [m_x, m_y, m_z]^T \tag{2.28} $$

$$\boldsymbol{n} = [n_x, n_y, n_z]^T \tag{2.29}$$

$$\boldsymbol{o} = [o_x, o_y, o_z]^T \tag{2.30}$$

$$\boldsymbol{p} = [p_x, p_y, p_z]^T \tag{2.31}$$

(F) Transformation equations

To describe or control the motion of a robot, it is necessary to establish the motion relationship between various rigid members of the robot and between the robot and the environment. It is necessary to establish a kinetic coordinate system on each member, as well as a reference coordinate system according to the environmental conditions, and determine the transformation relationship between various coordinate systems. According to the vector conservation theory (the vector from one point to another is constant, regardless of its motion path), the position and orientation of any rigid body in a space can be described in several methods, but the result should be the same regardless of the method adopted. Transformation equations in robotics are also established based on this theory.

Take the welding robot as an example, as shown in Fig. 2.6. During welding, the position of the welding torch coordinate system relative to the workpiece coordinate system directly determines the welding quality, which is the ultimate goal of robot planning. The welding torches cannot be directly controlled by the robot controller but needs to be controlled indirectly through the varous links and joints. To describe the position and orientation ${}^G_TT$ of the welding torch coordinate system relative to the workpiece coordinate system, several coordinate systems should be established, including $\{B\}$, $\{S\}$, {G}, $\{W\}$ and $\{T\}$. $\{B\}$ is the reference coordinate system; $\{S\}$ is the bench coordinate system; $\{G\}$ is the workpiece coordinate system (target coordinate system); $\{W\}$ is the wrist the coordinate system; $\{T\}$ is the welding torch coordinate system.

Generally, space vector diagrams are used to control the position and orientation of the welding torch coordinate system relative to the workpiece coordinate system, as shown in Fig. 2.7.

The position and orientation of the welding torch coordinate system relative to the reference coordinate system can be described either by the *B*-*W*-*T* path or by the *B*-*S*-*G*-*T* path, as follows:

$${}^B_TT = {}^B_WT{}^W_TT \tag{2.32}$$

$${}^B_TT = {}^B_ST{}^S_GT{}^G_TT \tag{2.33}$$

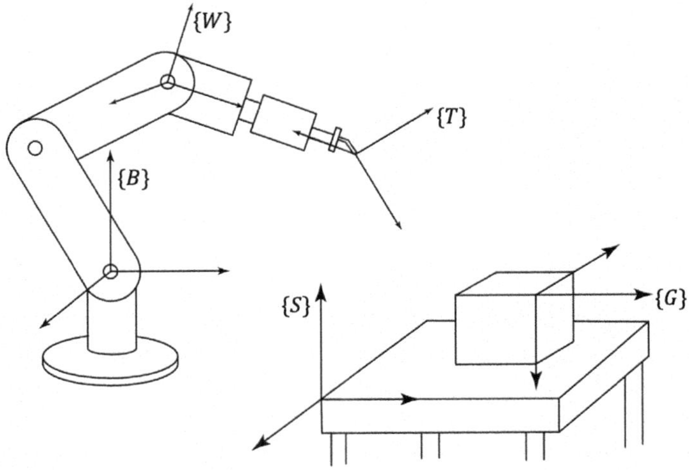

Fig. 2.6 Transformation equation of welding robot

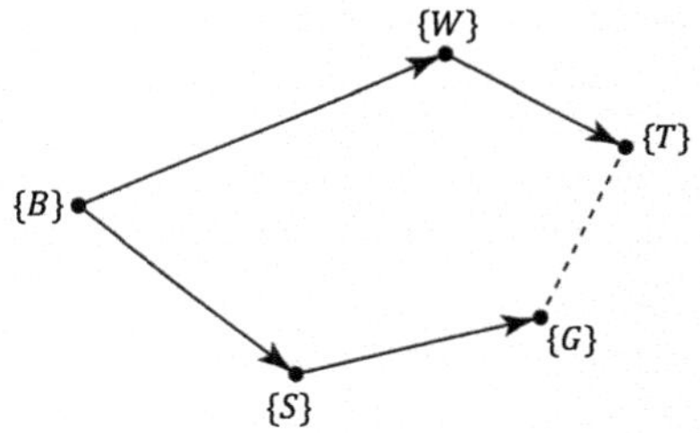

Fig. 2.7 Welding robot dimension chain

Since the starting and ending points of these two transformations are the same, therefore,

$$ {}^{B}_{W}T\,{}^{W}_{T}T = {}^{B}_{S}T\,{}^{S}_{G}T\,{}^{G}_{T}T \tag{2.34}$$

Either of the above matrices can be calculated from other known matrices. For example, if ${}^{B}_{S}T$, ${}^{S}_{G}T$, ${}^{G}_{T}T$ and ${}^{B}_{W}T$ are given, ${}^{W}_{T}T$ can be obtained through the following formula.

$$ {}^{W}_{T}T = {}^{B}_{W}T^{-1}\,{}^{B}_{S}T\,{}^{S}_{G}T\,{}^{G}_{T}T \tag{2.35}$$

and

$$ {}^{B}_{T}T = {}^{B}_{W}T\,{}^{W}_{T}T = {}^{B}_{W}T\,{}^{B}_{W}T^{-1}\,{}^{B}_{S}T\,{}^{S}_{G}T\,{}^{G}_{T}T \tag{2.36}$$

2.2 Robotic Kinematics

Robot kinematics is a discipline that studies the displacement, velocity, and acceleration of various parts of a robot body under static conditions, as well as their relationships. Since the first-order derivative of a displacement or angle is velocity, and the second-order derivative is acceleration, robot kinematics focuses on the study of robot's position & orientation and its relationship with the vectors of each joint. The robot body is actually a linkage mechanism consisting of a number of rigid rods connected by joints, and these rigid rods are called links. The joint to which the link is connected is usually a single degree-of-freedom joint. According to the mode of motion, the single degree-of-freedom joints of robot body mainly include prismatic joints and rotary joints. In very rare cases, robots also use multi-DOF joints. A n-DOF joint is generally regarded as a composite joint consists of n single-DOF joints and $n - 1$ links with zero length.

To perform kinematic analysis of robots, the first step is to establish the robot motion equation, also known as position equations. Robot motion equation is the basis for robot motion analysis. The basic procedure is to establish the robotic motion equation by establishing a link coordinate system on each link, and then use a homogeneous transformation to describe the position & orientation relationship between these coordinate systems. Then, the robotic motion equation can be established following the homogeneous transformation equation described in the previous section.

2.2.1 Link Motion Parameters

In the kinematic analysis of robots, the cross section, rigidity and strength of the link as well as the type, shape, weight and rotational inertia of the joint bearing are usually ignored. The link is regarded as a rigid body that limits the position relationship between two adjacent joint axes and the joint axis is regarded as a straight line or vector. This makes it easy to describe the status of any link and the relationship between adjacent links.

For the convenience of description, the various links of the robot body are numbered, its base is named as link 0, the first movable link is link 1, and so on, the ith link is called link i, and the link at the extreme end of the robot arm is called link n.

(1) Description of A link

The status of link itself can be described by two parameters, namely, link length and twist angle. For example, the intermidate rod $i - 1$ can be described by two of the following parameters, as shown in Fig. 2.8.

(a) The link length l_{i-1} refers to the distance between the two joint axes at both ends of the link, i.e. length of mutual perpendicular line length l_{i-1} between the two joint axes, as shown in Fig. 2.8. In the kinematic analysis, the mutual

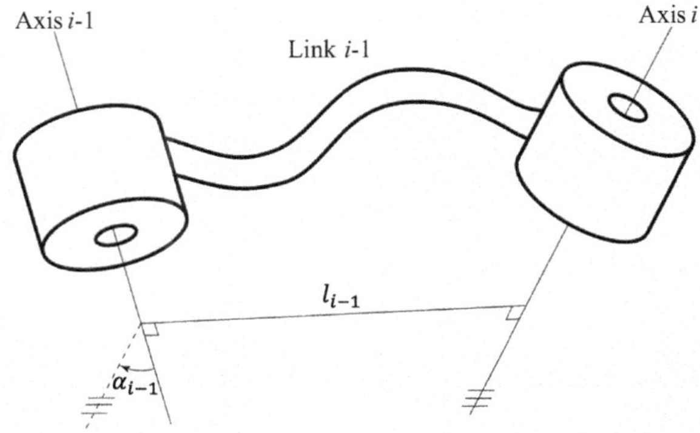

Fig. 2.8 Description of a link

perpendicular line l_{i-1} represents the link $i - 1$, and the link direction is defined as the direction from the joint axis $i - 1$ to the joint axis i. When joint axis i intersects with joint axis $i - 1$, the length of link i $l_{i-1} = 0$.

(b) The link twist angle α_{i-1} refers to the included angle between the joint axes at both ends of the link, as shown in Fig. 2.8. When the joint axis i is parallel with $i - 1$, $\alpha_{i-1} = 0$.

For the first and the end links, it is agreed as follows:

$$l_0 = l_n = 0 \ \alpha_0 = \alpha_n = 0.$$

(B) Link connection description

Two neighboring links are connected by a joint, and the connection between the two links can be described by two parameters of the joint, for example, links $i - 1$ and i are connected by joint i. The connection between the two links can be described by the following two parameters of the joint axis i:

A) Link offset d_i along joint axis i refers to the distance between the common vertical lines l_{i-1} and l_i of two neighboring links, as shown in Fig. 2.9.
B) Joint angle θ_1 of joint axis i refers to the included angle between the mutual perpendicular lines l_{i-1} and l_i of two neighboring links, as shown in Fig. 2.9.

When the joint i is a prismatic joint, the link spacing d_i is a variable; When the joint i is a rotary joint, the joint angle θ_i is a variable.

(C) Link motion parameters

The movement of each link can be described by the above four parameters, e.g. Link $i - 1$ can be described with $(l_{i-1}, \alpha_{i-1}, d_i, \theta_i)$. For a prismatic joint, d_i is a variable, the other three parameters are fixed. For rotary joints, θ_i is the joint variable, the other three parameters are fixed. This method of description is called Denavit-Hartenberg

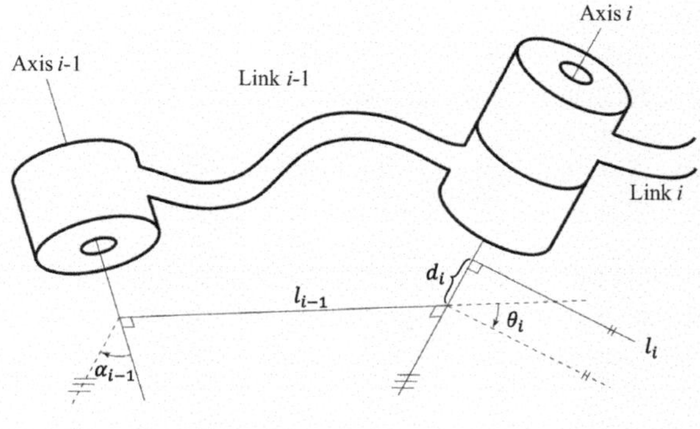

Fig. 2.9 Description of relationship between two adjacent links

method, which is the most widely used method in robotics. A 6-joint robot needs to be described with 24 parameters, of which 18 are fixed parameters and 6 are variable parameters. That is, 6-degree-of-freedom robot has 6 variables in its description parameters, so it is called 6-degree-of-freedom robot.

2.2.2 *Link Coordinate System*

For every link of the robot, a coordinate system shall be attached so that the motion relationship between the links can be described by homogeneous coordinate transformation to determine the position & orientation and motion of the end effector. The attached coordinate system of link i is called coordinate system $\{i\}$. The base coordinate system (link 0) is a fixed coordinate system, and other coordinate systems are moving coordinate systems. Normally, coordinate system 0 is usually used as the reference coordinate system to describe the position of other coordinate systems. Theoretically, the coordinate system attached to each member can be set arbitrarily. For convenience, the convention as shown in Fig. 2.10 is followed.

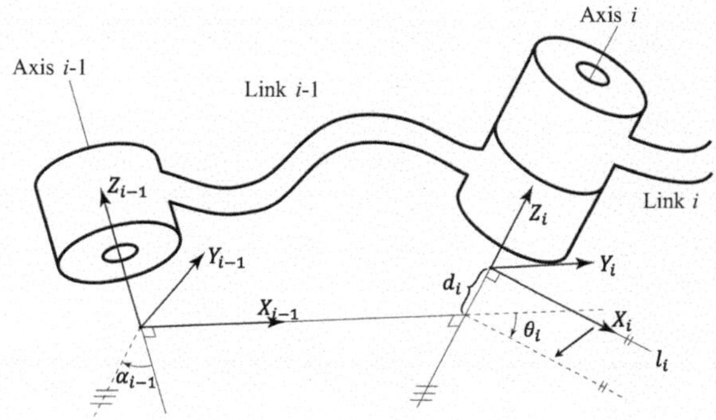

Fig. 2.10 Establishment of link coordinate system

(a) Z-axis Zi of the coordinate system $\{i\}$ coincides with the joint axis i with the direction pointing to joint i;
(b) Its origin is established at the intersection of the common vertical line l_i and the joint axis i.
(c) Axis X_i is coincident with l_i, with direction pointing towards joint $i + 1$;
(d) The axis Y_i is determined according to the right-hand rule.

Generally, when the joint vector of joint 1 is 0, the coordinate system $\{1\}$ is generally defined as the reference coordinate system $\{0\}$, so there will always be $l_0 = 0.0$ and $\alpha_0 = 0.0$ and if joint 1 is a rotary joint, $d_1 = 0.0$; and if joint 1 is a prismatic joint, $\theta_1 = 0.0$.

For the end link coordinate system $\{n\}$, if joint n is a rotary joint, X_n shall be set in the same direction as X_{n-1} (i.e. $\theta_n = 0.0$), and the origin of its coordinates should be set so that $d_n = 0$. If joint n is a prismatic joint, the origin of the coordinate system $\{n\}$ should be located at the intersection between X_{n-1}, and the joint axis n (i.e. $d_n = 0$), and the direction of the X_n axis should be set so that $\theta_n = 0$.

After establishing the coordinate system according to the above rules, the link parameters (l_{i-1}, α_{i-1}, d_i, θ_i) can be defined according to the following rules:

l_{i-1} is the distance from Z_{i-1} to Z_i in the direction of X_{i-1};
α_{i-1} is the angle of rotation from Z_{i-1} to Z_i around the axis X_{i-1};
d_i is the distance from X_{i-1} to X_i in the direction of Z_i;
θ_i is the angle of rotation from X_{i-1} to X_i around axis Z_i;

For arc welding robots, the tool coordinate system is established at the wire end of the torch. Generally, the end of welding wire is taken as the origin, and the direction in which welding wire feeds into the workpiece is taken as Z axis. Looking at the robot body, the horizontal left is Y axis and the direction away from the robot body is X axis.

2.2.3 Link Transformation Matrix

In order to correlate the motions of individual links, it is not enough to simply establish the link coordinate system, and it is necessary to determine the transformation relationship between individual links. The transformation of the coordinate system $\{i\}$ relative to $\{i-1\}$ is denoted as ${}^{i-1}_{i}T$. To transform any vector ${}^{i}p$ in the coordinate system $\{i\}$ into $\{i-1\}$, firstly rotate it around the X_{i-1} axis for α_{i-1}, then translate it for l_{i-1} along the X_{i-1} axis, then rotate around Z_{i-1} for θ_i, and finally translate for d_i along Z_{i-1}.

$$ {}^{i-1}_{i}T = Rot(x, \alpha_{i-1})Trans(l_{i-1}, 0,0)Rot(z, \theta_i)Trans(0,0, d_i) \quad (2.37) $$

When converted into matrix, it can be expressed as:

$$ {}_{i}^{i-1}T = \begin{bmatrix} \cos\theta_i & -\sin\theta_i\ 0 & & l_{i-1} \\ \sin\theta_i\cos\alpha_{i-1} & \cos\theta_i\cos\alpha_{i-1} & -\sin\alpha_{i-1} & -d_i\sin\alpha_{i-1} \\ \sin\theta_i\sin\alpha_{i-1} & \cos\theta_i\sin\alpha_{i-1} & \cos\alpha_{i-1} & d_i\cos\alpha_{i-1} \\ 0 & 0 & 0 & 1 \end{bmatrix} \tag{2.38} $$

Generally, each joint has only one degree of freedom. For the rotary joint, only θ_i is a variable in the link parameters (l_{i-1}, α_{i-1}, d_i, θ_i), therefore, ${}_{i}^{i-1}T = f(\theta_i)$. For a prismatic joint, there is only one variable d_i, so ${}_{i}^{i-1}T = f(d_i)$.

2.2.4 Robotic Kinematic Equations

For a n-DOF robot, the kinematic equation of the robot can be obtained by multiplying the transformation matrices of n links:

$$ {}_{n}^{0}T = {}_{1}^{0}T{}_{2}^{1}T \cdots {}_{n}^{n-1}T \tag{2.39} $$

where, ${}_{n}^{0}T$ is the transformtion matrix of the end effector coordinate system $\{n\}$ relative to the reference coordinate system $\{0\}$ of the robot. The position & orientation of the end effector relative to the reference coordinate system of the robot can be determined if the variables of each joint of the robot are known, i.e. all ${}_{i}^{i-1}T$ are known. This is a forward kinematic problem, and the solution of which is unique. After the vector of each joint is determined, the position & orientation of the end effector is uniquely determined.

For a given manipulator, it is a inverse kinematic problem to calculate the vector ${}_{i}^{i-1}T = f(\theta_i)$ of each joint using the given expected position & orientation ${}_{n}^{0}T$ of the end effector in the reference system. Robot inverse kinematic problem is more important in engineering application, and it is the basis for motion planning and path control of robots. The solution to the inverse kinematic problem is not unique and may not exist. The inverse problem is still solved using the above kinematic equations. The solution method is to continuously left multiply the inverse matrix of each link matrix at both sides of the equation, and then solve for $\theta_1, \theta_2 \ldots$ and θ_n successively.

By left-multiplying $\left({}_{1}^{0}T\right)^{-1}$, we get:

$$ \left({}_{1}^{0}T\right)^{-1}{}_{n}^{0}T = {}_{2}^{1}T \ldots {}_{n}^{n-1}T \tag{2.40} $$

There is only one variable θ_1 at the left side of the above formula. The joint variable θ_1 can be obtained by performing matrix transformation on the both sides and comparing the corresponding elements of the both sides. Substituting θ_1 into the above equation, the same method can be used to calculate θ_2 in this way.

Whether it is the computation of forward problem of robotic kinematics, the computation of inverse kinematic problems, or the path planning of end effector, all these operations are completed by the robot system itself. For example, the user only needs to give the desired position & orientation of the end effector, briefly describe it by means of some interaction, and the robot system itself determines the exact path and speed to reach the desired position & orientation. Many inverse kinematic problems have multiple solutions, but in most applications, it is not necessary to compute all solutions to save time.

For teaching-playback robot, the teaching-playback point is the point expected to be reached by the end effector. During the teaching process, when the operator controls the end effector to reach this point through a teach pendant, the robot records the Cartesian coordinates of the teaching-playback point and interpolation point under the reference coordinate system, and the robot controller only needs to perform simple inverse kinematic calculation to calculate the angle of each joint.

2.3 Robotic Motion Control

Welding robots are a multi-joint robot, making it difficult to directly control the end effector. The motion of the end effector can only be realized by synthesizing the motions of various joints. The motion of each joint is controlled by a servo control system, and all the servo systems work in collaboration to synthesize the motion of the end effector of the robot. Therefore, robot control requires two levels of control, as shown in Fig. 2.11. The first level is the servo control of the individual joint motor. The second level is the coordinated control of the motion of various joints, which is usually realized by the host computer. The host computer not only coordinates the motion of various joints to achieve the expected path, but also achieves human-machine interaction and completes other management tasks.

(1) Servo control of joints

For welding robot, whether the point-to-point motion control or continuous path motion control are achieved by controlling the position at individual moment. Therefore, the motion control system of each joint is actually a position control system.

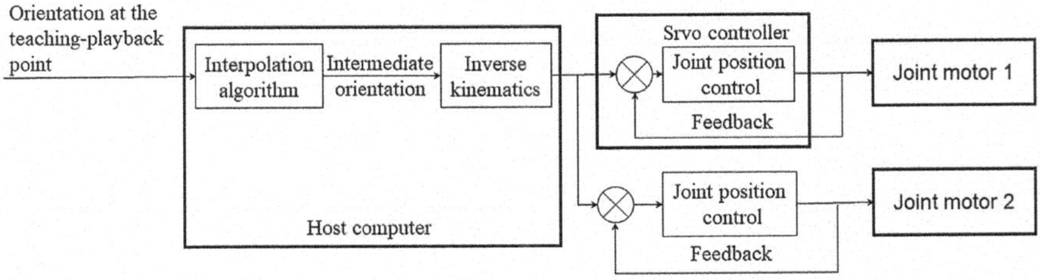

Fig. 2.11 Block diagram of robot motion control system

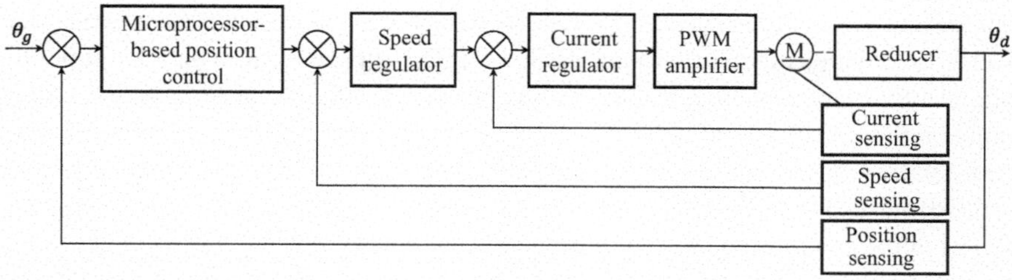

Fig. 2.12 Block diagram of a typical robot closed-loop control system of joint angle position

On this basis, the robot also uses sensors to detect the actual position or motion in real time, and the feedback control can improve the motion accuracy, so that the end effector of the robot can accurately achieve the desired position & orientation and path. All joints of welding robot are rotary joints and their position shall be controlled by angular position. Figure 2.12 shows a block diagram of a typical robot closed-loop control system for the joint angle position. θ_g is the given value of joint angle, which is calculated by the host computer through inverse motion kinematics. The three-loop joint angle closed-loop control system uses a certain algorithm to calculate the actual control quantity θ_d for each joint angle, and controls the movement of the drive elements to accurately achieve the desired position.

The control algorithms adopted by the position controller include PID control, variable structure control and adaptive control. The servo controller of welding robot mainly adopts PID control, i.e. proportional, integral and derivative control, which is the most commonly used control algorithm. Variable structure control means that there are more than one controller in the control system, and different controllers are used in different situations according to certain rules. Self-adaptive control means that the system automatically makes adjustments according to a certain control strategy upon detection of uncertain interference, automatically adapts to changes of external environment conditions, so as to achieve and maintain optimal performance indexes of system output.

(B) Synthesized control of joint motion

The servo control system of each joint controls each joint to make its joint angle reach the desired value. How is the desired joint angle of each joint of the robot determined? This is a problem to be solved by the host computer. The following illustrates the working principle using the teaching-playback robot as an example.

Before welding, the trajectory that the welding torch should follow during the welding process should be taught to the robot using a teaching pendant. First, the operator needs to decompose the complex trajectory curve into multiple straight lines and arcs, and then teach these straight lines and arcs. Only two characteristic points are needed to be taught for the straight line, i.e., starting and ending point. Three points need to be taught for a circular arc. Once the teaching is completed and the teaching program is stored, the robot remembers these taught points. The coordinates of the taught points recorded by the robot are those under Cartesian coordinate system (reference coordinate system). The robot controller carries out inverse kinematic calculation to get the joint angle of each joint, which is output to the closed-loop position control system as the given value θ_g of joint angle. Other points on the trajectory are obtained by the host computer using linear interpolation or circular arc interpolation algorithm. The coordinates of the interpolated points are also those under Cartesian coordinate system (reference coordinate system). These interpolation points also require inverse kinematics calculations by the robot controller. By repeating this calculation, the joint angles corresponding to all interpolated points on the trajectory is found, and the desired angular displacement of each joint is realized under the control of the joint angle closed-loop position control system to achieve the desired position & orientation and the desired path of the end effector. Interpolation is a process in which the host computer control system calculates point sequence and generates trajectories. Obviously, the continuity, smoothness and accuracy of actual motion path of robots depend on the distance between two interpolation points. The smaller the distance between two interpolation points, the closer the actual motion path is to the desired path, and the smaller the error. There are two commonly used interpolation methods: constant-spacing interpolation and constant-interval interpolation.

The so-called constant-spacing interpolation means that the distance between two adjacent interpolating points remains unchanged. As long as the interpolation distance is small enough, the trajectory accuracy can be guaranteed, so this interpolation algorithm is easy to ensure the stability of the motion. However, if the robot speed is not constant, the time interval T_s for different locations will be different. Therefore, it is relatively difficult to implement this method.

Constant-interval interpolation is a inerpolating process in which the interpolation points are calculated at a constant time interval T_s, i.e., the time interval between any two adjacent interpolated points is fixed. The length of this time interval T_s has an important influence on the accuracy of the motion path and the stability of the motion. Generally, T_s of welding robot shall not exceed 25 ms to ensure the stability of motion. The smaller the T_s, the better. However, as the robot has to perform interpolation and inverse kinematic calculation within T_s, its minimum vaule is limited by the calculation speed of the host computer. The space distance between two interpolation points P_i, P_{i+1} equals the motion speed of robot multiplied by *Ts*. Therefore, the motion speed of robot also has an important influence on the accuracy of motion path. Under certain T_s, the larger the motion speed, the larger the distance between two interpolation points, and the worse the accuracy of the motion path and stablity of the motion. Therefore, constant-interval interpolation is not suitable for robots moving at high speeds. This interpolation method is easy to realize and is widely used on welding robots with moderate motion speed.

(C) Interpolation algorithm

The basic motion path of robots includes linear path and circular arc path, and other non-linear or non-circular arc paths can be approximated using these two paths. Therefore, the algorithms for calculating interpolation points include linear interpolation method and circular arc interpolation method.

A) Linear interpolation algorithm

The linear interpolation algorithm is shown in Fig. 2.13. Given that the coordinates of the starting point P_o of a spatial straight line are (x_o, y_o, z_o), the coordinates of the ending point P_e are (x_e, y_e, z_e), and the number of interpolations is N, the increment of the interpolation point on each axis is:

$$\Delta x = \frac{(x_e - x_0)}{N + 1} \tag{2.41}$$

$$\Delta y = \frac{(y_e - y_0)}{N + 1} \tag{2.42}$$

$$\Delta z = \frac{(z_e - z_0)}{N + 1} \tag{2.43}$$

where, the number of interpolations can be calculated using the length L from the starting point to the ending point and the spacing d between interpolation points. The calculation formula is as follows:

$$N = \text{int}\left(\frac{L}{d}\right) \tag{2.44}$$

$$L = \sqrt{(x_e - x_o)^2 + (y_e - y_o)^2 + (y_e - y_o)^2} \tag{2.45}$$

For constant-spacing interpolation, d is the interpolation spacing; For constant-interval interpolation:

$$d = \nu T_s$$

where, ν is the motion speed of robot; T_s is the time interval for constant-interval interpolation.

The coordinates of interpolation points can be calculated using the following formula:

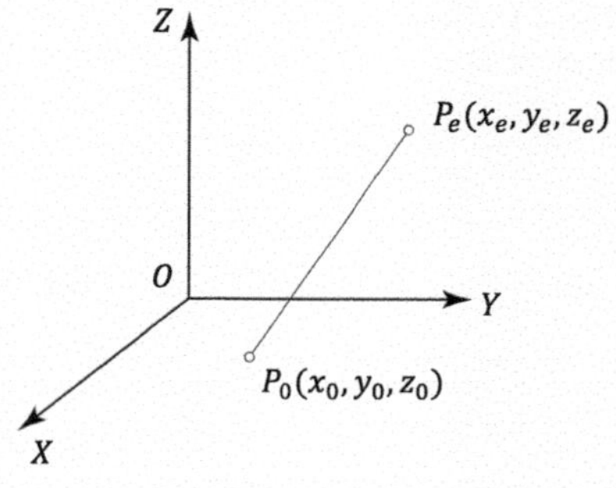

Fig. 2.13 Linear interpolation

$$\begin{cases} x_{i+1} = x_i + \Delta x \\ y_{i+1} = y_i + \Delta y \text{ (i = 1, 2, ..., } N) \\ z_{i+1} = z_i + \Delta z \end{cases} \tag{2.46}$$

B) Plane arc interpolation

Plane arc interpolation is performed when the plane of the arc is parallel or coincident with with one of the three reference planes (i.e., *XOY* plane, *YOZ* plane or *ZOX* plane) of the reference coordinate system. The *XOY* plane is taken as an example for illustration below. Given that three points $A\ (x_A, y_A, z_A)$, $B\ (x_B, y_B, z_B)$ and $C\ (x_C, y_C, z_C)$ on the arc with A-B-C arranged in clockwise, the position of the center, starting angle, radius of the arc and central angle of the arc can be calculated, as shown in Fig. 2.14.

Assuming that the coordinates of the center O_0 are (x_0, y_0, z_0), since $AO_0 = BO_0$, then

$$\sqrt{(x_A - x_0)^2 + (y_A - y_0)^0 + (z_A - z_0)^0} = \sqrt{(x_B - x_0)^2 + (y_B - y_0)^2 + (z_B - z_0)^2} \tag{2.47}$$

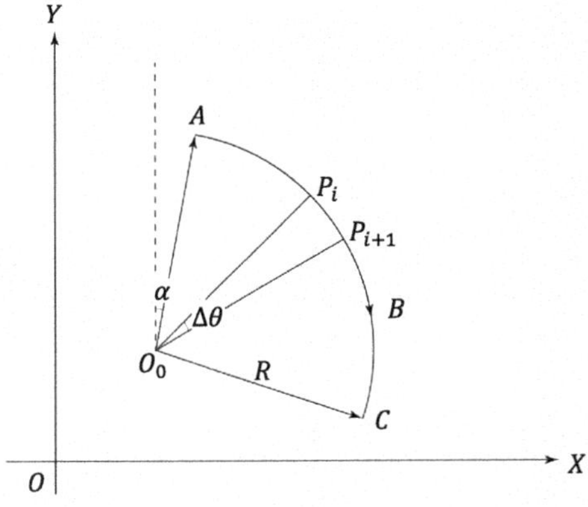

Fig. 2.14 Plane arc interpolation

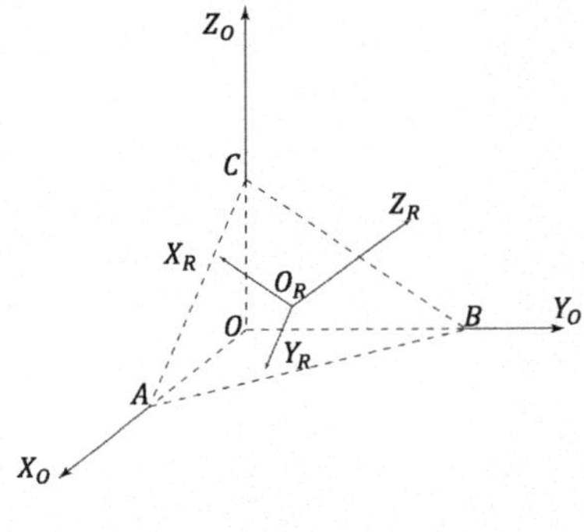

Fig. 2.15 Space arc interpolation

Since $AO_0 = CO_0$, then

$$\sqrt{(x_A - x_0)^2 + (y_A - y_0)^2 + (z_A - z_0)^2} = \sqrt{(x_C - x_0)^2 + (y_C - y_0)^2 + (z_C - z_0)^2} \tag{2.48}$$

Since $A\ (x_A, y_A, z_A)$, $B\ (x_B, y_B, z_B)$ and $C\ (x_C, y_C, z_C)$ are in different lines, then:

$$\begin{bmatrix} x_0 & y_0 & z_0 & 1 \\ x_A & y_A & z_A & 1 \\ x_B & y_B & z_B & 1 \\ x_C & y_C & z_C & 1 \end{bmatrix} = 1 \tag{2.49}$$

(x_O, y_O, z_O) can be calculated from the above equations, R can be obtained from the following formula:

$$R = |AO| = \sqrt{(x_A - x_0)^2 + (y_A - y_0) + (z_A - z_0)^2} \tag{2.50}$$

The starting angle α can be calculated via (2.51):

$$\alpha = \arcsin\left(\frac{x_A - x_0}{R}\right) \tag{2.51}$$

The central angle θ of the arc is:

$$\begin{aligned} \theta =& \text{acrcos}\left[\frac{(x_B - x_A)^2 + (y_B - y_A)^2 - 2R^2}{2R^2}\right] \\ &+ \text{acrcos}\left[\frac{(x_C - x_B)^2 + (y_C - y_B)^2 - 2R^2}{2R^2}\right] \end{aligned} \tag{2.52}$$

For constant-spacing interpolation, $\Delta\theta$ is the given angular displacement; For constant-interval interpolation, $\Delta\theta = \frac{T_s v}{R}$, (v is the moving speed along the circular arc).

The number of interpolations $N = \frac{\theta}{\Delta\theta}$

After the above data are obtained, the coordinates of starting point A can be expressed as:

$$\begin{cases} x_A = x_0 + R\sin\alpha \\ y_A = y_0 + R\cos\alpha \\ Z_A = z_0 \end{cases} \tag{2.53}$$

The coordinates of the endpoint C can be expressed as:

$$\begin{cases} x_C = x_0 + R\sin(\alpha + \theta) \\ y_C = y_0 + R\cos(\alpha + \theta) \\ Z_C = z_0 \end{cases} \quad (2.54)$$

The coordinates of any other interpolation points on the arc AC can be obtained from the following formula:

$$\begin{cases} x_i = x_0 + R\sin(\alpha + i\Delta\theta) \\ y_i = y_0 + R\sin(\alpha + i\Delta\theta)\ (\mathrm{i} = 1, 2, \ldots, \mathrm{N}) \\ z_i = z_0 \end{cases} \quad (2.55)$$

C) Space arc interpolation

Spatial arc interpolation shall be perfomed when the plane of the arc is not on or parallel to any of the three reference planes (i.e. *XOY*, *YOZ* or *ZOX* planes) of the reference coordinate system. In this case, an intermediate coordinate system needs to be established to convert spatial arc into plane arc, then the coordinates of each interpolation point are obtained by means of plane arc interpolation, and finally they are converted into the coordinates of the reference coordinate system through coordinate transformation.

The three points on a given arc are $P_1\ (x_1, y_1, z_1)$, $P_2\ (x_2, y_2, z_2)$, and $P_3\ (x_3, y_3, z_3)$. The center and radius of the arc can be determined by the similar method as that for a plane arc.

Assuming that the coordinates of the center O_R are (x_0, y_0, z_0), since $P_1O_0 = P_2O_0$, then

$$\sqrt{(x_1 - x_0)^2 + (y_1 - y_0)^2 + (z_1 - z_0)^2} = \sqrt{(x_2 - x_0)^2 + (y_2 - y_0)^2 + (z_2 - z_0)^2} \quad (2.56)$$

Since $P_1O_0 = P_3O_0$, then

$$\sqrt{(x_1 - x_0)^2 + (y_1 - y_0)^2 + (z_1 - z_0)^2} = \sqrt{(x_3 - x_0)^2 + (y_3 - y_0)^2 + (z_3 - z_0)^2} \quad (2.57)$$

Since $P_1\ (x_1, y_1, z_1)$, $P_2\ (x_2, y_2, z_2)$, and $P_3\ (x_3, y_3, z_3)$ are different from each other, then:

$$\begin{bmatrix} x_0 & y_0 & z_0 & 1 \\ x_1 & y_1 & z_1 & 1 \\ x_2 & y_2 & z_2 & 1 \\ x_3 & y_3 & z_3 & 1 \end{bmatrix} = 1 \tag{2.58}$$

From the above three equations, R can be obtained through the following formula:

$$R = |P_1O| = \sqrt{(x_1 - x_0)^2 + (y_1 - y_0)^2 + (z_1 - z_0)^2} \tag{2.59}$$

An intermediate coordinate system $X_RY_RZ_R$ is established with O_R as the coordinate origin, and the outer normal direction of the plane determined by the three points of P_1 (x_1, y_1, z_1), P_2 (x_2, y_2, z_2), and P_3 (x_3, y_3, z_3) is the Z_R axis, and the angle between the X_R axis and X_O is set as θ, as shown in Fig. 2.15. The following can be obtained:

$$\cos\theta = \frac{B}{\sqrt{A^2 + B^2}} \tag{2.60}$$

$$\sin\theta = \frac{-A}{\sqrt{A^2 + B^2}} \tag{2.61}$$

where, A, B and C are respectively the intercepts of the plane in which the arc is located on the three axes of the reference coordinate system.

Using the coordinates of P_1 (x_1, y_1, z_1), P_2 (x_2, y_2, z_2) and P_3 (x_3, y_3, z_3), the unit vector of axis Z_R in the reference coordinate can be obtained, and the sine and cosine of the angle between axis Z_R and axis Z_O can be obtained.

$$\cos\alpha = \frac{C}{\sqrt{A^2 + B^2 + C^2}} \tag{2.62}$$

$$\sin\alpha = \frac{\sqrt{A^2 + B^2}}{\sqrt{A^2 + B^2 + C^2}} \cdot \frac{A}{|A|} \tag{2.63}$$

Translate, the origin O_R of the intermediate coordinate system to the origin O of the reference coordinate system, and then rotate the intermediate coordinate around the axis Z_R for θ, and finally rotate around axis X_R for α, so that $X_RY_RZ_R$ coincides with $X_OY_OZ_O$. Therefore, the transformtaion matrix from $X_RY_RZ_R$ to $X_OY_OZ_O$ is as follows:

$${}^{O}_{R}T = \text{Trans}(x_0, y_0, z_0)Rot(z, \theta)Rot(x, \alpha) = \begin{bmatrix} \cos\theta & -\sin\theta\cos\alpha & \sin\theta\cos\alpha & x_0 \\ \sin\theta & \cos\theta\sin\alpha & -\cos\theta\sin\alpha & y_0 \\ 0 & \sin\alpha & \cos\alpha & z_0 \\ 0 & 0 & 0 & 1 \end{bmatrix} \tag{2.64}$$

Before interpolation of plane arcs, it is necessary to convert the coordinates in the reference coordinate $X_O Y_O Z_O$ to the coordinates in the intermediate coordinate system $X_R Y_R Z_R$. Therefore, the matrix ${}_O^R T$ shall be given.

$$\begin{aligned} {}_O^R T &= {}_R^O T^{-1} \\ &= \begin{bmatrix} \cos\theta & \sin\theta & 0 & -(x_0\cos\theta + y_0\sin\theta) \\ -\sin\theta\cos\alpha & \cos\theta\cos\alpha & \sin\alpha & -(x_0\sin\theta\cos\alpha + y_0\cos\theta\cos\alpha + z_0\sin\alpha) \\ \sin\theta\sin\alpha & -\cos\theta\sin\alpha & \cos\alpha & -(x_0\sin\theta\sin\alpha + y_0\cos\theta\sin\alpha + z_0\cos\alpha) \\ 0 & 0 & 0 & 1 \end{bmatrix} \end{aligned} \tag{2.65}$$

In this way, space arc interpolation can be performed as follows.

Firstly, the coordinates of the three teaching-playback points in the reference coordinates are converted to the coordinates in the intermediate coordinate system;

$$P_{Ri} = \begin{bmatrix} x_{Ri} \\ y_{Ri} \\ 0 \\ 1 \end{bmatrix} = {}_O^R T \begin{bmatrix} x_i \\ y_i \\ z_i \\ 1 \end{bmatrix} (i = 1, 2, 3) \tag{2.66}$$

Then, the plane arc interpolation is performed in the $X_R Y_R Z_R$ coordinate system according to the preceding description;

Finally, the coordinates of interpolation point P_{j} in the intermediate coordinate are converted to those in the reference coordinates.

$$P_J = \begin{bmatrix} x_j \\ y_j \\ z_j \\ 1 \end{bmatrix} = {}_R^O T \begin{bmatrix} x_{Rj} \\ y_{Rj} \\ 0 \\ 1 \end{bmatrix} (\mathrm{j} = 1, 2, 3) \tag{2.67}$$

2.4 Brief Introduction to Robotic Dynamics

The motion of each joint of a robot is achieved by the driving force of the drive. The discipline that studies the relationship between force and motion of each joint or link of the robot is called robotic dynamics, which is usually expressed as the relationship between position, speed and acceleration of each joint and the force or moment of this joint. The forces on the link or joint include the driving force or moment applied by the robotic driver, external force and moment, etc.

There are two problems in robotic dynamics that need to be study. The first one is the forward dynamics problem, in which the force or moment of each joint is known and its displacement, velocity and acceleration are to be found. The second one is the inverse dynamic problem, where the displacement, velocity and acceleration of each joint are known, and the force or moment is to be found.

Forward dynamics equation is mainly used for robot dynamics simulation, for example, to determine the structural parameters and drive scheme of robot manipulator through simulation. Generally, the structure, dimensions and transmission mechanism characteristics of each link are first selected based on the kinematic and dynamic parameters and the rated load, and then continuously optimized through dynamic simulation calculation. For robots with high speed motion, off-line programming also requires robotic dynamics simulation to evaluate dynamic loads and path deviations due to high speed motion. Welding robot is a low-speed robot, and generally no such simulation calculation is required.

The inverse problem of robotic dynamics mainly solves the problems related to the real-time control of robot motion to achieve the expected motion trajectory. The solution to the inverse dynamics problem are also used to improve the dynamic and static characteristics, control accuracy, resolution, stability and repeatability of the robot.

As a robot has multiple links and joints, its dynamic equation is a non-linear problem with multiple inputs and multiple outputs, and has intricate coupling relationships. Therefore, solving the robot dynamics problem is very complex. There are many methods to establish and solve robotic dynamic equations, such as Lagrangian method, Newton-Euler method, Abel method and Kane method. The commonly used method is to use Lagrange method to establish equations, and use Lagrange functional equilibrium method or Newton-Euler dynamic equilibrium method to solve.

The motion speed and acceleration of a welding robot are generally not high, so only some simple dynamic control is required.

Chapter 3
Robot Body and Control System

The robot body itself is also called robot arm, manipulator, mechanic arm or robot manipulator. It is the actuator of robot system, which can perform various operations in lieu of human arm. This chapter mainly introduces the robot body and control system by taking 6-axis articulated robot commonly used in industry as an example.

3.1 Robot Body Structure

The robot body consists of rigid links, drives, transmission mechanisms, joints and internal sensors (such as encoder disk), etc., as shown in Fig. 3.1. Its main task is to control the end effector (such as a welding torch) to achieve the desired position, orientation and to ensure that the end effector moves along the desired path at a certain speed. Generally, the robot body can be considered to be composed of rigid links and joints, while the drive, transmission mechanism and internal sensors can be considered as parts of the joints. In this way, the robot body can be seen as an open-loop joint chain, as shown in Fig. 3.2. The robot body can be divided into the following parts: Base, upper arm, forearm, wrist, elbow joint, shoulder joint and waist joint. For welding robots, there are 2–3 joints at the wrist, each with one degree of freedom.

3.1.1 Robot Base

A robot base, also called base, pedestal, or frame, is a component that directly connects and supports the arm and traveling mechanism. The base may be fixed or walking type. The fixed base is fixed to the ground or to a platform. The walking base can travel along the track; or it is mounted on a gantry and travels along the track with the gantry.

M. Chen et al., *Technologies of Robotic Welding*, Advanced and Intelligent Manufacturing in China, https://doi.org/10.1007/978-981-97-3149-7_3

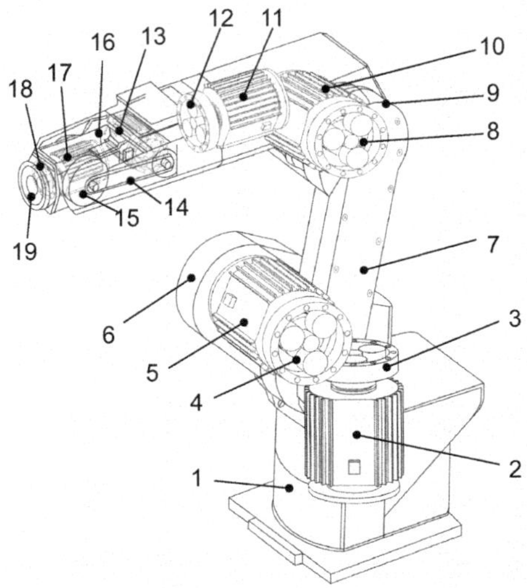

1-Base; 2- Drive motor of waist joint axis (J1 axis motor); 3- J1 axis reducer; 4- J2 axis reducer; 5- Driver motor of shoulder joint axis (J2 axis motor) ; 6- Shoulder; 7- Upper arm; 8- J3 axis reducer; 9- Elbow; 10- Drive motor of elbow joint axis (J3 axis motor); 11- Drive motor of wrist joint axis (J4 axis motor); 12- J4 joint reducer; 13- Drive motor of wrist joint axis (J5 axis motor); 14- J5 axis timing belt; 15- J5 joint reducer; 16- Forearm; 17- Driver of wrist joint axis (J6 axis motor) 18- J6 axis reducer; 19- Wrist joint housing

Fig. 3.1 Basic structure of a typical 6-axis robot body

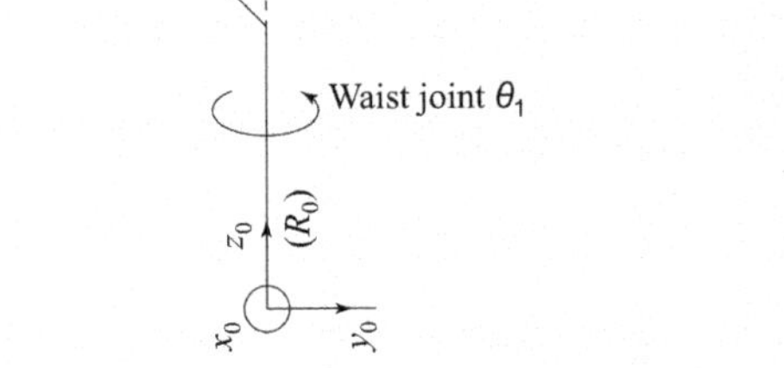

Fig. 3.2 Sketch of the geometry of a 6-axis manipulator

The base bears all the weight of the robot, and its performance has a great impact on the load capacity and motion accuracy of the robot. Generally, the base shall have sufficient strength, rigidity, accuracy, and stability.

The stiffness of robot base refers to the ability of the body to resist deformation under external force. Generally, it is measured by the amount of deformation along the direction of external force under certain external force. The smaller the deformation, the greater the stiffness. For the robot base, the stiffness is more important than the strength. In order to improve the rigidity, the links are usually made of usually made of closed hollow section cast iron or cast steel with high flexural and torsional rigidity. This kind of structure can not only improve stiffness, but also allow for installation of the driving device, drive mechanism, and pipeline inside the link, making the overall structure compact and neat in appearance. The support stiffness of base and the contact stiffness between the base and its bearer also have important effects on robot performance. Generally, appropriate bearer and base-plate connections are adopted to improve the support stiffness. The contact stiffness is guaranteed by ensuring the machining accuracy and surface roughness of the mating surfaces. In the case of rolling rails or rolling bearings, the contact stiffness should also be increased during assembly by the proper application of preload.

The position accuracy of the robot base affects the position accuracy of the wrist. The factors that affect the position accuracy of the base include not only stiffness, but also machining and assembly accuracy, connection methods, motion guidance devices, and positioning methods. The guidance accuracy, rigidity and wear resistance of the guidance device have a great influence on the accuracy and other performance of the robot.

Since the base has a large mass, if its motion speed and load are also high, the rapid change of motion state will cause shock and vibration. This will affect the precision of the position & orientation of end effector and even lead to abnormal operation. In order to prevent shock and vibration, welding robots are usually provided with effective buffering devices to absorb energy. In addition, the moving parts of the base, including drive transmission mechanism, pipeline systems and motion measuring elements, shall be designed in a compact and light-weight structure so as to reduce inertia force and improve the drive accuracy and efficiency.

The base is equipped with a waist joint, which is used to lift, rotate or pitch the arm. The waist joint is the most loaded motion axis and has the greatest influence on the position & orientation and motion accuracy of the end effector. Therefore, it is required to have high design and manufacturing accuracy, especially the support accuracy. There are two support modes commonly used. The first is ordinary bearing structure, as shown in Fig. 3.3. The advantages of this structure are easy installation and adjustment, but it has high waist height. The second support structure is the circular-crossover roller bearing support structure, as shown in Fig. 3.4. The advantages of this structure include high stiffness, high load capacity and easy assembly, but its manufacturing cost is relatively high. Figure 3.5 shows the typical structure of waist joint drive mechanism. Servo motor 1 drives the rotation of the waist joint by means of bevel pinion 2, bevel gear 3, drive shaft 5, small spur gear 4 and big spur gear 6.

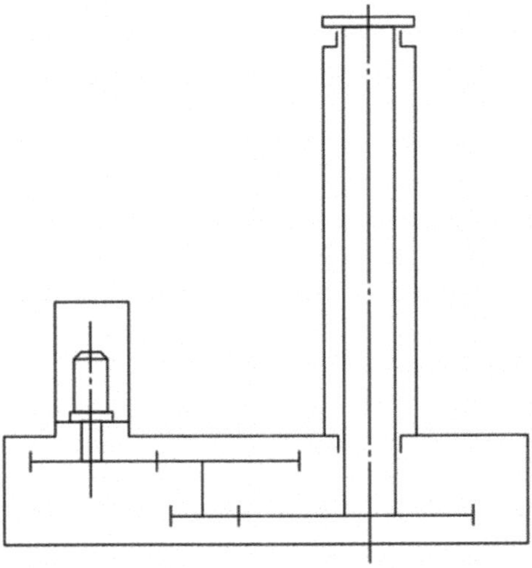

Fig. 3.3 Robot base using ordinary bearings to support waist joint

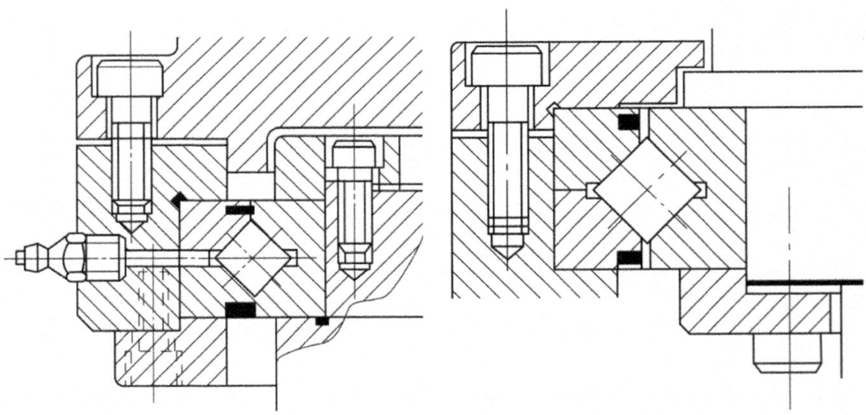

a) Bearing with outer ring rotating b) Bearing with inner ring rotating

Fig. 3.4 Circular crossover roller bearing support structure

3.1.2 Robotic Arm

The operating arm of robot (arm for short) is used to connect the base and wrist and is the main actuating component of robot. Its main function is to support and drive the wrist and end effector. Each joint of the arm is equipped with a corresponding drive and driving mechanism.

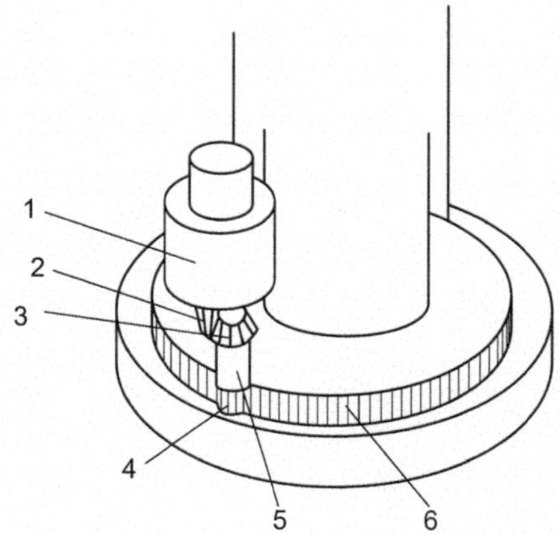

1-Servo motor; 2-Bevel pinion; 3-Bevel gear; 4-small spur gear; 5-Drive shaft; 6-big spur gear

Fig. 3.5 Sketch for structure of waist joint drive mechanism

The arm directly bears static and dynamic loads from the wrist, end effector and workpiece during operation, and it moves frequently and in a complicated state. Therefore, it is subject to complicated forces, both bending force and torsion force. To ensure the precision of movement, the arm shall be made of hollow rod with high rigidity in bending and torsion resistance. In addition, the internal hollow space can accommodate the drive, transmission mechanism and pipelines to make the overall structure compact and neat. The support stiffness of arms and the contact stiffness between the arm and its bearer also have important effects on robot performance. Generally, reasonable bearer structures and base-plate connections are used to improve the support stiffness. The contact stiffness is guaranteed by ensuring the machining accuracy and surface roughness of the mating surfaces.

The machining and assembly accuracy, connection method, motion guide and positioning method of arms also affect their position and motion accuracy. The accuracy, rigidity and wear resistance of arm guide devices have a great influence on the accuracy and other performance of the robot.

Rapid changes in the movement of the arm can also cause shock and vibration. This may affect the precision of the position & orientation of the end effector and, in severe cases, may lead to abnormal operation. In order to prevent shock and vibration, welding robots are usually provided with effective buffering devices to absorb energy. In addition, the moving parts of arm, including drive devices, transmission mechanism, pipeline system and motion measuring elements, shall be of compact and lightweight structure so as to reduce inertia force and improve drive accuracy and efficiency. In addition, the axes of joints shall be parallel as far as possible, and the axes of joints that are perpendicular to each other shall meet at a point as far as possible.

The arms consist of both upper and fore arms and are usually made of high-strength aluminum alloy, with thin-walled and enclosed section. Gear drive is used to ensure high drive rigidity. The transmission mechanism is mounted inside the thin-walled enclosed arm. Figure 3.6 shows the typical structure of the upper arm transmission mechanism. Figure 3.7 shows the typical structure of the fore arm transmission mechanism.

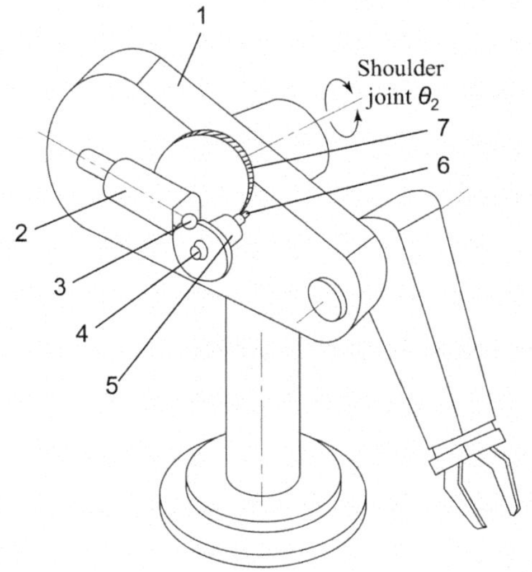

1-Upper arm 2-Upper arm motor 3-Bevel pinion 4-Bevel gear 5-Eccentric sleeve 6-Pinion 7-Big gear

Fig. 3.6 Structure of upper arm transmission mechanism

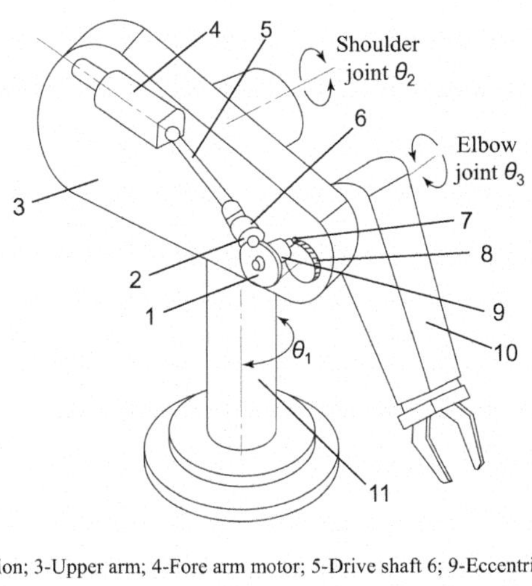

1-Bevel gear; 2-Bevel pinion; 3-Upper arm; 4-Fore arm motor; 5-Drive shaft 6; 9-Eccentric sleeve; 7-Pinion 8-Big gear; 10-Fore arm 11-Base

Fig. 3.7 Structure of fore arm transmission mechanism

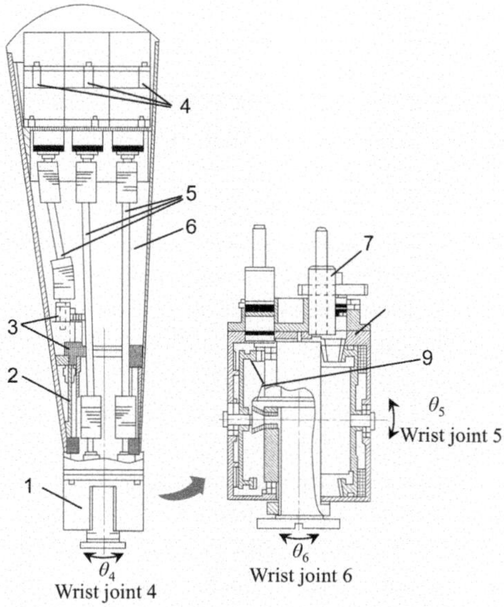

1-Wrist; 2-Bracket of joint 4; 3-Gear of joint 4; 4-Servo motor; 5-Drive shaft; 6-Fore arm; 7, 8-Gear of joint 5; 9-Gear of joint 6

Fig. 3.8 Wrist with 3 DOFs

3.1.3 Wrist and Its Joint Structures

The wrist is the end of the robot body and is used to connect the end effector. Its main function is to determine the position and orientation of the end effector based on the arm motion. The wrist section generally has 2–3 degrees of freedom. The wrist section is designed with compact structure, good rigidity and low weight. Each motion axis is equipped with an independent drive motor and drive systems. See Fig. 3.8 for the typical structure. The three drive motors and drive systems of the wrist are installed at the rear of the fore arm of the robot, which not only reduces the size of the wrist, but also allows the weight of the motor to be used as counterweight for a certain balance. Three motors use flexible couplings and transmission shafts to drive the transmission gears of the wrist joints, and after being decelerated by the transmission gear, the joint shaft is driven to realize joint movement. The movement of joint 4 is wrist rotation, the movement of joint 5 is wrist swing, the movement of joint 6 is a wrist twist.

3.2 Robot Joints and Transmission Mechanism

The motion of the end effector of a robot is synthesized by the motions of all joints of the robot. Each joint has a driving mechanism.

3.2.1 Joints

Joints are motion pairs located at the junctions of links. According to the motion mode, robot joints are classified into rotary joints and prismatic joints. The joint of welding robot is generally a rotary joint. It serves as a connection between the base and arm, the upper and fore arms, and fore arm and wrist, and transfers motion between various parts. The rotary joint consists of rotor shaft, bearing and transmission mechanism. According to the arrangement of driving mechanism and rotor shaft, joints can be classified into coaxial joints, orthogonal joints, external joints and internal joints, as shown in Fig. 3.9. The rotary axis of coaxial joint is coincident with the rotating axis of transmission mechanism, which has the advantage of high positioning accuracy, but requires small speed reducer and increased arm rigidity. This is the common joint type of multi-joint robots. The rotary shaft of orthogonal joints is perpendicular to the rotating shaft of the driving mechanism. The reduction mechanism of these joints can be installed on the base and driven through gears or chains. It is suitable for robots with compact arm structure. The transmission mechanism of the externally mounted rotary joint is installed outside the joint and is suitable for heavy duty robots. The drive motor and reduction mechanism of the internally mounted rotary joint are mounted inside the joint.

There are two types of rotary joints, i.e. cylindrical joints and spherical joints, depending on the joint structure. Spherical joints are commonly used by welding robots. Ball bearings in spherical joints can accommodate both radial and axial loads, and have the advantages of small friction factors and low requirements on bearings and bearing seats. Figure 3.10 shows several typical spherical bearings. Both plain centripetal ball bearings and centripetal thrust ball bearings, which have two contacts between each ball and raceway, must be used in pairs. The raceway of four-point contact ball bearing is a pointed-arch-shaped semicircle with four contact points between the ball and raceway. Preload can be achieved by controlling the magnitude of interference between the two raceways. It has such advantages as compact structure, small clearance, high load carrying capacity and rigidity, but high cost.

3.2.2 Drive Unit

(1) Drive classification

According to the type of power source, robot drives can be classified into hydraulic drive, pneumatic drive and electric drive. The joints can be driven directly, but in most cases indirectly through mechanical transmission mechanisms such as timing belts, chains, wheels, harmonic gears and RV reducers, etc.

The hydraulic drive is powered by an motor-driven high pressure fluid pump, such as plunger pump, and vane pump, etc. The advantages include high power, no need for reduction device, compact structure, high rigidity, quick response and high precision,

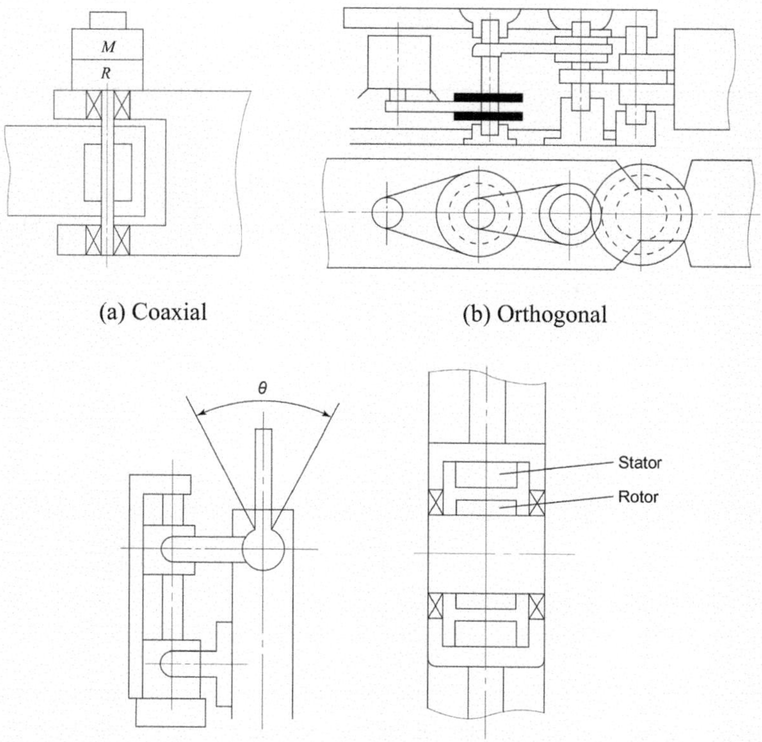

(a) Coaxial (b) Orthogonal

(c) Externally mounted (d) Internally mounted

Fig. 3.9 Types of rotary joints

etc.; while, the disadvantages include the possibility of hydraulic oil leakage, high maintenance cost and samll working temperature range (30–80 °C), etc. They are generally used for large-diameter heavy-duty robots, and not for welding robots.

The pneumatic drive is powered by the cylinder, which has the advantages ofsimple structure, easy usage, low cost, cleanliness and quick response, etc. Disadvantages include low power, poor rigidity, low precision, being difficult to control speed, and high noise, etc. This driving mode is mostly used in point-to-point control robots with low precision requirement or in some situations where electric driving devices cannot be used for safety reasons, such as handling robots and stamping robots, but rarely used in welding robots.

Electric drive uses the force or moment generated by various electric motors to drive the robot joint, achieving control of the position, speed and acceleration of the end effector. Welding robots usually use this drive mode, which has the advantages of quick start-up speed, wide adjustment range, high overload capability, high precision and low maintenance cost, etc. However, it cannot be driven directly, and a gear is required to slow down the speed.

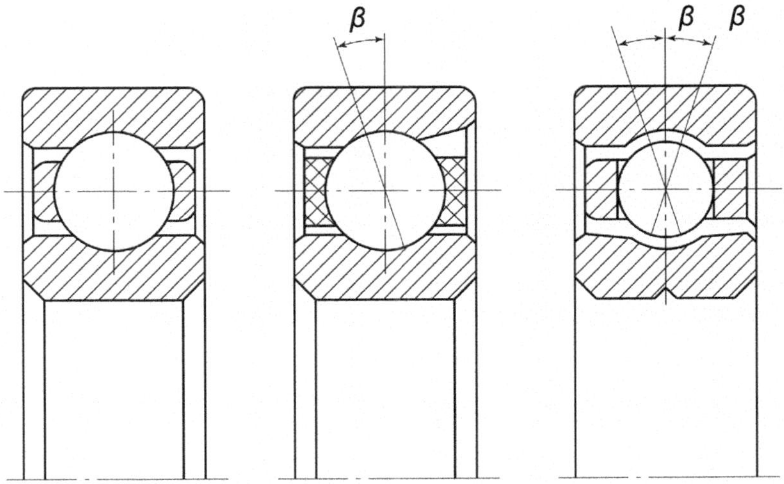

(a) Centripetal ball bearing (b) Radial-thrust ball bearing (c) Four-point contact ball bearing

Fig. 3.10 Basic patterns of joint ball bearings

(2) Drive motor

The driving elements used for electric drives include DC servo motor, AC servo motor or stepper motor. Table 3.1 compares the performance characteristics of several different motors.

The joint drive motor is generally required to have a large power-to-mass ratio, torque-inertia ratio, large starting torque, large torque, low inertia, high response speed, wide speed regulation range and large short-term overload capacity. Therefore, servo motors are mostly used in the welding robots. The stepped motor drive system is mostly used on small simple robots with low precision and speed requirements.

(3) Working Principle of Servo Motor and Drive

1) Working principles of servo motor

The servomotor reduce the rotation error through comparing the received and emitted electrical pulses. It rotates by a certain angle when receiving one pulse. At the same time, a pulse is generated for every rotation of this angle. In this way, the control system detects and adjusts the error by comparing the number of pulses sent and the number of pulses received, forming a closed-loop control and realizing more accurate positioning, with an accuracy up to 0.001 mm. Servo motors are divided into two types, i.e., DC servo motors and AC servo motors.

The structure and working principle of DC servo motors are similar to that of ordinary small-capacity separately-excited DC motors. There are mainly two differences: One is that the armature current of the DC servo motor is very small, making

Table 3.1 Performance comparison of common electric drive components

Motor type	Stepper motor	DC servo motor	AC servo motor
Basic principle	Electrical pulses are used to control the movement. One pulse corresponds to a certain step angle, the displacement is controlled by the number of pulses, and the movement speed is controlled by the pulse frequency	Pulses are used to control the displacement. When one pulse is received, the control system will rotate the corresponding angle and send one pulse. In this way, the control system can accurately control the rotation angle by comparing the number of pulses sent and received, that is, the position and speed are accurately determined through feedback	The principle is similar to the DC servo motor. The difference is that the torque is less pulsating with sine wave control
Structural features	Simple structure and small volume	Simple structure, large volume and weight	Smaller volume and weight than DC servo motor; without brushes and commutators, reliable operation and low maintenance requirements
Overload capability	None	Large	Maximum
Output speed range	Small	Large	Maximum
Speed response time	Long (it takes 200–400 ms to accelerate from standstill to rated speed)	Short (it takes only a few ms to accelerate from standstill to its rated speed)	Minimum
Torque-frequency characteristics	The output torque decreases with the increase of speed, and drops sharply when the speed is above 600 r/min. The maximum operating speed is generally below 600 r/min	Rated torque output at speeds below 2000 r/min	Basically constant torque output, capable of delivering rated torque at 3000 r/min
Low frequency characteristics	Low frequency vibration is likely to occur	With resonance suppression function; stable operation, and very stable even at low frequency	

(continued)

Table 3.1 (continued)

Motor type	Stepper motor	DC servo motor	AC servo motor
Control accuracy	Depending on the number of phases and beats. The higher the number of phases and beats, the higher the accuracy. Generally, the step angle is 1.8° and 0.9°	Depending on the internal encoder scale, the finer the scale, the higher the accuracy. For example, for a servo motor with a 17-bit encoder, the motor rotates once for every 131072 pulses received by the driver, resulting in a pulse equivalent of 360°/131072=0.0027466°, which is only 1/655 of the pulse equivalent of a stepper motor with a step angle of 1.8°	
Performance	Open loop control, too high starting frequency, or too high load will lead to loss of synchronization or rotor locking; it is also susceptible to overshoot at standstill	Closed loop control, not prone to loss of synchronization, stalling and overshoot	

the commutation easy and eliminating the need for commutation poles. The other is that the DC servo motor has a slender rotor and small air gap, which leads to a high armature resistance and unsaturated magnetic circuit. There are two types of DC servo motors: Brushed and brushless. Brushed motors are characterized by simple structure, low cost and large starting torque. However, the brushed motors are mainly used in unimportant situations due to the small power-to-volume ratio, need for maintenance, low speed, high thermal inertia and electromagnetic interference. Nonetheless, brushless motors have the advantages of small size, larger torque, small inertia, quick response and smooth and stable rotation, but a relatively complicated control circuit. Before the mid-1980s, DC servo motors were widely used in robots, but they are currently rarely used in welding robots.

The AC servo motors are further classified into permanent magnet synchronous type and asynchronous type. Table 3.2 compares the characteristics of the two types of AC servo motors. Permanent magnet synchronous motor can operate stably with excellent low-speed servo performance, which is the most commonly used type for welding robots at present.

The permanent magnet synchronous AC servo motor mainly consists of a stator and a rotor, as shown in Fig. 3.11. The stator core is composed of stacked silicon steel sheets. The slot of the stator is equipped with an excitation winding and a control winding, with the two winding having a 90° directional difference in space. The former is powered by an AC excitation power source while the latter is used for accessing control voltage signals. The rotor is a permanent magnetic pole made of a high-coercivity rare-earth magnetic materials. The non-load end cover of the servomotor is provided with an optical encoder, which is used to output feedback pulses. The servomotor driver compares the number of pulses received with the number of pulses emitted, based on which it adjusts the rotation angle of the rotor and controls the displacement accuracy. When no control voltage is applied to the

Table 3.2 Comparison between permanent magnet synchronous AC servo motor and asynchronous AC servo motor

	Permanent magnet synchronous AC servo motor	Asynchronous AC servo motor
Motor structure	Relatively simple	Simple
Maximum torque limit	Demagnetization of permanent magnet	Magnetic circuit saturation
Heat generation	Low	High
Electrical power conversion rate	High	Low
Response speed	Fast	Relatively fast
Rotational inertia	Small	Large
Speed range	Large	Small
Braking	Easy	Complex
Reliability	Good	Good
Environmental adaptability	Good	Good

stator control winding, the air gap in the stator only contains the pulsating magnetic field generated by the excitation winding and the rotor remains stationary. When a control voltage is applied to the control winding and the control winding current is out-of-phase with the field current, a rotating magnetic field is produced in the stator air gap, driving the rotor to rotate synchronously in the direction of the rotating magnetic field. The motor speed increases with control voltage under certain loads; the rotor stops immediately upon loss of control voltage; and when the control voltage is inverted, the servo motor will reverse.

2) Working principle of permanent magnet synchronous AC motor drive

AC servo drive, also known as AC servo controller, consists of a servo control unit, a power drive unit, various interfaces and a feedback system, as shown in Fig. 3.12. The new type AC servo drive uses a digital signal processor (DSP) as the control center, an intelligent power module (IPM) as the power drive unit, and an incremental photoelectric encoder as the speed measuring and position sensor. The triple-closed-loop (torque (current), speed and position loop) control mode is adopted to ensure high precision positioning and stability of the servo motor, as shown in Fig. 3.13. The adoption of DSP facilitates the implementation of vector-based closed-loop control algorithm of current, speed and position, and the digitalization, networking and intelligence of robots. AC servo drive integrates a drive circuit, a fault detection circuit and several protection circuits (including over-voltage, over-current, over-temperature, and under-voltage, etc.), and a soft start circuit, etc. In addition to driving the servo motor, AC servo drive can provide various protections and reduce the shock during startup. The drive circuit first rectifies the incoming three-phase

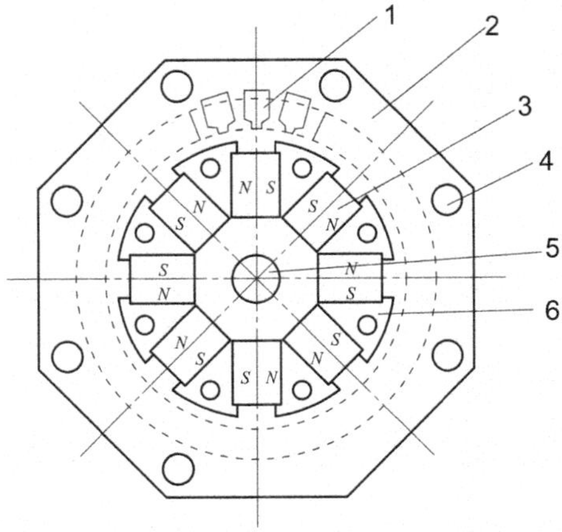

1- Stator winding (three-phase); 2- Stator core; 3- Permanent magnet (rotor); 4- Axial venting holes; 5- Rotor shaft; 6- Soft magnetic pole shoe

Fig. 3.11 Structure of permanent magnet synchronous AC servo motor

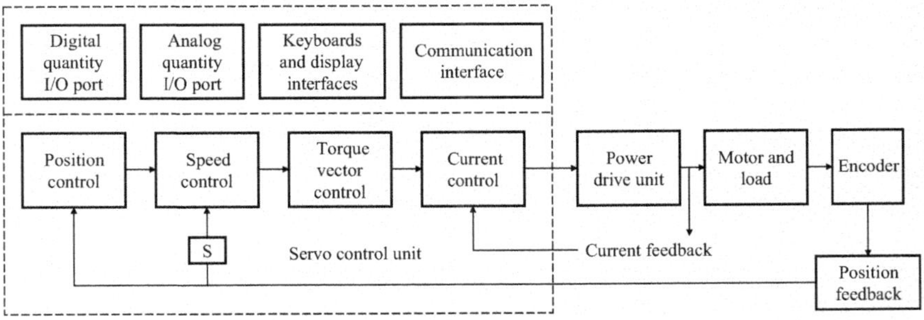

Fig. 3.12 Structure of AC servo controller

mains supply using a three-phase full bridge rectifier circuit, and then drives the three-phase permanent magnet synchronous AC servo motor using a three-phase sinusoidal pulse width modulation (PWM) voltage type inverter frequency conversion.

3.2.3 Transmission Device

The function of the transmission device is to transmit power from the drive to the actuator. Common transmission mechanisms include linear drive and rotary drive. The gearing usually has a fixed gear ratio.

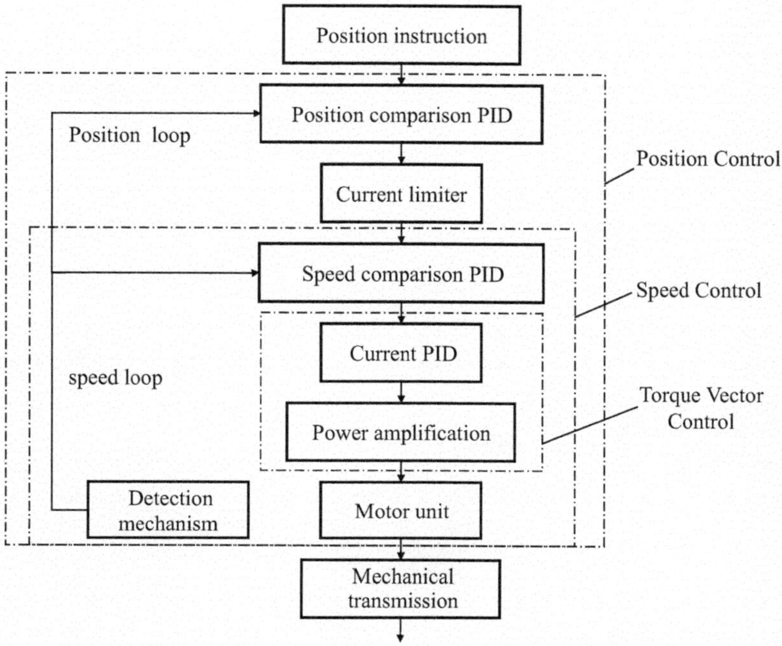

Fig. 3.13 Closed loop control of AC servo controller

(1) Linear transmission device

Linear transmission device may be used for *X*, *Y* and *Z* direction drive of robot with Cartesian coordinates, radial drive and vertical lift drive with cylindrical coordinates, and radial expansion drive with spherical coordinates.

There are several types of linear transmission devices, such as rack and pinion transmission devices and ball-screw transmission devices. The rack and pinion transmission devices are usually fixed, and the rotating motion of the gear is converted into a linear motion of the support plate, as shown in Fig. 3.14. The advantages of this device include simple structure, high transmission dynamics and power, large transmission ratio, high accuracy, stability and reliability. However, high installation accuracy is required and the return difference is relatively high. The ball-screw pair transmission device consists of a screw and a nut. Embedded in the spiral grooves of the screw and the nut, the balls are continuously recirculated through the guide groove in the nut, as shown in Fig. 3.15. Its advantages include small friction, high drive efficiency and high precision, while the disadvantages include high manufacturing cost and complex structure.

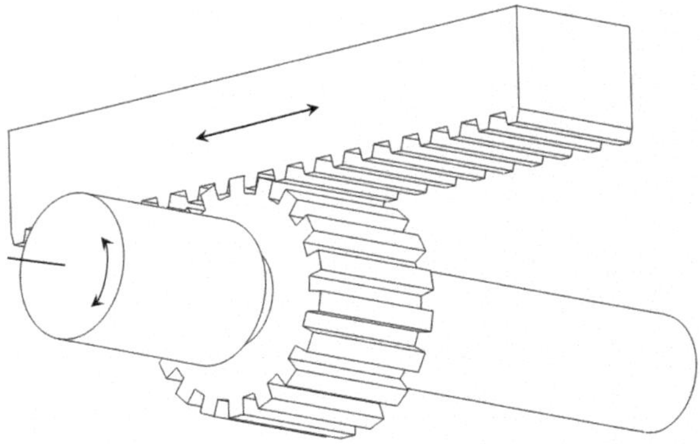

Fig. 3.14 Rack and pinion transmission device

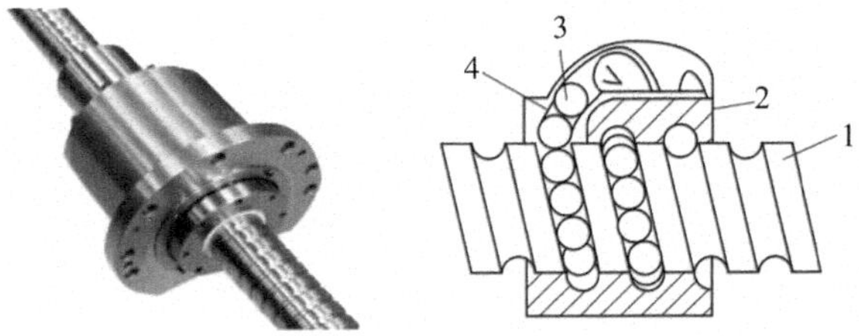

1-Screw 2-Nut 3-Ball 4-Guide groove

Fig. 3.15 Ball-screw pair transmission device

(2) Rotary transmission device

The rotary transmission devices commonly used by robots include gear chain, timing belt, harmonic gear, and RV cycloidal pin-wheel reducer.

1) Gear chain transmission device

The movement of the driving sprocket is transmitted through the chain to the driven sprocket as shown in Fig. 3.16. This transmission mode has the advantages such as large energy transmission, high overload capacity and accurate average transmission rate, but has poor stability, easy wear of transmission elements and easy slippage. Therefore, it is rarely used in welding robots.

2) Timing belt transmission device

Timing belt transmission is a meshing type belt transmission that uses the meshing between the transverse teeth equally spaced on the inner surface of the belt and the corresponding slots on the pulley to transmit motion, as shown in Fig. 3.17. This kind of transmission combines the advantages of the friction belt transmission and chain transmission, such as non-sliding, stable transmission, good buffering effect, low noise, low cost, high repeat positioning precision, large transmission ratio and high transmission speed, etc. Therefore, it is widely used in the welding robots. However, it requires high installation accuracy and has a certain degree of elastic deformation.

Fig. 3.16 Gear chain transmission

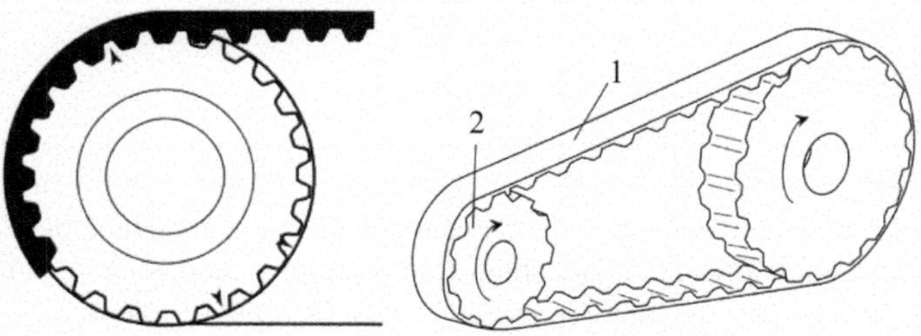

1- Timing wheel ; 2- Timing chain

Fig. 3.17 Timing belt transmission device

3) Harmonic gear reducers

The harmonic gear consists of a rigid inner gear (rigid wheel for short), a harmonic generator, and a flexible outer gear (flexible wheel). The harmonic generator is installed inside the flexible wheel and consists of an elliptical cam and a flexible rolling bearing on the outer ring, as shown in Fig. 3.18. Generally, the rigid wheel is fixed, and the harmonic generator serves as the active part to drive the flexible wheel to rotate. During the rotation, the flexible wheel may experience certain radial elastic deformation. Since the number of inner teeth of the rigid wheel is more than that of the outer teeth of the flexible wheel, the inner teeth of the rigid wheel in the direction of the long axis of cam will fully engage with the outer teeth of the flexible wheel when the harmonic generator rotates. However, in the short-axis direction, the outer and inner teeth are completely disengaged. In this way, the outer teeth of the flexible wheel can repeatedly enter, mesh and exit out of the inner teeth of the rigid wheel. The radial displacement at any point on the flexible wheel periphery varies according to the sine wave pattern. Therefore, this type of transmission is called harmonic transmission, and this kind of gear combination is named as harmonic gear reducer. The transmission ratio is equal to the number of teeth on the flexible wheel divided by the difference between the number of teeth on the rigid wheel and the flexible wheel.

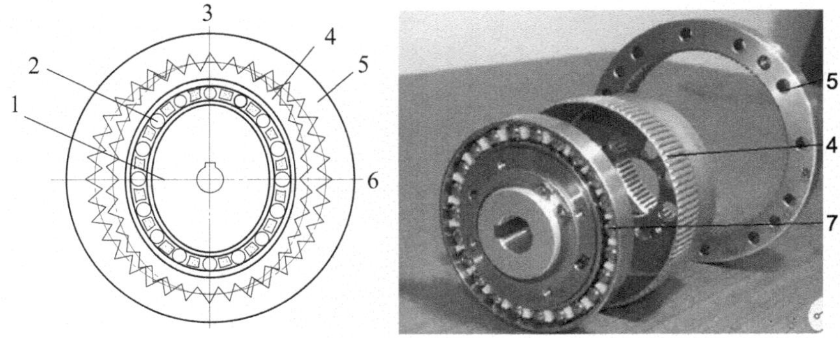

1-Elliptical cam ; 2- Flexible bearing 3- Long axis of the cam; 4- Flexible outer gear. 5- Rigid inner gear; 6- short axis of the cam; 7- Harmonic generator

Fig. 3.18 Harmonic gear

The advantages of this transmission mode include large transmission ratio (60–320 per single stage), small volume, stable transmission, low noise, large load-carrying capacity, high efficiency (70–90%), high transmission precision (45 times that of ordinary gear transmission), small return difference (less than $3'$), and easy sealing, as well as convenient repair and maintenance. The disadvantage is that the intermediate output cannot be obtained, and the flexible wheel has a low rigidity. Harmonic gear reducer is widely used in industrial robots, mainly at the wrist joints of robots.

4) RV cycloidal pin-gear speed reducer

The RV cycloidal pin-gear speed reducer is a two-stage reduction mechanism. The Stage I reduction mechanism is an involute cylindrical gear planetary reduction mechanism, as shown in Fig. 3.19. The input gear transmits the rotation of the servomotor to the spur gear. The reduction ratio is the tooth ratio of the input gear to the spur gear. The Stage II speed reducer is a cycloidal pinwheel planetary reduction mechanism, as shown in Fig. 3.20. The spur gear connected directly to the crankshaft is the input of the Stage II reduction mechanism. Two RV gears are fitted in the eccentric part of the crankshaft through the needle bearing. On the inner wall of the casing, there is a needle gear with one more tooth than the RV gear as well as equally spaced slots, as shown in Fig. 3.21. Figure 3.22 shows a cross-sectional view of a RV cycloidal pinwheel reducer. For each crankshaft rotation, the RV gear makes an eccentric motion while in contact with the needle notch, causing the RV gear to rotate by one tooth along the direction opposite to the rotation direction of the crankshaft.

This speed reducer has a large number of teeth that mesh simultaneously, making it compact in structure, small in size, high in torque, good in rigidity, wide in transmission ratio range, high in accuracy, small in return clearance, small in inertia, and strong in resistance to overload impact loads. In addition, due to the small backlash and small inertia, it has good acceleration performance, stable operation and good positioning accuracy. This speed reducer is now widely used on various joints of robots.

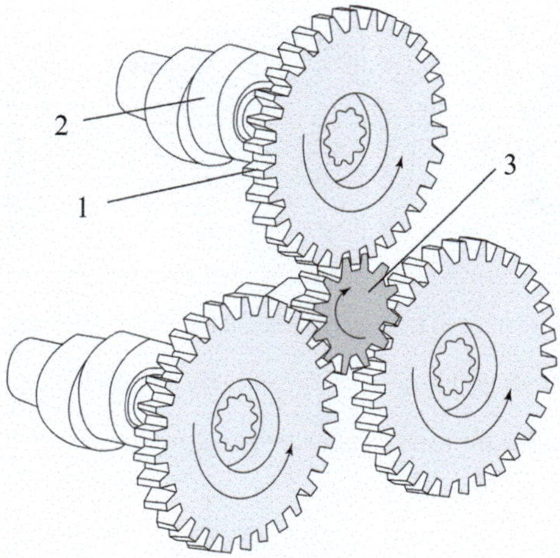

1- Spur Gear; 2- Crank Shaft; 3- Input Gear

Fig. 3.19 Cylindrical gear planetary reducer for RV cycloidal pinwheel planetary reducer

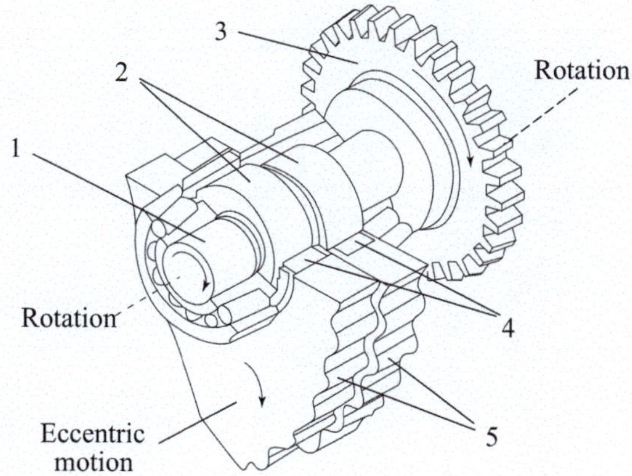

1- Crank Shaft; 2- Eccentric Shaft; 3- Spur Gear; 4- Needle Bearing; 5- RV Gear

Fig. 3.20 Cycloidal pinwheel planetary reducer

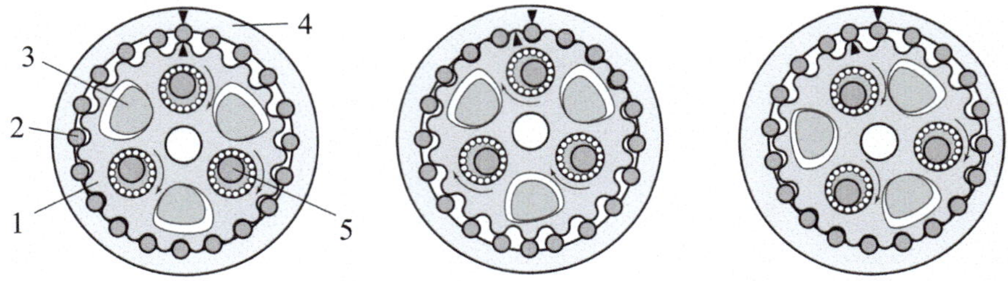

(a) Crankshaft rotated by 0° (b) Crankshaft rotated by 180° (c) Crankshaft rotated by 360°

1- RV Gear; 2- neelde gear; 3 - Drive shaft; 4- Housing; 5- Crankshaft

Fig. 3.21 The fitting between the teeth and slots of RV gear

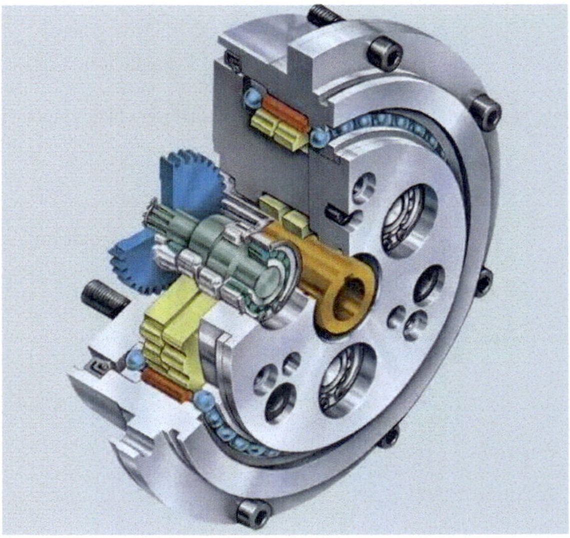

Fig. 3.22 RV reducer

3.3 Robotic Control Systems

3.3.1 The Role of Robotic Control Systems

If the robot body is compared to the arm of man, the robot control system is equivalent to the brain of man. It receives instructions and information from outside, processes and judges these instructions and information to form control instructions, and sends the instructions to the robot body to control the coordinated motions of various joints of the robot to achieve the expected motion position, orientation, path and operation sequence, etc. In addition, robotic control systems provide the following functions:

(a) Memory function: The robotic control system is equipped with a memory, which can store the planned task and the teaching program. During work, the memory also store the movement information of joints, position and orientation information as well as the operation information of control system.
(b) The teaching and program executing functions: The robot teaching, teaching program storing, and program executing are all implemented under the control of the robot control system.
(c) Communication with peripheral devices: For example, it communicate with the welding power supply to adjust the parameters in real time.
(d) Sensor interface: These interfaces are used to connect various sensors, such as position sensor, visual sensor, tactile sensor and force sensor, etc. The internal and external information collected by these sensors is sent, through the interface, to the control system.
(e) Position servo function: The motion control, speed and acceleration control are realized via position servo function. When using external axes, multi-axis linkage also requires servo control.
(f) Fault diagnosis safety protection function: The control system of the robot always monitors the operation status of the robot and stops the robot in case of fault to protect the robot.

3.3.2 Robotic Control System Software

The robotic control system consists of computer hardware and software. The control system software includes controller system software, robot-specific language compilation software, robotic kinematics software, robotic dynamics software, robot control software, robot self-diagnosis software and self-protection function software.

1) Controller system software

Similar to a computer operating system, this software mainly realizes hardware abstraction description, bottom-level driver management, execution of common functions, inter-program message transfer, and program package management, etc. In addition, it also provides some tools and functions for programming and program executing. Robots manufactured by different manufacturers have different controller system software.

2) Robot-specific language compilation software

The robot-specific language of robot is also called robot programming language and communication tool between robot and user. The user controls the robot to perform the task by programming with the robot-specific language. There is no a universal robot programming language, and each robot manufacturer has its own language.

There are mainly two types of robot-specific language:

(a) Robot-oriented programming language. Its main feature is the use of statements to describe the action sequence of the robot with each statement roughly equivalent to an action of the robot. The entire program controls the robot to complete a whole task. These languages include robot-specific language VAL, BASIC-based language, AR-BASIC and Robot-BASIC, as well as the computer general-purpose language AML added with the robot subroutine library, etc.
(b) Task oriented robot programming language. This kind of language allows the user to issue direct commands to control the robot to accomplish a specific task without specifying the action detail that the robot needs to make. This language is not yet mature.

At present, all robot manufacturers provide their own special programming languages for their robots. For example, ABB robot uses RAPID programming language, Yaskawa robot uses INFORM, KUKA robot uses KRL programming language, and FANUC robot uses KAREL programming language. These robotic languages are specific to the particular robot. Some welding robots are also equipped with off-line programming simulation software, which is used for off-line simulation. And the results can be downloaded directly to the robot controller to run after completion of the simulation.

3) Robotic kinetics software

In the process of teaching or operation, this software can solve the forward and inverse kinematics problems as required, control the coordinated movement of various joints, and synthesize the motion orientation of the end effector through the movement of various joints. For example, during the teaching process, the inverse kinematics problem is calculated based on the teach position and orientation to determine the angle vector of each joint.

4) Robotic dynamics software

It is used to solve the forward and inverse dynamics problems and control motion accuracy. For example, based on the path planning results (position, speed and acceleration of joints) of various robot, the driving moment can be calculated using the robot dynamic model to realize servo control of various joints.

5) Robotic control software

The software provides coordinated the control of the servo systems of various joints to achieve the desired position or motion. By comprehensively analyzing the given information and feedback information and according to the control strategy, control signals are generated to effectively control the individual parts.

6) Robot self-diagnosis software

This software can perform startup test in-process test, and prepare and display the test reports based on the test results.

7) Self-protection software

The software can make specific response based on the detected dangerous conditions, send out control signals, and stop the equipment in time to realize self-protection.

3.3.3 *Robot Control System Hardware*

It consists of a controller, an actuator, a controlled object and a detecting unit, as shown in Fig. 3.23. The controller receives the input signal and the feedback signal from the detecting unit, and generates a control signal through comparison and judgment, and sends it to the actuator for execution. The actuator receives the control signal from the controller and forms a signal that can directly manipulate the controlled object. The controlled objects are the driving mechanisms of various joints of the robot. The detecting unit is used to detect the controlled variable and convert the detected controlled variable into a feedback signal and send it back to the controller so that the controller can make adjustment decision.

Robot control modes include PTP control, path control, program control, adaptive control and artificial intelligent control. The PTP control only controls the position and orientation of the starting and ending position of the end effector of, and adopted by most spot welding robots. The path control mode requires the robot to move along the taught path. Most of the arc welding robot belongs to this control mode. In the program control mode, each degree of freedom of robot is controlled according to certain patterns, so that the robot can achieve the desired space path. The self-adaptive control robot adjusts the parameters of non-linear model based on the detection and judgment of the status of manipulator and servo error until the error disappears. It has high control quality and certain intelligent performance. The artificial intelligence control is relatively complicated. It detects the surrounding status information at any time during the motion of robot, and makes real-time judgment and decision based on certain intelligent algorithm.

The controller is the core of the control system. According to the handling method of the control algorithm, it can be classified into serial and parallel structures.

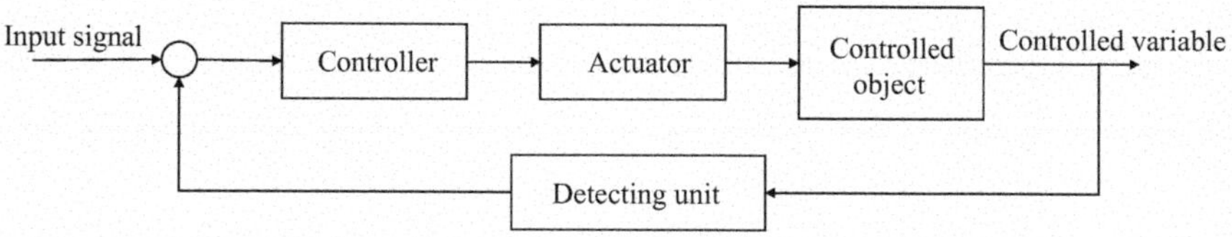

Fig. 3.23 Hardware structure of industrial robot control system

1) **Serial structural controller**

A serial structure controller is actually a controller in which the robot control algorithm is processed by a serial computer, and it can be classified into three types according to the number of CPUs used: Single CPU controller, dual-CPU controller and multiple-CPU controller.

(a) Single CPU controller uses one CPU to perform all control functions and adopts centralized control mode. Since many mathematical calculations are involved in the control process of robots, it is difficult for this type of controller to meet the production requirements due to its low control speed. Therefore, this type of controller is basically not used in industrial robots.
(b) The dual-CPU controller uses two CPUs to control in host-slave mode. The two CPUs share one memory where the exchanged data are stored. The main CPU is responsible for system management, robotic language compilation, man-machine interface management, kinematic calculation and path planning, and regularly transmits the calculation results as the increments of joint motion to the common memory. The slave CPU retrieves the computed data of the main CPU from the common memory and digitally control the positions of the individual joints.
(c) Multi-CPU controller uses multiple CPUs to perform two-level distributed control via host-slave mode, as shown in Fig. 3.24. This is the most commonly used controller in industrial robots today. The host computer is responsible for organizing and managing the whole system. It can be connected to display, keyboard, display box, and floppy disk drive, etc. are all connected to the host CPU to form the host copmuter. The host computer converts the expected tasks into the joint motion through kinematic computation and path planning, and also tests the motion and operation of each part of the robot at any time to deal with unexpected events. The slave computer is composed of several CPUs and is responsible for real-time control. Each CPU controls one joint and uses digital position control. It carries out motion interpolation and servo control of joints according to the dynamic characteristics and current motion of the robot, and drives individual robot joints to perform specified movement and operation. Each CPU of the slave computer communicates with the host computer and the other slave CPUs through the bus. Since each CPU undertakes fixed tasks and communicates quickly and conveniently, the control speed and accuracy of this type of controller are significantly improved.

The serial controller performs various calculations sequentially. However, the dynamic equations of the articulated robot are a set of second-order differential equations for the non-linear strong coupling problems, which are very complex and involve a large amount of computation and time-consuming. As a result, the serial control has poor real-time performance, and it is difficult to ensure high-speed motion accuracy in the case of interference. In actual production, off-line planning and feedforward compensation decoupling method may be used as required to solve the heavy burden of compensation calculation.

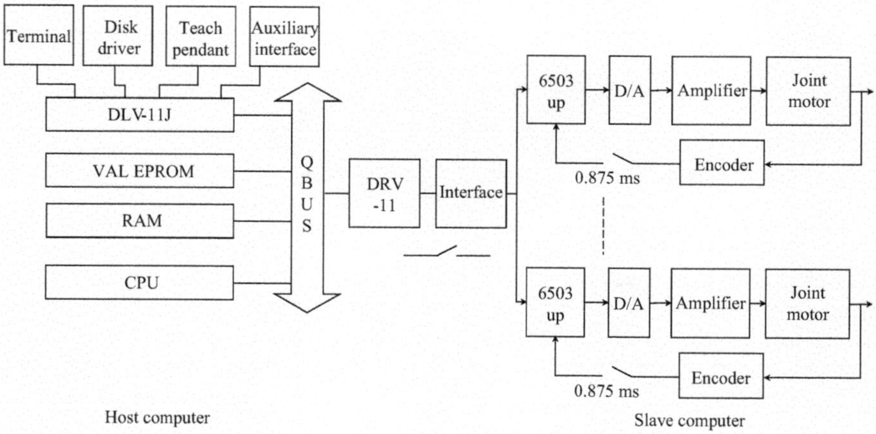

Fig. 3.24 Controller with two-level distributed control

2) **Parallel structure controller**

The parallel structure controller maps the algorithms to the the parallel CPUs for the parallel computation of robotic kinematics and dynamics. There are two implementation methods. The first way is to develop parallel algorithms based on the existing parallel processor structure. The second way is to first develop the parallelism of the algorithm, and then design the parallel processor structure that supports the algorithm to achieve the best parallelism efficiency. At present, parallel structure controllers are not widely used.

3.3.4 Microcomputers for Robotic Controllers

The controller of an industrial robot usually consists of multiple microcomputers. The host computer generally adapts a single-chip microcomputer or embedded microcomputer, while the slave computer may adapt a single-chip microcomputer or embedded computer or STM32 and AVR control series chips.

(1) **Single-chip microcomputer**

Single-chip microcomputer, also known as microcontroller unit (MCU), is a complete microcomputer system integrated on a single chip, as shown in Fig. 3.25. The chip adopts super-large-scale integration technology which integrates the central processing unit (CPU), RAM, ROM, necessary I/O interfaces, interrupt system, A/D converter, and timer/counter, etc..

Like the CPU of a PC, the CPU of a single-chip processor is composed of an arithmetic unit, a controller and a register. The arithmetic unit is responsible for data processing and calculation, the controller is responsible for issuing control signals, and the register is used to temporarily store data and control information. Compared

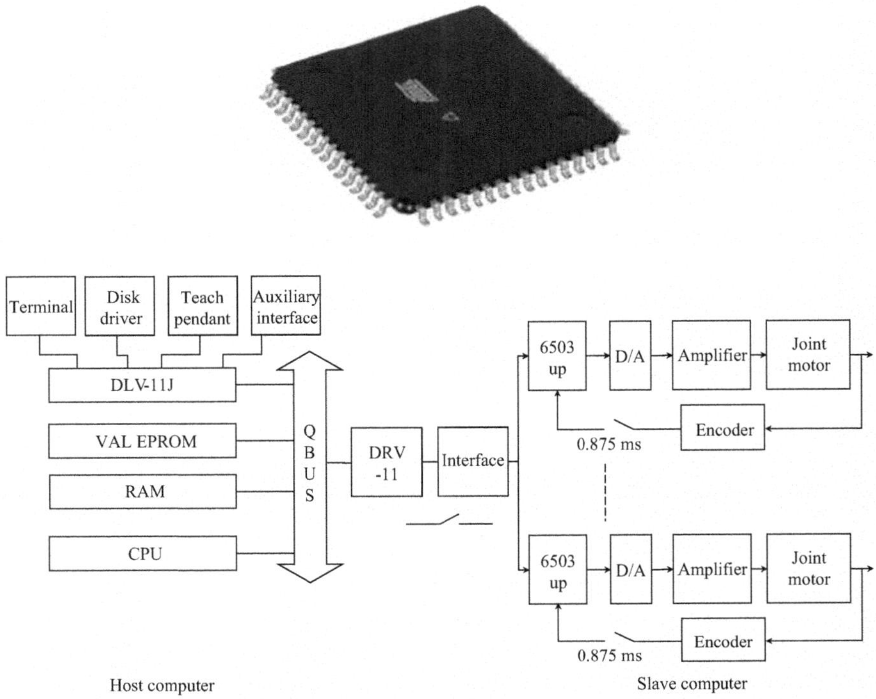

Fig. 3.25 Single-chip microcomputer and its structure block diagram

with the Random Access Memory (RAM) and Read Only Memory (ROM), the registers can exchange date with the arithmetic units and controllers at a higher transmission speed.

The ROM and RAM of the single-chip microcomputer respectively correspond to the hard disk and memory of the PC for data storage. The RAM memory exchanges data with the register. The data to be processed by the CPU is input into the register by the RAM. This operation is called a read operation. The data that has just been processed by the CPU is output into the RAM by the register, which is called a write operation. The data in the RAM can be updated quickly, and the CPU will read and write the data at the specific memory area of the RAM according to the executed program. RAM requires power to maintain the stored data. Once the power is cut off, all data in the RAM will lose. The ROM has the ability to store the data after power failure (loss of power protection), but it has slow read-write speed. The user-written programs are stored in the ROM, and some data that needs to be stored on a long-term basis are also stored on the ROM.

The PC is connected to the mouse and keyboard via USB, to the display screen via HDMI or VGA port (there are other methods, such as Bluetooth, for connection), and to the Internet via network port or wireless network; and the single-chip microcomputer are connected to external equipment through the pins on the periphery of it. The mode in which the data is transmitted simultaneously through several pins is called parallel communication. In this mode, the theoretical transmission rate is high, but the actual transmission rate is is usually much lower, and susceptible to the interference due to the need for all the pins to keep pace during transmission. Multiple single-chip robots usually use serial communications with higher communication efficiency. Although serial communication has a lower theoretical transmission speed due to the use of more data transmission channels (pins), it will be less time-consuming in actual operation because there is no need to maintain the timing relationship of individual pins.

The single-chip processor is provided with a clock circuit, a timer/counter and an interrupt system, which work together with the controller to ensure the orderly operation of the whole single-chip microcomputer. The clock provides a reference signal, which is used to control the individual components of a single-chip microcomputer to work in an orderly manner at a certain pace. It is like that when a group of people are walking in a queue, someone shouts "one-two-one", everyone will walk at the rhythm of the shouting. When executing a task, both the PC and the single-chip microcomputer decompose the task into several basic operations, each of which is accomplished by one instruction. The number of clock cycles required to complete a specific instruction is a constant, and the samller the clock cycle number, the faster the processing speed of the single microcomputer. Therefore, the clock frequency is an important measure of the single-chip microcomputer performance.

When a special situation that requires processing or a task with higher priority arises during the running of a program, the processor will automatically stop the current operation and switch to the new emergent task. After completion of the new task, it will return to the breakpoint in the previous program to continue running. This control method is called interrupt control. The interruption control mode improves the system efficiency and meets the requirements of real-time processing while maintaining reliable operation of the system. Due to the high processing speed of single-chip microcomputer, the duration of interruption process is usually very short, which is non-perceivable. Therefore, it can be considered that the robot performs all operations in real-time control mode.

The single-chip microcomputer completes the assigned task by executing the program, while the program of single-chip microcomputer is composed of a sequence of instructions. Each single-chip processor has its own instruction system, i.e., a collection of all instructions that can be executed. A instruction corresponds to a basic operation. The instruction system is generally different for different types of single-chip microcomputers. However, the robot specific language is usually used for robot programming, instead of the instruction system of single-chip microcomputer. When a completed program is saved, the robotic language compiling program will convert the program into a instruction sequence of the single-chip microcomputer, which is stored in the ROM of the single-chip microcomputer. The memory consists of a number of memory units (one unit consists of 8 logic switches for 8-digit microcomputer), with each unit similar to a room. Each room in the building is assigned a unique room number and, likewise, each memory unit is also assigned a unique address number, which is called memory unit address or memory address.

The CPU accesses a memory unit by using its address. When executing a programme, the CPU takes the instructions that make up the programme out of the memory units one by one in order. Therefore, an address tracking component must be provided to track the address of the memory units where the instruction to be executed is located. This component is called program counter, abbreviated as PC (included in the CPU). When the program is executed, the address of the first memory unit for storing the first instruction will be assigned to the PC (PC points to the first instruction). When the instruction is read out, the address number in the PC will be automatically increased by the number of memory units occupied by this instruction, and after that, PC points to the next instruction.

Single-chip microcomputer is mainly intended for real-time control while PC is mainly intended for data processing. As the real-time control involves less data types and data volume and the processing method is relatively simple, the data processing function of the single-chip microcomputer is weaker than that of PC, and its calculation speed and accuracy is also lower. The organizational structure of the memory in the single-chip microcomputer is relatively simple, the storage chip is directly connected to the bus. The CPU addresses a memory units directly by its physical address for reading and writing, resulting in faster accessing speed.

(2) **ARM embedded computer**

The ARM embedded computers are a type of single-chip microcomputer developed and produced by chip manufacturers using ARM's core architecture. ARM does not produce single-chip microcomputers itself, and it authorizes other chip manufacturers to manufacture single-chip microcomputer on the basis of the ARM core.

ARM means Advanced RISC Machine (ARM). RISC is a 32-bit reduced instruction set. ARM core architecture can be further divided into ARM7, ARM9, ARM11, Cortex-M, Cortex-R and Cortex-A series, etc. Cortex-M, Cortex-R and Cortex-A are the latest architectures. Figure 3.26 shows the structure block diagram of Cortex-M3 embedded computer chips.

The ARM core adopts the Reduced Instruction Set code (RISC). The so-called reduced instruction is a simple and efficient instruction that can be operated within a single cycle at a high clock frequency. Simple instruction means that the hardware used to execute the instruction in the processor is also simple. Compared with the Complex Instruction Set Computer (CISC), the RISC instruction sets and associated decoding mechanisms are much simpler. When using this instruction set, the complex tasks can be accomplished by calling more instructions through software design. As software is more flexible than hardware, microcomputers with this architecture have such advantages as small size, low power consumption, low cost and good control performance. It supports both 16-bit and 32-bit instruction sets and is compatible with 8-bit instruction devices. The structure of the RISC is characterized by fixed instruction length, uniform format of instructions, large quantities of general purpose registers and simple addressing method. The processor only processes the data in the registers and does not directly access the memory.

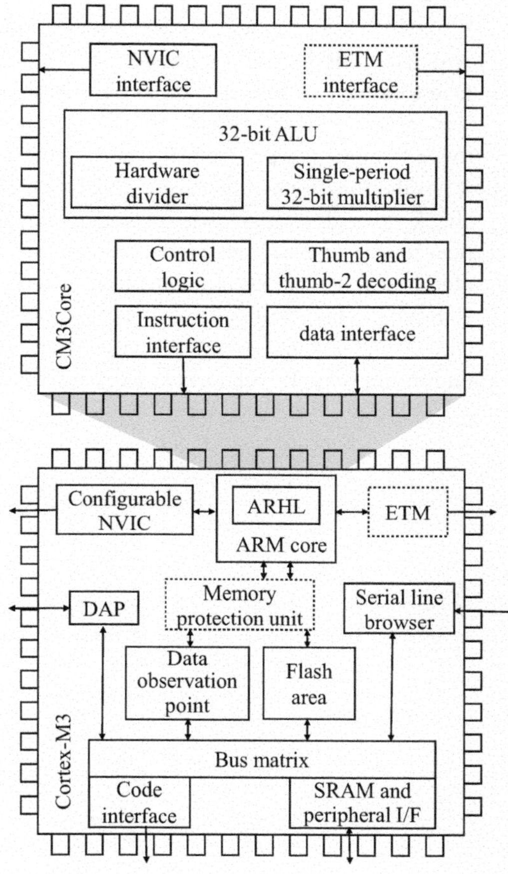

Fig. 3.26 Structure block diagram of Cortex-M3 embedded computer chip

(3) **ARV single-chip microcomputer**

The ARV single-chip microcomputer is a high-speed 8-bit single-chip microcomputer based on the RISC's reduced instruction set. Its ruduced instruction uses the word length as the unit of instruction length, which means that the operand and opcode are arranged in one word. Most instructions are single-cycle instructions and can be pre-retrieved for flow process, resulting in extremely fast execution and high reliability. The features are as follows:

(a) Simple hardware and software structure, which ensures quick execution, high reliability at low software/hardware cost, a well as high cost-performance ratio.
(b) Embedded flash program memory, which is easy to erase and write, making it convenient for users to coding and debugging.
(c) Embedded EEPROM for long-term storage of critical data to avoid loss due to power failure.

(d) Embedded high-capacity RAM which is extendable, enabling the programming with using high-level languages.
(e) Due to the short service interruption duration and strong serial port functions (provided with high-speed synchronous/asynchronous serial port TWI, SPI), the communication speed is fast and multi-machine communication is possible.
(f) It has the functions of hardware generation check code, hardware detection and check error detection, two-stage receiving buffer, automatic adjustment and positioning of baud rate (at the time of receiving), and data frame shielding, etc.. It also has high communication reliability.
(g) Equipped with multiple reset functions (automatic power on/off reset, external reset, watchdog reset, BOD reset), watchdog circuit, and power failure detection circuit improve the reliability of the embedded system.

Chapter 4
Sensing Technology of Welding Robots

Robots use sensing systems to achieve precise motion control and sense changes in external conditions. The robot must ensure the smooth movement of the end effector and the precise movement path, speed and acceleration etc. The motion of welding robots is synthesized by the motions of individual joints (individual axes), therefore, the servo system of each joint adopts the three-loop control mode of position, speed and acceleration. This closed-loop control requires the use of sensors for measuring position, velocity and acceleration. Therefore, high-precision angular displacement sensors and speed sensors are installed in the manipulator drives. In addition, robots need to collect a large amount of information about the surrounding environment just like other robots, so that they can work efficiently and with high quality. For example, a robot's manipulator must keep clear of various obstacles during its movement in space and approach the working object at a certain speed, which requires the robots to make identification and decision. Another example is that arc welding robot is required to move along the seam on the weldment. If the robot is not capable of sensing, error may occur in its motion path, which will affect the welding quality. Therefore, the arc welding robot shall also be provided with a sensing system, such as an automatic seam tracking system.

4.1 Internal Sensors

Internal sensor is a sensor used by the robot for internal feedback control, mainly for detecting the robot joint signals. The signals to be detected include joint side position and speed, motor side position and speed, motor armature current, and joint torque, etc. Position detection is generally performed by means of photoelectric encoder and resistance displacement sensor, while speed detection is usually achieved by position difference. Motor current can be detected by converting it into voltage via

M. Chen et al., *Technologies of Robotic Welding*, Advanced and Intelligent Manufacturing in China, https://doi.org/10.1007/978-981-97-3149-7_4

a Hall sensor or resistor. The joint torque can be detected by means of strain gauge based torque sensors.

4.1.1 Position Sensors

Position sensors are used to measure the angular or linear displacement of joints as well as the speed. The commonly used displacement sensors in robots include potentiometer and encoder.

(1) Resistive displacement sensor

The resistive displacement sensor converts position change into resistance change, and its basic principle is shown in Fig. 4.1. These sensors can be either linear (Fig. 4.1a) or rotary (Fig. 4.2b). The resistance displacement sensor is actually a high-precision sliding resistor. The displacement of the measured object causes the sliding contact to move, and the moving distance of the contact is proportional to the moving distance of the measured object. In this way, the ratio of the resistance difference caused by the change of contact to the total resistance of the sensor reflects the change of position, with the sign of resistance difference indicating the direction of displacement. Generally, a certain power supply voltage is applied across the potentiometer to measure the change of resistance by measuring the change of voltage signal on the sliding end, as shown in Fig. 4.2. The output signal V_{out} of resistive displacement sensor may be calculated using the following formula:

$$V_{out} = V_p \frac{R_i}{R} = V_p \frac{x}{x_p}$$

where: Vp is the input voltage of power supply; R is the total resistance of sensor; R_i is the resistance difference caused by the displacement of contact; x_p is the maximum displacement, and x is the displacement of contact.

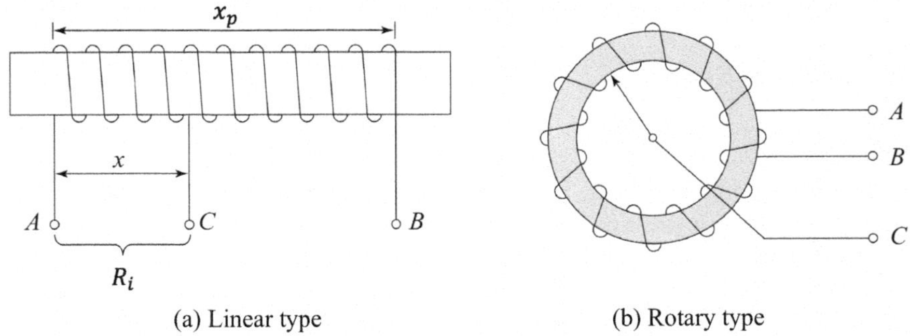

Fig. 4.1 Principle of resistive displacement sensor

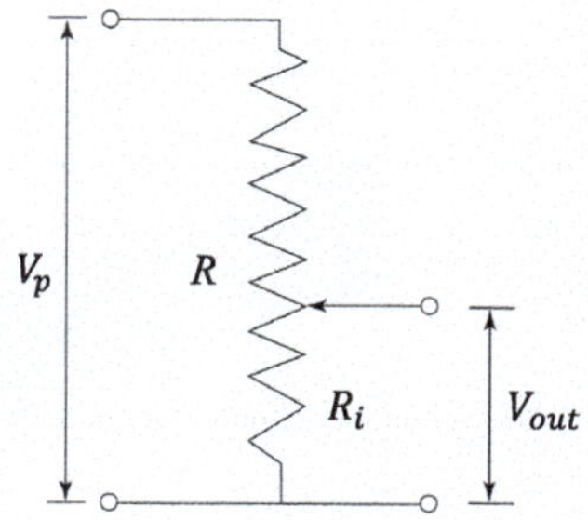

Fig. 4.2 Output Signal of resistive displacement sensor

Sliding resistors used as displacement sensors are usually conductive plastic resistors (also known as spray-plated film type, usually carbon film-plated). Such resistors have the advantages of small size, low cost, continuous output, high precision, low noise, and retention of position information in case of power failure. In addition, the sensor can differentiate the output voltage to detect the velocity. While the resistance of ordinary wound-type resistors varies discontinuously, with low measuring accuracy is low, so it is impossible to make differential calculation.

Resistive displacement sensors can be used to detect the position of joints or links. Due to the abrasive effect of the sliding contact on the surface of the resistive coating, these resistive displacement sensors have limited reliability and service life and have been gradually replaced by photoelectric encoders with decreasing cost. The combined use of resistive displacement sensors and encoders can significantly improve detection accuracy and reduce input requirements. The resistive displacement sensor is responsible for detecting the starting position, while the encoder is responsible for detecting the current position during motion.

Figure 4.3 shows two types of resistive displacement sensors.

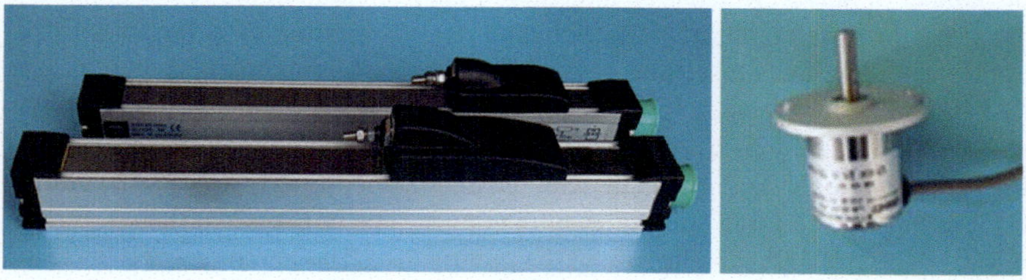

(a) Linear type (b) Rotary type

Fig. 4.3 Picture of resistive displacement sensor

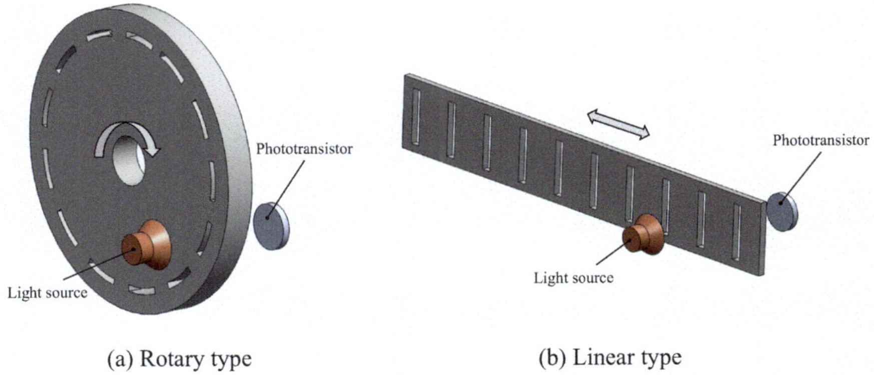

Fig. 4.4 Schematics of photoelectric encoder

(2) Photoelectric encoders

The photoelectric encoder is the most commonly used position sensor for welding robot joints. It is a device that converts the displacement into electrical pulse signals through photoelectric conversion. There are two types: rotary and linear. The former is also called encoder, as shown in Fig. 4.4a. The latter is also referred to as a code scale, as shown in Fig. 4.4b. Only the rotary coding disk is used on welding robots. The rotary type can be classified into incremental rotary type and absolute rotary type according to the working principles.

The incremental photoelectric encoder is composed of a grating disc, an indicating disc, luminescent devices and photo sensors, as shown in Fig. 4.5. A grating disc is a metal disc with a long radial hole, or a transparent plastic or glass disc with a radial strip-shaped light-blocking coating. The spacing between two adjacent light transmitters is called a grid section, and the total number of grid sections is called the number of pulses, which is an indicator of the resolution of the grid section. The encoder is provided with a housing for mounting the grating disk, indicating disc, luminescent devices and sensing devices, etc. Infrared luminous tubes are usually used as luminescent devices; high frequency photo-sensitive elements such as silicon photocell and photo-sensitive transistor are usually used as photo sensors. The grating disc is installed on the same spindle with the motor, and the motor drives the grating disc to rotate at the same speed during operation. The luminescent device emits a bundle of infrared light on one side of the grating disc. The photo sensor detects the light passing through the grating hole and outputs the same number of pulse signals as the number of transparent holes. The angular displacement and speed of the motor can be obtained by measuring the number of pulses output from the photoelectric encoder per second.

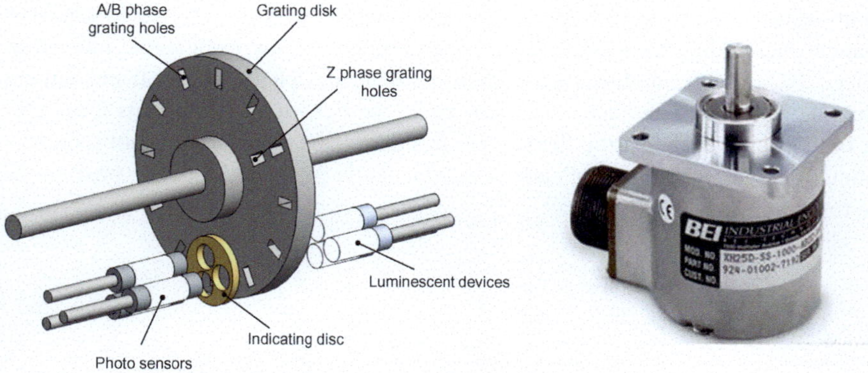

(a) Structure diagram (b) Physical photo of rotary encoder

Fig. 4.5 Rotary encoder

The encoder uses two photoelectric converters to output square-wave pulses of phases A and B with a phase difference of 90° for determining the rotation direction. When rotating clockwise, the phase A leads the phase B by 90°; when rotating counter-clockwise, the phase A lags behind the phase B by 90°. In addition, the incremental encoder is provided with a single pulse for phase Z (single-turn pulse, with one pulse per revolution for reference point positioning). The advantages of incremental encoder include simple structure, long service life (average mechanical service life can be more than tens of thousands of hours), high anti-interference capability, high reliability and accuracy. The disadvantage is that it is unable to output the absolute position of the motor rotation. Usually a proximity switch is used to detect the mechanical zero position. If the motor-driven mechanical device activates the proximity switch, the system identifies that the zero position has been reached and uses this position as a reference to calculate the position at each instant of time.

Absolute encoders directly output digital quantities representing the magnitude of displacement. It consists of multiple light sources (usually light-emitting diodes), a photoelectric encoder and photosensitive elements. The encoder disk and the servo-motor are installed on the same shaft. The light source is arranged on one side of the encoder and a photo-sensitive element is arranged on the other side for every code track. The typical structure is shown in Fig. 4.6. The coaxial code disc is provided with several concentric code tracks, each of which is alternately arranged with light-transmitting and opaque sectors. The more the code tracks, the higher the accuracy. For a binary encoder with N bits, the encoder shall have N bar code tracks. The number of sectional areas in a circumferential direction is called the number of sectors. The larger the number of sectors, the higher the resolution. Figure 4.7 show a sixteen-sector, four-code-track code disc and its arrangement with photo-electric elements. When the encoder is driven to rotate to a certain position, each photo-sensitive element will output corresponding level signal. The photo-sensitive element facing the transparent area is on and outputs a low level signal indicating binary "0", while the photo-sensitive element facing the opaque area is off and outputs a high level signal. The binary signals from all the light-sensitive outputs form a N-bit binary number indicating the current position. This encoder does not require a counter, and an unique digital code is automatically generated for each position of the rotating shaft.

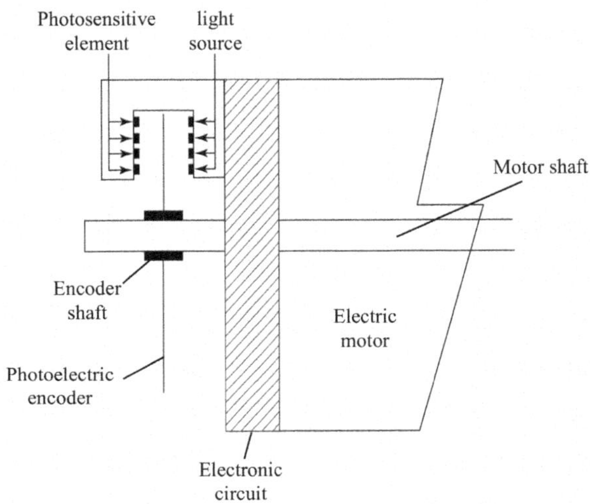

Fig. 4.6 Structure of absolute encoder

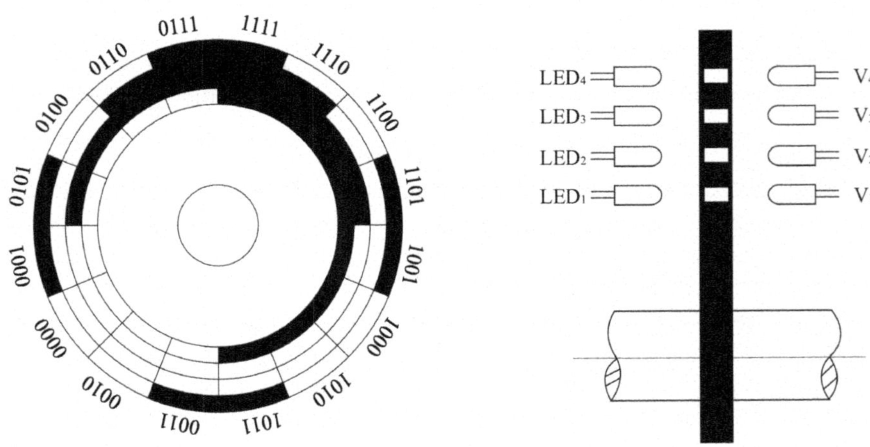

(a) Code disk (b) Element arrangement

Fig. 4.7 Code disc of absolute encoder

Absolute encoders have the advantage that they can directly output the number indicating the position without accumulated error and the position information will not be lost in case of power failure.

(3) Magnetic encoder

Magnetic encoders usually consist of a magnetic disc or bar, a sensor and a control circuit. When the magnetic disk or strip moves, the sensor detects the change in magnetic field and converts it into an electrical signal that reflects the displacement. In contrast to optical encoders, magnetic encoders have high immunity to interference, and are robust and durable. The air gap in optical encoders must be kept clean and transparent, while magnetic encoders are not sensitive to contaminants such as dust, dirt, grease and vibration.

1) Magnetic rotary encoder

The magnetic rotary encoder consists of a disk, a reading head and a control circuit, as shown in Fig. 4.8. A certain number of magnetic poles are installed on the circumference of the disk. When the disk rotates, the spatial leakage magnetic field generated by the magnetic pole will change, which can reflect the change of displacement. The greater the number of magnetic poles, the higher the resolution. The reading head is generally made of magneto-resistive device that can sense the changes in the magnetic field, or a Hall effect device which can sense voltage change. The reading head converts the detected magnetic field change into an electrical signal change. The control circuit amplifies, divides, or interpolates the signal to produce a digital signal output reflecting the amount of displacement.

Like the photoelectric encoder, magnetic rotary encoders can be classified into incremental encoders and absolute encoders. An absolute rotary encoder is assigned

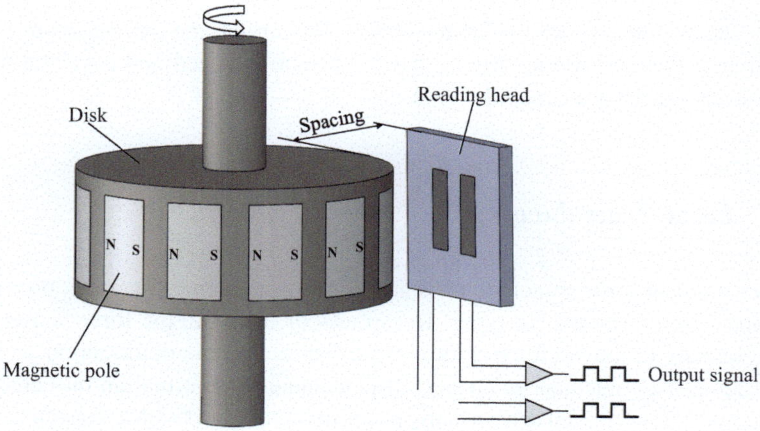

Fig. 4.8 Structure of magnetic encoder

a unique binary code to each measuring position, which tracks the exact position of the encoder even when the encoder is powered off.

2) Linear magnetic encoder

Linear magnetic encoders use linear magnetic strips instead of discs. The reading head detects the magnetic field change caused by the change of magnetic pole position during the movement of magnetic strip to reflect the position information. Each position of the magnetic strip of an absolute linear magnetic encoder corresponds to a unique binary digit for the reading head to determine the position. Incremental linear encoders are provided with one or more reference marks on the magnetic strip, which are used for resetting after power failure.

4.1.2 Velocity and Acceleration Sensors

Velocity sensors are used to detect the motion speed of robotic joints. Velocity sensors are usually not necessary when using photoelectric encoders as position sensors, since the robot's motion speed can be calculated directly from angular displacement per unit time interval. The shorter the time interval, the closer the calculated speed is to the real instantaneous speed. However, if the motion speed is very slow, the speed measurement accuracy will be very low. In this case, a tachometer generator is generally required as a speed sensor.

The tachometer generator is a device that converts speed into voltage signal, as shown in Fig. 4.9. The output voltage μ is proportional to the measured speed n. The rotation speed of the robot joint can be measured by connecting the rotor of the tachometer generator coaxially with the drive shaft of the robot joint servo drive motor.

The acceleration sensor is a device for measuring the acceleration of the robot joints. Welding robots usually have low motion speed and acceleration, so acceleration sensors are generally not used.

4.1.3 Force Transducers

Force or moment sensors include resistance strain gauge, piezoresistive, piezoelectric, capacitive and inductive types. Regardless of the type, the force sensors use elastic elements to convert force or torque into displacement or deformation, and then use certain sensing element to convert displacement or deformation into electrical signal output. A robot joint drive usually needs to be equipped with a resistance strain force sensor or piezoresistive force sensor to measure the output force and moment of the drive itself.

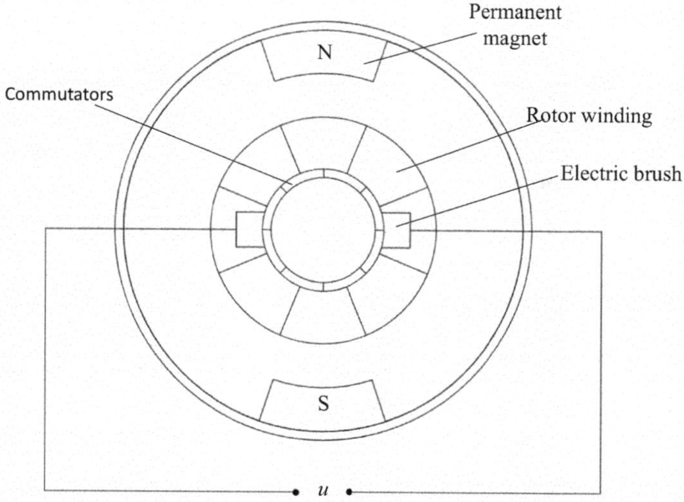

Fig. 4.9 Schematic diagram of tachometer generator

The resistance strain gauge and compression-resistance force transducers have similar working principles. Both make use of the change in resistance of sensitive element under the action of force or moment to cause the change in electric signal, and the magnitude of force or moment is determined through the change in electric signal. The sensitive element of resistance strain gauge sensor is made of a metal foil or wire, and the change in its geometrical dimensions under the effect of force will cause the change of resistance. The industrial sensitive element of piezoresistive sensor is a semiconductor, and the change in its resistivity under the effect of force will cause the change of resistance. As the resistance change of sensitive elements is small under the effect of pressure, the sensitive elements of these two types of sensors are usually connected to form a bridge, and the unbalanced output of the bridge under the effect of external force can reflect the magnitude of external force and improve the detection accuracy. The materials of the sensitive elements (usually called substrates or diaphragms) of piezoresistive sensors are mainly silicon and germanium. The piezoresistive sensor has high accuracy, quick response and high reliability, but it highly susceptible to the influence of temperature, which requires temperature compensation.

Piezoelectric sensors measure forces using the piezoelectric effect of crystal materials such as quartz and barium titanate. When these piezoelectric crystals are subjected to pressure, they generate opposite charges on their opposite sides, which, after being measured and amplified by the measuring circuit and amplification circuit, will form electrical signals in direct proportion to the applied force. Advantages of these sensors include simple structure, large measuring range, high sensitivity, high signal-to-noise ratio, and high reliability, etc. The disadvantage is that it can only be used for pulsating pressure measurement and not for static pressure measurement.

4.2 External Sensor

The external sensor is used to detect the environment and conditions of a robot. Welding robot uses such sensors to detect changes in external environmental conditions, and adjust or correct welding process parameters according to such changes to ensure welding quality.

4.2.1 Proximity Sensors

Proximity sensors are used to sense the proximity between welding robots and peripherals to avoid obstacles and prevent collisions. Non-contact sensors are usually used, mainly including eddy-current sensors, photoelectric sensors, and ultrasonic sensors.

(1) Eddy current proximity sensor

The eddy current proximity sensor implements proximity measurement based on the fact that the magnitude of the induced eddy currents generated by the alternating magnetic field in the metal body varies with the distance between the surface of the metal body and the coil, as shown in Fig. 4.10. This sensor consists of a probe coil, an oscillator, a detection circuit and an amplifier, as shown in Fig. 4.11. The high-frequency current flowing through the probe coil excites a high-frequency alternating magnetic field inside the metal conductor, which induces eddy currents in the metal conductor, The eddy currents generate an alternating eddy current magnetic field in the opposite direction to the high-frequency alternating magnetic field near the metal surface. The alternating eddy current magnetic field changes the amplitude and phase of the high-frequency current flowing through the probe coil, that is, the effective impedance of the probe coil changes, and the change reflects the distance from the probe coil to the surface of the metal conductor. The detection circuit converts the detected change in coil impedance into an electrical signal, which can reflect the distance between the probe and the surface of the measured metal conductor.

Eddy current proximity sensors can be used as either proximity switches or distance sensors. When used as a proximity switch, a magnetic metal inductor is usually installed on the object under detection. When the metal inductor approaches the sensor by a certain distance, the sensor will send a trigger signal, to initiate the relay and the robot will stop moving. When used as a distance measuring sensor, an analog signal is output, the magnitude of which varies linearly with the distance.

(2) Optical proximity sensor

Optical fiber sensors can be classified into laser proximity sensors, infrared proximity sensors and natural light proximity sensors according to the light source used.

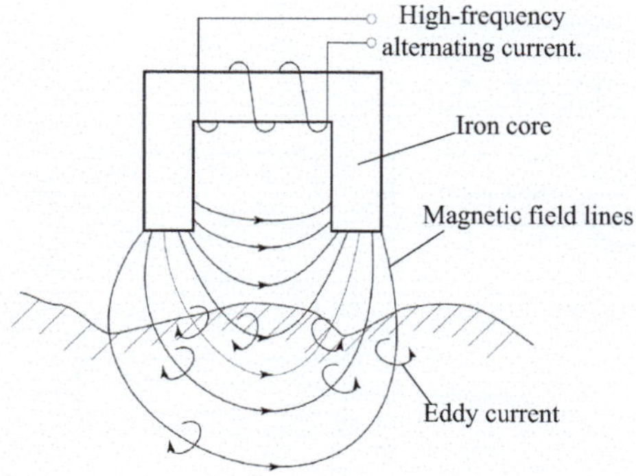

Fig. 4.10 Measurement principle for eddy current proximity sensors

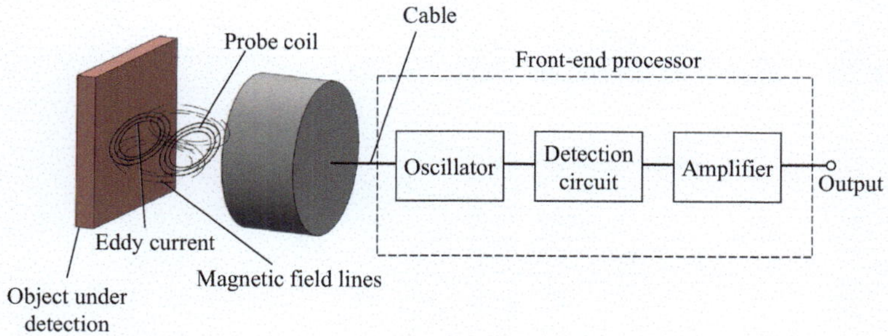

Fig. 4.11 Block diagram of eddy current proximity sensors

1) Laser proximity sensor

The laser proximity sensor consists of a laser transmitter and a laser receiver, as shown in Fig. 4.12. This sensor has the advantages of long-distance contact-free measurement, high speed, high precision, large range and high resistance to photoelectric interference.

The laser proximity sensor can be used as either a proximity switch or a distance sensor, and it is mainly used as proximity switch in welding robots.

Laser proximity sensors can be classified into direct-reflective, mirror-reflective, reciprocal-reflective and optical-fiber types. Figure 4.13 shows the schematic diagram of direct-reflective laser proximity sensor. The signal from the transmitter is reflected by the object under detection, and the receiver judges the object under detection according to the change of the reflected beam received. Figure 4.14 shows the schematic diagram of a mirror-reflective laser proximity sensor. The light emitted from the transmitter is reflected back to the receiver through a mirror. When the object under detection appears, it blocks the light, and the reflected signal received by the receiver changes. Figure 4.15 shows the schematic diagram of a reciprocal-reflective

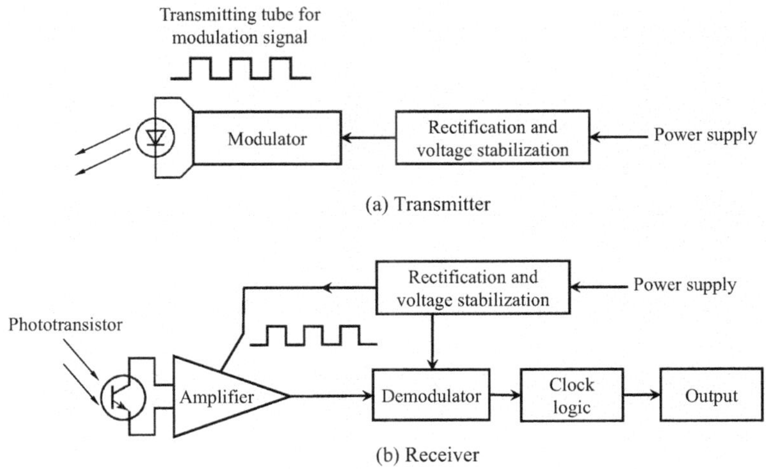

Fig. 4.12 Block diagram of laser proximity sensor structure

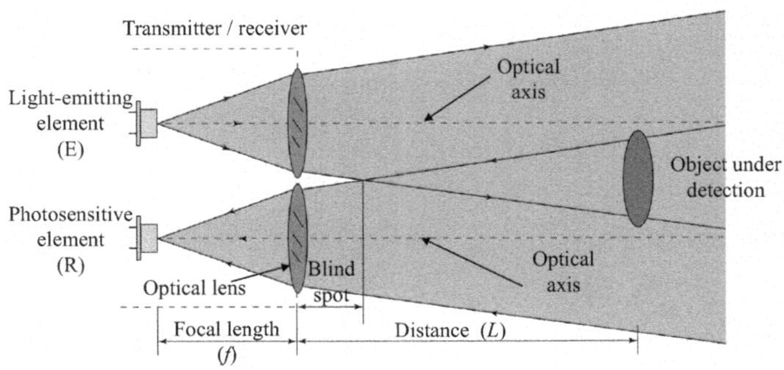

Fig. 4.13 Schematic diagram of direct-reflective laser proximity sensor

laser proximity sensor. The receiver and transmitter are installed on the same axis and they receive the light from the transmitter directly. If the object under detection is present, the object under detection will block the light between the transmitter and receiver, and the light signal received by the receiver will change. An optical fiber laser proximity sensor uses plastic or glass fiber to conduct laser beams and it can carry out long-distance detection. See Fig. 4.16 for its schematic diagram.

2) Infrared proximity sensors

An infrared proximity sensor consists of an infrared light emitter and an infrared light-sensitive receiver. The infrared emitter is used to emit modulated infrared light signals. If the infrared light signals meet the detected object, the reflected energy changes according to the distance of the object under detection. The infrared photo-sensitive receiver can judge the received signals and obtain the position information

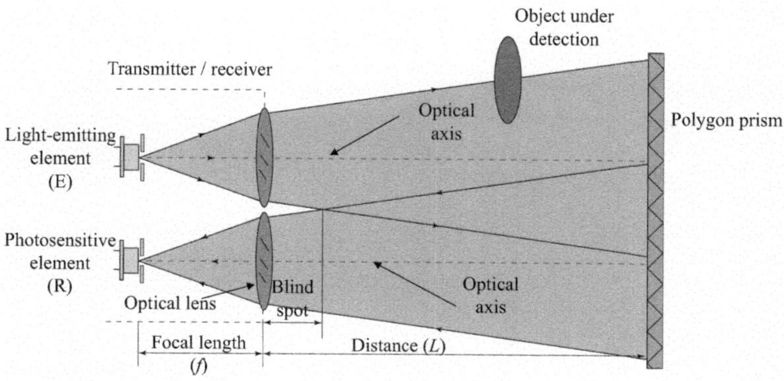

Fig. 4.14 Schematic diagram of mirror-reflective laser proximity sensor

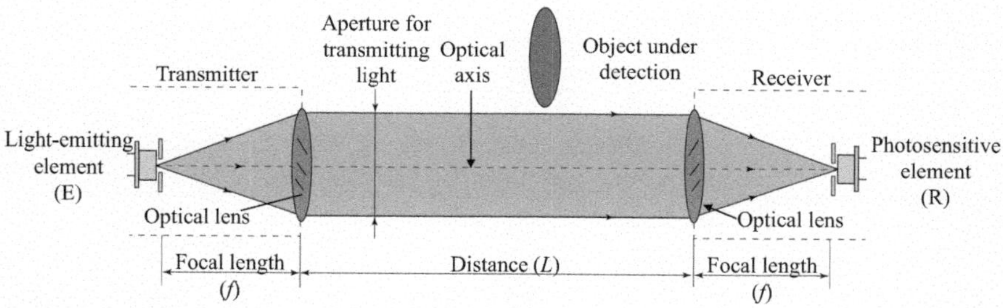

Fig. 4.15 Schematic diagram of a reciprocal-reflective laser proximity sensor

of the object. This sensor has the advantages of high sensitivity, quick response and compact size. This kind of sensor is usually attached to the robot's end effector to detect the foreign object in its working space.

(3) Ultrasonic proximity sensor

Ultrasonic wave is a sound wave with a vibration frequency higher than 20 kHz. Due to the high frequency and short wave length, ultrasonic wave is characterized by small diffraction, good directionality, and strong reflection echo, etc. The ultrasonic proximity sensor consists of an ultrasonic generator, a receiver, and a controller, etc. According to the arrangement of transmitters and receivers, it is classified into reflective and reciprocal reflective ultrasonic sensors. The reflective ultrasonic sensors are widely used. Figure 4.17 shows the structure and schematic diagram of a reflective ultrasonic transmitter.

The ultrasonic generator emits ultrasonic waves in a certain direction and starts the timer simultaneously. When the ultrasonic wave encounts the object under detection during transmission, a reflected wave is returned, and the ultrasonic receiver stops the timing immediately upon receiving the reflected wave. The microprocessor in the receiver calculates the transmission and reception total time t, and then calculates

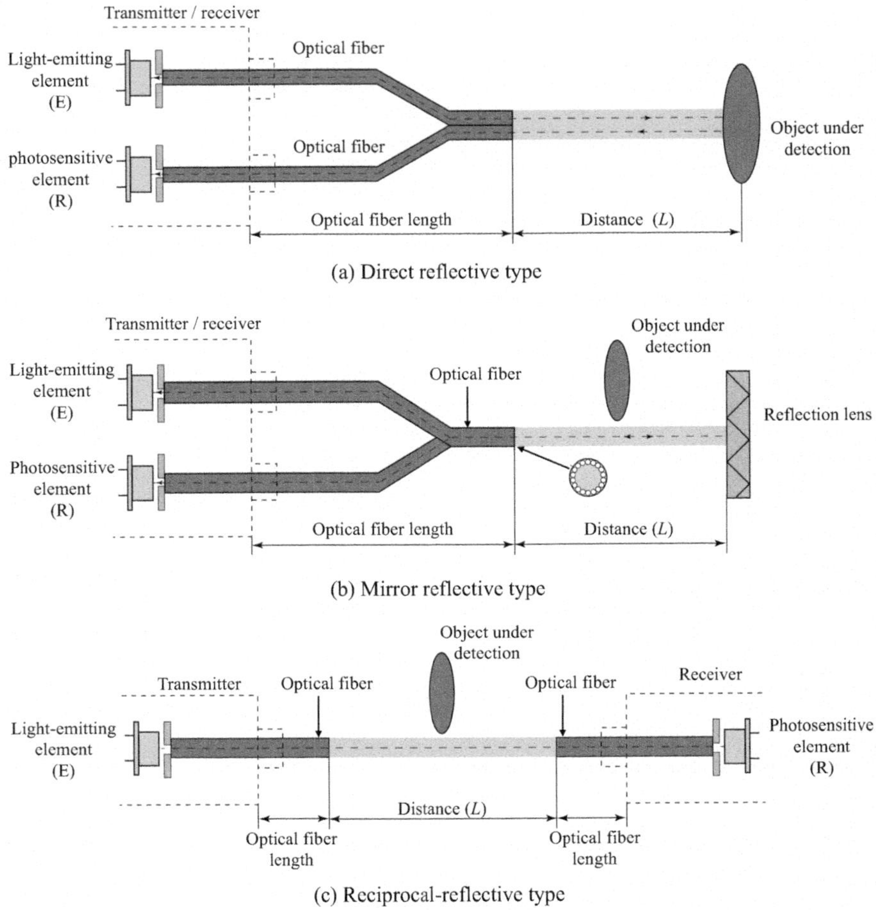

Fig. 4.16 Schematic diagram of optical fiber laser proximity switch

Fig. 4.17 Structure and basic principle of reflective ultrasonic proximity sensor

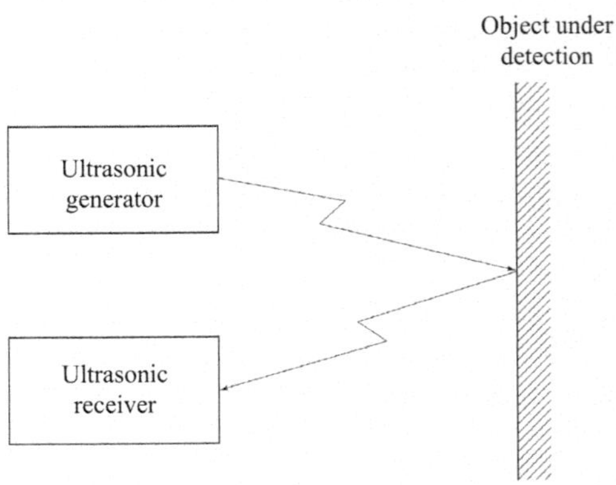

Table 4.1 Comparison of performance characteristics of several proximity sensors

	Eddy current proximity transducer	Optical proximity sensor	Ultrasonic proximity sensor
Detection distance	A fraction of millimetre to tens of millimetres	Up to 60 m	A fraction of meter to several meters depending on wavelength
Response frequency	Up to 10 kHz	1–2 kHz	10–40 Hz
Error	$\leq$5%	Up to 0.005%	$\leq$0.6 mm
Cost	Low	High	High
Environmental requirements	Operable in air, oil and water and suitable for a wide temperature range. No external electromagnetic field is allowed in the surrounding environment	There shall be no obvious dust in the surrounding environment, otherwise the accuracy will be affected	Used in air medium, adjustment is needed for other media; not suitable for use when the wind speed is higher than 10 m/s; humidity and temperature can affect the accuracy

the distance $S = (vt/2)$ of the object to be detected based on the propagation speed v in the medium. After that, the distance is displayed or a digital signal is issued. Ultrasonic proximity sensors can be used either as proximity switches with digital output or as distance sensors with analog output.

The sensor is characterized by fast detection speed, high measurement accuracy, simple structure, easy use and wide application. In arc welding robot, the ultrasonic proximity sensor is usually used for distance measurement and is not used as proximity switch.

The advantage of ultrasonic proximity sensor is that it is insensitive to the dust in the environment, transparency of the measured object, surface color and oil stains on the surface, and its disadvantages include slow response speed and the high sensitivity of measurement accuracy to the ambient wind speed, temperature and pressure. Ultrasonic transducers are generally not recommended if high measurement accuracy is required. Table 4.1 compares the performance characteristics of several proximity sensors.

4.2.2 *Arc Parameter Sensors*

The arc parameters mainly include the welding current and arc voltage, which directly determine the stability and quality of welding process. The robots usually need to measure and regulate these two parameters in real time during welding. Due to the complexity and variability of welding process, the welding current and arc voltage sensors shall have strong isolation and anti-interference capability. Hall current sensor and voltage sensor are usually used in welding robot. The data acquired by the sensor is received and processed by the host control computer via the data acquisition card.

Hall sensors are magnetoelectric conversion components made of semiconductor materials. The principle of Hall closed-loop current sensor is shown in Fig. 4.18. The magnetic field generated by the primary side current I_n (measured current) is compensated by the magnetic field generated by the secondary side current I_m, so that the Hall elements are always at zero flux. When the magnetic field generated by the primary and secondary side current reaches an equilibrium, the following relationship exists:

$$N_1 \times I_n = N_2 \times I_m \tag{4.1}$$

where, N_1 and N_2 are the number of coils on the primary side and secondary side respectively. As the data acquisition card usually acquires voltage signals, a measuring resistor R_{m} shall be connected to the secondary current output terminal and the voltage U_{m} across the measuring resistor shall be used as the voltage input of the data acquisition card. The installation method of Hall current transducer is very simple, all that need to do is just to pass the welding cable through the center hole. The sensor is of non-contact type and does not interfere with the current output from the arc welding power supply. The measurement frequency is up to 100 kHz and the conversion speed is up to 50 A/s.

The working principle of closed-loop Hall voltage sensor is basically the same as that of closed-loop Hall current sensor. The only difference is that the Hall voltage sensor first converts the measured voltage into current, which requires a current-limiting resistor at the input terminal, as shown in Fig. 4.19. The ratio of primary side current to measured voltage is determined by the current-limiting resistor R_i. The current output at the output end of the voltage sensor is converted into a voltage signal via a resistance R_m and transmitted to the data acquisition card. Generally, the current-limiting resistance R_i is high, the input current of voltage sensor is low and the magnetic field intensity generated is also low, so its measuring accuracy is sensitive to the surrounding magnetic field. During use, the Hall voltage sensor shall be kept away from the cable carrying welding current as far as possible so as to prevent the magnetic field caused by welding current from affecting the measurement accuracy.

Fig. 4.18 Schematic diagram of hall current sensor

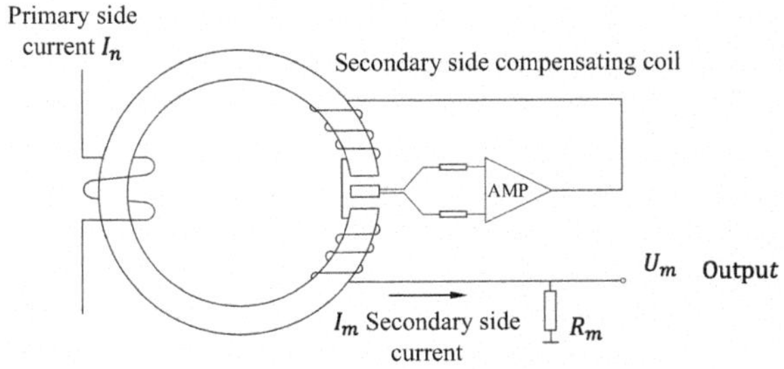

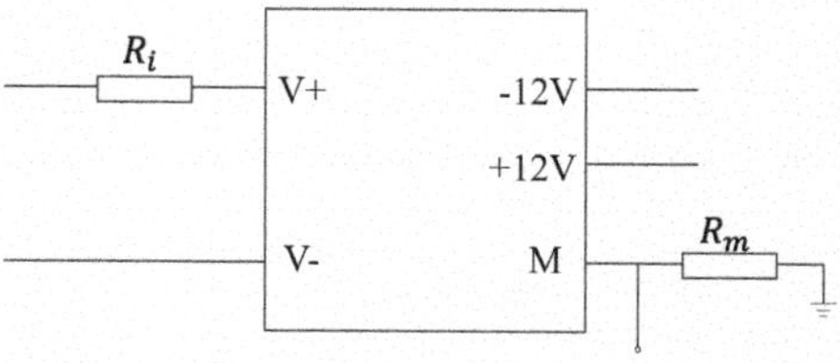

Fig. 4.19 Schematic diagram of hall voltage sensor

The input end of Hall voltage sensor shall be connected to the contact tip of the of welding torch and the workpiece respectively.

4.2.3 Seam Tracking Sensors

Errors in groove size and assembly of to-be-welded workpieces are inevitable, and unpredictable changes may occur during welding due to thermal effects. If the error or change exceeds the allowable range, the robot cannot perform its task efficiently and with high quality. However, the error can be indirectly eliminated by the use of sensors that monitor the geometric parameters in real time and adjust the welding parameters accordingly in a timely manner. This kind of control is classified into two categories. One is to keep the arc center-line always aligned with the groove angle bisector through deviation detection and correction, which is called seam tracking control. The other is to output electrical parameters fitting with the groove size and ensure the weld formation quality by detecting the change in groove size and making the arc welding power supply automatically adapt to such change. This kind of control is called self-adaptive control of weld quality. At present, seam tracking sensors are widely used, while self-adaptive weld quality control sensors are not very common. There are several types of seam tracking sensors, including arc type, mechanical contact type, laser vision type, ultrasonic type, and so on. Widely used are through arc sensors and laser visual sensors.

(1) Through arc sensor

The through arc sensor calculates the deviation between the TCP (abbrivation of tool center point, for arc welding robot, it is the tip of the welding wire) path and the expected path by detecting the electric signal of the welding arc itself. This deviation is used to control the arc weaving device to make correction, as shown in Fig. 4.20. In the Fig. 4.20a, *x*-*y*-*z* is the coordinate system of the workpiece, and *x* direction refers to the welding direction; *o*-*n*-*a* is the tool coordinate system, and TCP is the weaving center and also the origin of the tool coordinate system. On the one hand, the torch travels in the *x* direction, and on the other, it swings in a sine wave path in the tool coordinate system. Figure 4.20b shows the expected position of the weaving center during tracking. Ideally, the axis *a* of the tool coordinate system is located on the bisector of the groove angle, with the two vertices of the weaving path are equidistant from the workpiece.

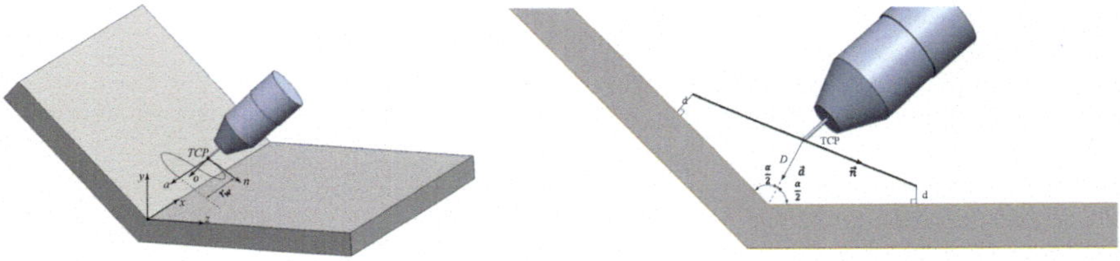

(a) Definition and weaving path of TCP (b) Expected TCP Position during seam tracking

Fig. 4.20 Schematics of through-arc seam tracking

The through arc sensor is based on the principle that the welding current changes with the distance l from the contact tip to the workpiece, which can be expressed by the following formula:

$$U = \beta_1 I + \beta_2 + \beta_3/I + \beta_4 l \tag{4.2}$$

where, U, I, l are respectively the arc voltage, the welding current and arc length, β_1, β_2, β_3 and β_4 are all constants. In the case of GMAW, the power supply with flat characteristics is usually used and the arc voltage remains unchanged. Therefore, it can be seen from Formula 4.2 that the welding current decreases with the increase of arc length. Figure 4.21 illustrates the sensing principle. The arc swings transversely (perpendicular to the travelling direction of the welding torch) and the welding current changes periodically. If the weaving center is on the desired path, i.e. the torch travels without deviation, the current varies in a sine wave pattern, with the average value of the current in each 1/4 period being constant $\left(I_{\mathrm{L}}^{*} = I_{\mathrm{R}}^{*} = I^{*}\right)$, as shown by the dotted line in Fig. 4.21b. Therefore the average current deviation between two adjacent 1/4 period can be used to determine the deviation. If there is any deviation, the current variation curve will not follow a sine wave pattern, and the average value of the current within each 1/4 period will deviate from I^{*}. In the Fig. 4.21, $I_{\mathrm{L}} > I_{\mathrm{L}}^{*}$, $I_{\mathrm{R}} < I_{\mathrm{R}}^{*}$, indicating that the distance between the contact tip and the workpiece becomes lower when the welding torch swings to the left, and higher when the welding torch swings to the right, and the weaving center of the welding torch deviates to the left. Using this deviation as an input signal, the position deviation of the weaving center point can be corrected, allowing the arc to align with the groove center-line.

The through arc sensor is the most commonly used real-time seam sensor in welding robots today. This kind of sensor has the following advantages:

(a) A non-contact sensor with sensing accuracy unaffected by the condition of the workpiece surface;
(b) Implementing control by detecting the variation of current of the arc itself and therefore immune to the influence of arc light and flue gas;
(c) Being able to implementing vertical and horizontal tracking;

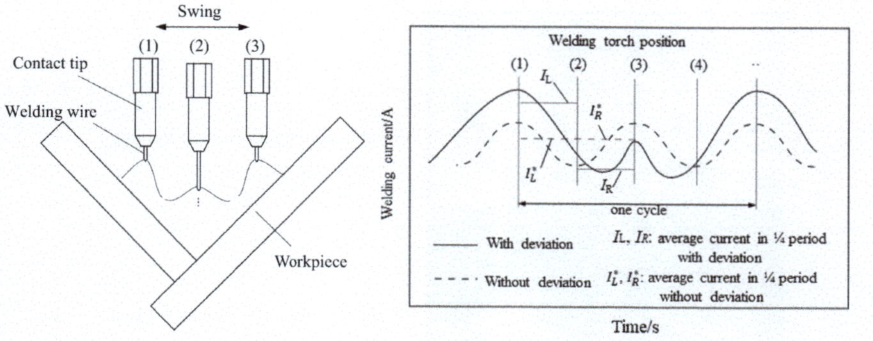

(a) Weaving path of welding torch (b) Variation of welding current in different conditions

Fig. 4.21 Principle of through-arc seam tracking sensors

(d) No additional devices required, with low space requirements and high accessibility of the welding gun;
(e) Low cost.

The through-arc seam tracking sensors are mainly used for gas metal arc welding. This kind of sensor is only suitable for corner joints, thick plate lap joints (workpiece thickness greater than 2.5mm), or butt joints with symmetrical grooves (V, U, and Y-shaped grooves, etc.), and not for I-type groove butt joints. The disadvantage is that the starting point of the weld cannot be found before arcing. Moreover, the accuracy of the sensor is subject to the interference factors affecting arc stability. For CO_2 gas shielded welding operating with short-circuit transfer, the welding current changes with the arc status, therefore measures shall be taken to ensure the stability of detection signal.

Figure 4.22 demonstrates the effectiveness of through-arc seam tracking sensor for the T-joint fillet welding. The vertical plate deviates from the centerline of the flat plate, while the preset trajectory of the welding gun coincides with the centerline of the flat plate. When the through-arc seam tracking sensor is not used, the welding gun deviates from the weld seam (top image). When the sensor is used, the arc always travels along the centerline of the weld with the position deviations well compensated.

The through-arc tracking sensor have a low weaving frequency, usually less than 5 Hz, and it therefore cannot be used for high-speed welding and lap welding of thin plates. To solve this problem, the high-speed rotating arc method was proposed. Figure 4.23 shows the principle of the high-speed rotating arc method in TIG welding. The torch is fixed to an eccentric gear, which is rotated by an electric motor, so that the arc will rotate at a high speed, with a rotation frequency of up to 100Hz. TIG welding uses a constant-current power supply. When the distance from the contact tip to the workpiece changes, the welding current does not change, but the arc voltage

Fig. 4.22 Effectiveness of through arc seam tracking sensor for the T-joint fillet welding

changes. Figure 4.23b shows how the arc voltage varies during the torch weaving under different conditions. If the weaving centerlies on the expected path (in the swing, the center-line of the tungsten electrode is coincidental with the bisector of the groove angle, as shown in Fig. 4.20b), i.e, the torch travels without deviation ($\Delta x = 0$), the arc voltage variation curve will be regular, as shown by the dotted line in Fig. 4.23b. The arc length is the longest and the arc voltage is the highest when the torch swings to the front edge C_f and rear edge C_r of the molten pool; when the torch swings to the left edge L and right edge R of the molten pool, the length and voltage of the arc are the shortest. The waveform of arc voltage undergoes one cycle for each turn of the torch swing, with its two half-cycle are the same. If the torch weaving center deviates from the ideal path, i.e., there is a deviation (e.g., to the right), then $\Delta \neq 0$, and the voltage variation curve will be regular, as shown by the solid line in Fig. 4.23b. The maximum value of arc voltage moves forward relative to the front edge of the molten pool C_f and backwards relative to the tail edge of the molten pool C_r. In addition, the two half-cycels of the arc voltage variation cycle are asymmetrical relative to point C_f during one rotation of the welding torch. The deviation can be determined by integrating the arc voltage between the left and right sides of the C_f point within the same time interval, and the deviation can be used for corrective control.

The high-speed rotating arc sensor improves the detection sensitivity for welding torch positon, and significantly raises the tracking accuracy, making fast control possible.

(2) Laser visual sensor

The laser visual sensor is a sensor based on the triangulation principle, as shown in Fig. 4.24. A laser beam is irradiated to the surface of the object under detection, and its reflective beam passes through imaging lens to form a image point on the optical sensitive element. The connection line between the laser head and the imaging lens is named as a reference line. The distance between the two is s, the focal length of the lens is f, and the angle between the laser and reference line is β. The intersection

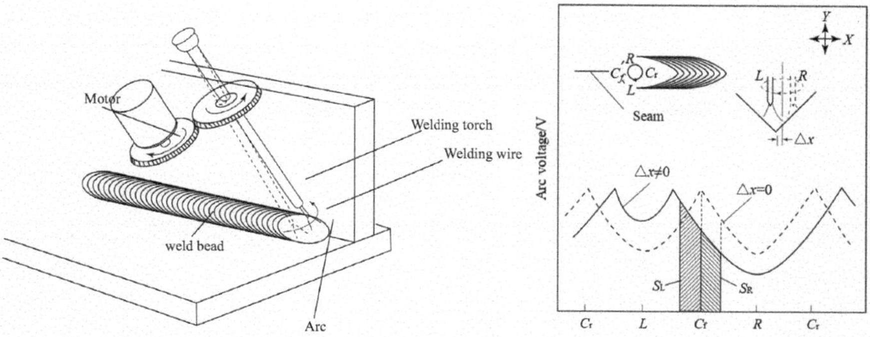

(a) Weaving path of the torch

(b) Variation of the arc voltage in different situations

Fig. 4.23 Principle of high-speed rotating arc tracking sensor

Fig. 4.24 Principle of triangulation

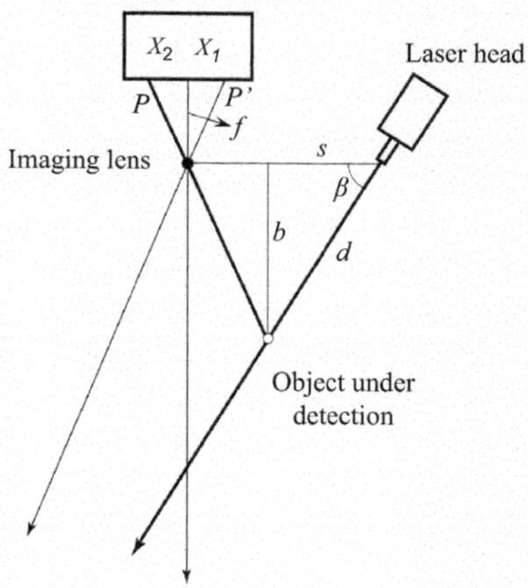

point between the laser beam reflected by the to-be-detected object and the imaging plane of the sensitive element is point P; a line parallel to the incident laser is made through the imaging lens, and its intersection with the imaging plane is $P^{'}$. The triangle formed by the optical head, imaging lens and the object under detection is similar to that formed by the imaging lens, the imaging point P and the auxiliary point $P^{'}$.

Assuming $PP^{'} = x$, then

$$b = f_{\mathrm{s}}/x \tag{4.3}$$

$$X = x_1 + x_2 = f/\tan\beta + PS \times P \tag{4.4}$$

where, PS is the pixel number per mm; P is the location of the imaging point relative to the imaging center in the pixel coordinates. The height b can be obtained from Eqs. (4.3) and (4.4), and the distance d between laser head and the to-be-detected object can be obtained using b:

$$d = b/\sin\beta \tag{4.5}$$

When the distance between the laser head and the object under detection changes, the position of image point on the optical sensitive element will change accordingly. Therefore, the height change can be calculated based on the triangle relationship of the object, as a result, this kind of sensor can be used for measuring height change. When a laser beam scans the surface of the object being detected either through moving the laser head along a certain trajectory or throuogh a scanning lens, a laser strip image reflecting the characteristics of the surface of the object being detected can be obtained on the arrayed photosensitive elements. When the laser visual sensor scans along the groove of a to-be-welded workpiece, it can not only obtain the profile information of the bevel, but also determine whether the scanning center-line is on the groove center-line. Therefore, the laser sensor can be used for seam positioning, seam tracking, measuring groove dimensions and inspection of weld formation.

According to whether the laser beam scans or not, laser visual sensors can be classified into structured-light sensors and scanning sensors. The structured-light sensor uses single- or multi-light-plane laser beams with large spot sizes and plane-type light-sensitive elements. Since the power of the laser used is generally lower than that of the arc, it is usually necessary to place the sensor in front of the torch to avoid the interference of the direct arc light, as shown in Fig. 4.25. The disadvantage of this sensor is that it is difficult to ensure even distribution of the intensity of the beam on the spot, resulting in low image quality. In addition, the surface of high-reflectivity metal such as aluminum alloy, stainless steel and galvanized plate will cause secondary reflective light which strongly interferes with the images, making subsequent image processing very difficult.

Scanning laser visual sensors use a beam of light with a small spot diameter for scanning imaging, which results in a high signal-to-noise ratio and easier reflection treatment. Such sensors generally use array CCD devices as imaging devices, as shown in Fig. 4.26. The laser emitted from the laser head is focused into a very small laser beam through the focusing lens, which is deflected by the deflection mirror to irradiate onto the workpiece. The deflection mirror rotates under the drive of the motor to make the laser beam scan the workpiece at a certain angle controlled by the angular displacement sensor. The reflection beam of the scanning beam enters the CCD array after passing the inspection mirror and imaging lens to form an image that can reflect the geometric dimensions and spatial position of the groove of workpiece. This scanning sensor has a large depth of field of up to 280 mm. As the size of the spots of laser beam cannot be very small, the lateral resolution is relatively low, usually >0.3 mm. In addition, as mechanical scanning is adopted, the scanning frequency is not high, usually only 10 Hz. Therefore, this sensor is mainly used for seam tracking

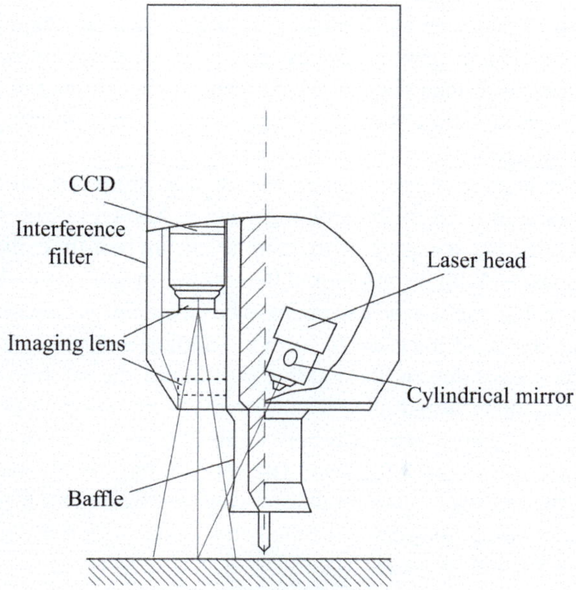

Fig. 4.25 Schematics of structured-light vision sensor

and self-adaptive quality control of thick workpieces. High-precision and high-speed tracking or detection mostly uses structured-light sensors.

If the image-sensitive elements of laser visual sensors are upgraded from analog CCD to digital CMOS devices, the frame rate for image acquisition can reach 3,000–10,000 frames/s, which can significantly improve the imaging quality, sensing speed and accuracy. With the aid of digital technology, the interference caused by secondary reflections from surface of high-reflectivity metal such as aluminum alloy, stainless

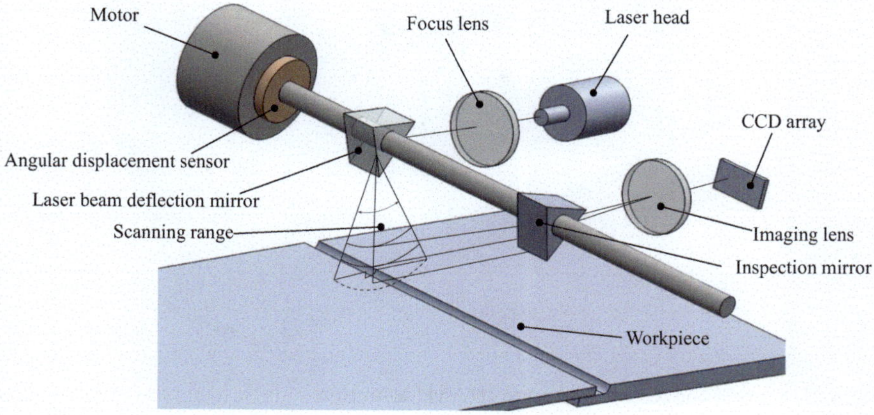

Fig. 4.26 Structure and principle of scanning laser visual sensor

steel and galvanized plate can be eliminated through proper image processing algorithm. This enables clearer identification of weld grooves, precise weld tracking, and accurate measurement of the geometric parameters such as joint gap, unfitness butt joint, and cross-sectional dimensions of the groove, etc., thus allowing for adaptive control of the weld quality.

Scanning laser visual sensors are usually mounted on a torch and should be located at a certain distance in front of the welding wire, as shown in Fig. 4.27. Welding robots need to utilize one a degree of freedom to ensure real-time alignment of the welding gun and sensor with the bisector of the groove. In addition to the mechanical connection to the robot, the sensor also needs to be electrically connected to the robot controller via an electrical interface to form a complete sensing system, as shown in Fig. 4.28. The sensor detects the relative position of the torch to the groove in real time and sends the information to the information processor for processing. The message processor is connected to the robotic controller via an electrical interface that transmits information in both directions. The robot controller receives the sensing information via the interface and then makes judgment based on other information, and sends corresponding commands to the robot body, which drives the welding torch to align the arc with the seam.

Laser visual sensors can be used not only for robotic welding process detection and control, but also for real-time or post-weld quality inspection. The sensor is installed behind the welding torch to scan the welds to obtain 3D images of the weld surface. The digital image processing algorithm can not only calculate the geometric

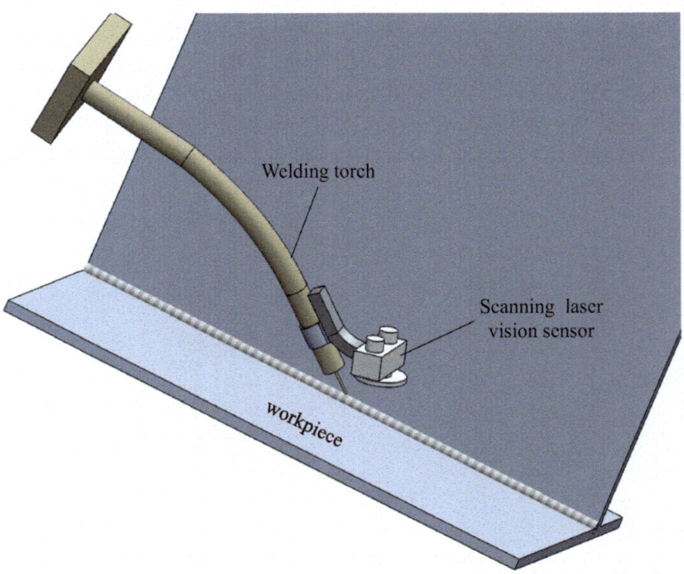

Fig. 4.27 Laser vision sensor mounted on the welding torch

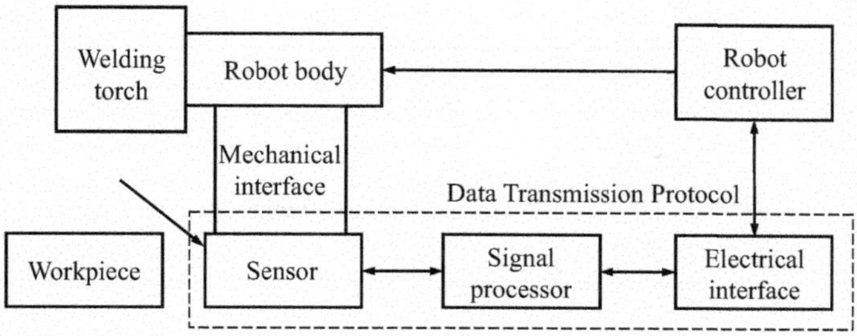

Fig. 4.28 Block diagram of seam tracking system

shape parameters of welds, such as weld width, reinforcement and weld toe angle, with high speed and precision, but also detect such defects as undercut, hump and surface porosity, etc.

Chapter 5
Welding Robot System

The basic welding robot system consists of a robot body, a robot controller (control system), a welding system, a positioner and clamping device, a positioner and clamping device, a safety protection device and a welding torch cleaning station, etc. The Robotic welding station or production line usually consists of one or more welding robots, several handling robots, several positioners, several welding systems and a unified robot control center.

The robot systems can be classified into arc welding robot systems, electric resistance welding robot systems, laser beam Welding robot systems and friction stir welding robot systems according to the welding process.

5.1 Resistance Spot Welding Robot system

5.1.1 Structure and Characteristic

The spot welding robot system generally consists of a robot body, a robot control system, a teach pendant, a spot welding tong, an air/water pipeline, an electrode grinder and relevant cables, as shown in Fig. 5.1. Suitable positioners and fixtures are usually required. Figure 5.2 shows the typical structure of the robot body and the welding tong. Spot welding robot workstations usually have several robots working at the same time. Figure 5.3 shows a spot welding robot workstation for a vehicle body production line.

Spot welding robot systems are widely used for welding of thin plate assemblies such as automobiles, electronics, instrumentation and household appliances. There are more than 4000 welding spots on a body-in-white (BIW). Manual welding is susceptible to the operator's technical level and fatigue degree, and it is therefore difficult to ensure the reliability and consistency of welding quality. Spot welding robot can successfully solve this problem with the following main advantages:

M. Chen et al., *Technologies of Robotic Welding*, Advanced and Intelligent Manufacturing in China, https://doi.org/10.1007/978-981-97-3149-7_5

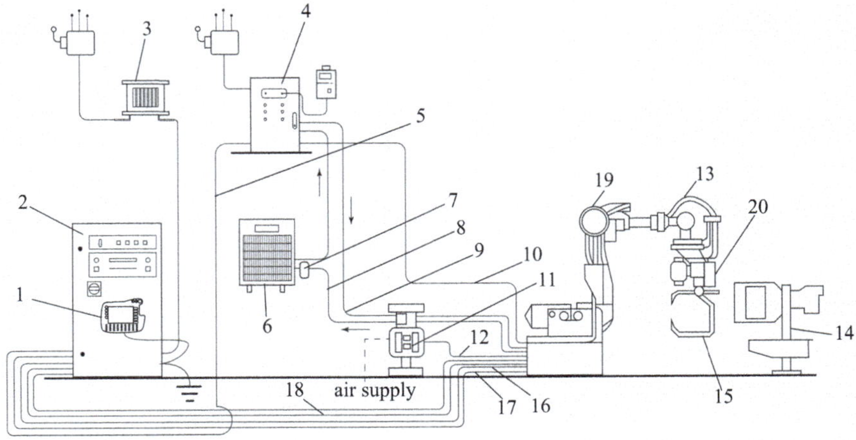

1-Robot teach pendant; 2-Robot controller; 3-Robot transformer; 4- Welding control box; 5-Control cable; 6-Water-cooling unit; 7-Cooling water flow switch; 8-Inlet pipe for cooling-water; 9-Outlet pipe for cooling-water; 10-Power cable for welding tong; 11-Air supply; 12- Inlet pipe for air; 13- Cable cluster; 14-Electrode grinder; 15-Welding tong; 16,18-Robot control cable; 17- Power cable for robot ; 19- Control cable for welding tong (pneumatic/servo); 20-Robot body

Fig. 5.1 Structure of spot welding robot system

Fig. 5.2 Welding robot body and welding tong

(a) Enabling complete automate of welding process with improved welding quality, and high quality and the quality consistance
(b) High welding productivity;
(c) Being able to works continuously for 24 h in a day;

Fig. 5.3 Spot welding robot workstation for automobiles

(d) Good working conditions and low labor intensity, significantly lowering requirements on worker's operation technology;
(e) Good flexibility, suitable for both large-scale and small-scale production;
(f) Easy to implement group control and be integerated in the production line to further improve productivity.

5.1.2 Robot Body and Control System

(1) Robot body

Robotic spot welding is usually performed with an electrically driven, full-articulated robot body. The requirements of resistance spot welding on the robot body are as follows:

(a) DOF. The robot used for spot welding is required to have no less than five degrees of freedom. Currently, robots with six degrees of freedom are mostly used, namely, waist axis, shoulder axis, elbow axis, wrist rotary axis, wrist swing axis and wrist twisting axis.
(b) Drive mode. Spot welding robot can be pneumatic, hydraulic or electric type. The electric robot is used most widely due to its advantages such as easy maintenance, low energy consumption, fast speed, high precision and good safety.
(c) Work space. Robots with work space that meets the actual working requirements shall be selected. Generally, the robots shall be selected based on the positions and quantity of welding spots. Generally, the work space shall not be less than 5 m^3.
(d) Motion range and maximum motion speed of each DOF. Table 5.1 shows the typical motion range and maximum motion speed of a spot welding robot at each DOF.
(e) Loading capacity. As the spot welding tong is heavy, the robot is required to have a large load capacity, generally no less than 50 –120 kg.

Table 5.1 Typical motion range and maximum motion speed of spot welding robots at various DOF

DOF	Motion range/(°)	Maximum moving speed/[(°)/s]
Waist axis	±135	50
Shoulder axis	Front 50, rear 30	45
Elbow axis	Upper 40, down 20	40
Wrist rotary axis	±90	80
Wrist swing axis	±90	80
Wrist twisting axis	±170	80

(f) Control mode and repetition accuracy. For spot welding, the positions of the welding spots shall can be controlled. Therefore, the point-to-point (PTP) control mode may be used, with the positioning accuracy being higher than ±5 mm.

(g) Spot welding speed. The spot welding speed of spot welding robots is relatively high, generally above 60 spots/min. When selecting the spot welding robot, attention shall be paid to matching the single-spot weld time with the logistics speed of the production line.

(2) Teach pendant

The teach system is the interaction interface between robots and humans. It is essentially a dedicated smart terminal. Figure 5.4 shows a typical spot-welding robot teach pendant.

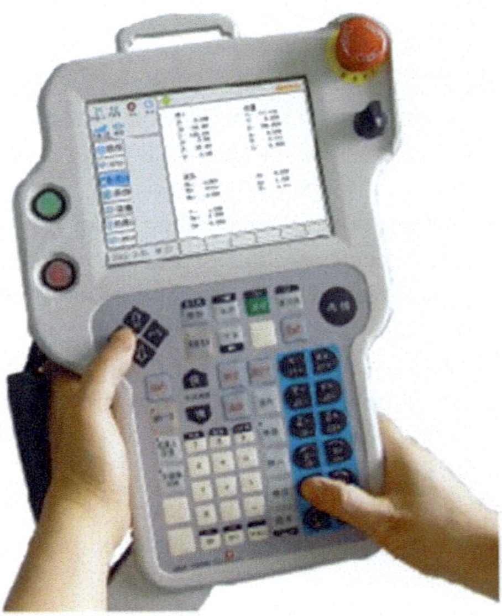

Fig. 5.4 Spot welding robot teach pendant

The teach pendant adopts a graphical interface. All the instruction operations can be realized through the touch screen and keys, including all the settings and status information display. The control mode of spot welding robot is usually PTP (point-to-point control type). Therefore, the teaching is relatively simple. Only the position & orientation of the welding tong at two end points and the welding parameters need to be taught. The path between the two points doesn't need.

(3) Robotic controller

The controller of the spot welding robot processes all the information during the working process and issues control commands to control all the action of the robot system and the welding parameters of the welding tong. In case of mis-operation by an operator during the teaching process or malfunction during welding, the controller will trigger the alarm system to sound an alarm, stop the manipulator, and display error or fault messages.

Generally, the point-to-point I/O mode is used for communication between the spot-welding robot controller and the resistance spot-welding controller.

5.1.3 Spot Welding System

The spot welding system consists of an welding tong, a spot welding power supply, a welding controller and auxiliary systems such as water, electricity and gas supplies etc.

(1) Welding tong and power supply

1) Structure of the welding tong

The welding tong is the end effector of spot welding robot, which can be classified into C-shape and X-shape according to its structure, as shown in Fig. 5.5. C-shaped welding tongs are used for welding spots in vertical and near vertical positions, and X-shaped welding tongs are used primarily for horizontal and near horizontal positions.

2) Driving method of the welding tong

Spot welding robot welding tongs can be classified into pneumatic welding tongs and servo welding tongs according to the driving method. The pneumatic welding tong is driven by a compressed air cylinder. The opening/closing operation is controlled by a reversing valve, and the welding controller sends an analog signal to drive a proportional valve to control the welding pressure. This kind of welding tong has the advantages of simple structure and easy maintenance. The main disadvantage is that the welding pressure cannot be adjusted in real time during welding, which is not conducive to improvement of welding quality. The movement speed of electrode cannot be controlled during pressurization, causing great impact on workpiece, making the workpiece prone to deformation and easily causing high noise. This

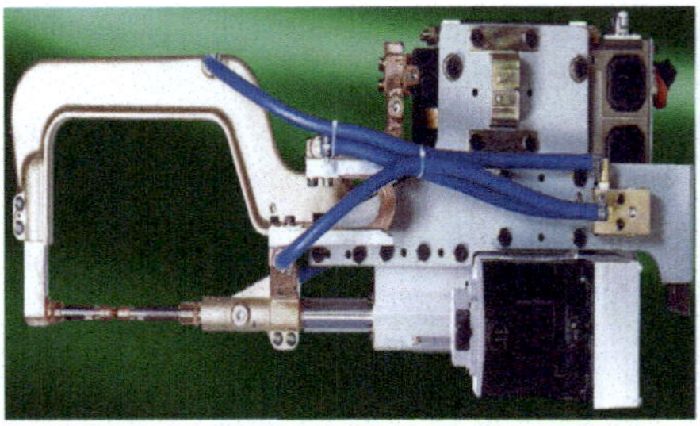

(a) C-shaped servo welding tong

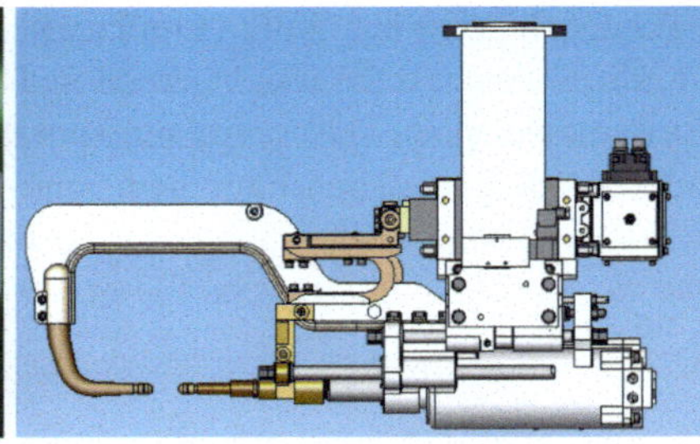

(b) C-shaped pneumatic welding tong

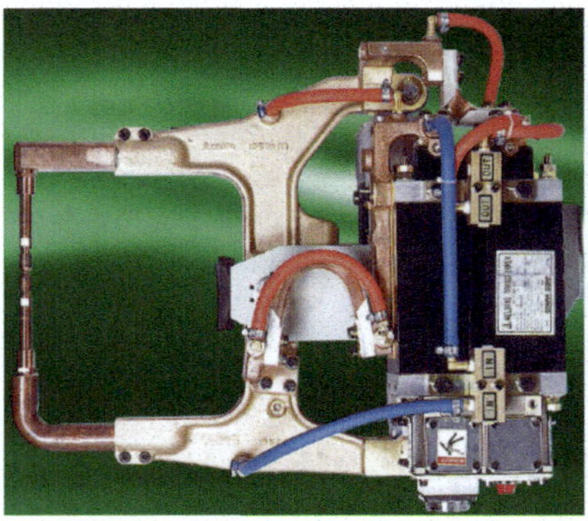

(c) X-shaped servo welding tong

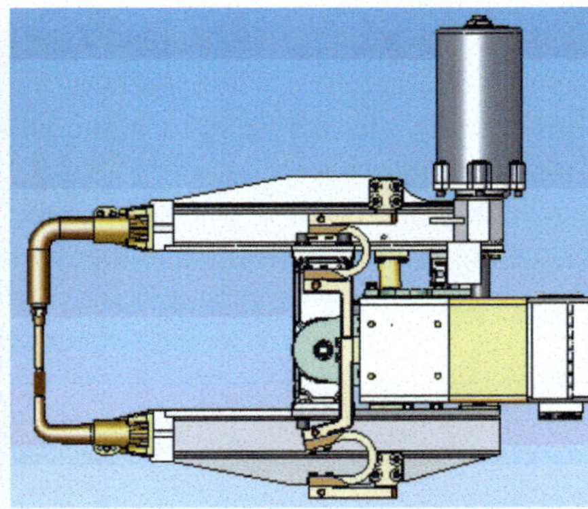

(d) X-shaped pneumatic welding tong

Fig. 5.5 Welding tong used for spot welding robot

driving method also has the downside of low control accuracy of welding pressure, which tends to cause large spatters. In addition, the electrode of this welding tong is susceptible to wear, which will also affect the welding quality.

The servo welding tong is driven by a servo motor. The opening, closing and welding pressure of the welding tong are controlled by a servo motor. The opening and welding pressure are subject to stepless adjustment with high precision. Some robots can also use the servo drive motor of the servo holder as a linkage axis. During the closing and pressurization process of the servo welding tong, the pressure can be adjusted in real time to ensure that the two electrodes are gently closed to achieve soft contact, thus minimizing the impact on the workpiece, and weakening the noise and spatter. As the opening of the welding tong can be steplessly adjusted and accurately controlled, the jaws can be closed gradually during the movement from one welding point to another, thus shortening the weld cycle and improving the production efficiency.

3) Connection between spot welding tong and power supply

A welding power supply is to be used to provide welding current. According to the connection between the welding tong and the power supply for spot welding, the

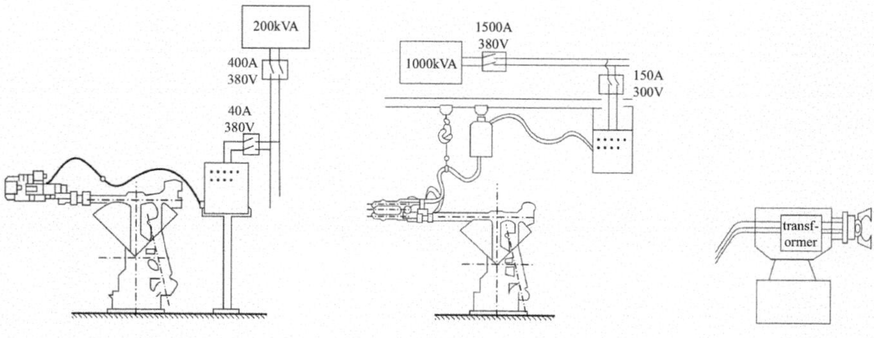

(a) Integrated type (b) Separated type (c) Built-in-type

Fig. 5.6 Schematic diagram of spot welding robot

welding tong can be classified into three types: Integrated welding tong, separated welding tong and built-in welding tong, as shown in Fig. 5.6.

For the separted type, the power supply and the tong are assembled together, and then the installed at the end of the manipulator arm of the robot, as shown in Fig. 5.9a. Its advantages include no need for large-diameter secondary cables and rack for mounting the transformer, simple structure, low maintenance cost and energy saving (2/3 energy saved compared with separated type). The disadvantage is that the end of the manipulator is subjected to a large load (generally 60 kg), resulting in poor accessibility of the welding tong. Figure 5.7 shows the structure of an integrated point holder.

For the separted type, the welding tong and the power supply for spot welding are installed independently. The former is installed at the end of the manipulator arm of the robot, while the latter is suspended on the suspension rail above the robot and can move along with the welding tong on the rail. The two are connected by cables, as

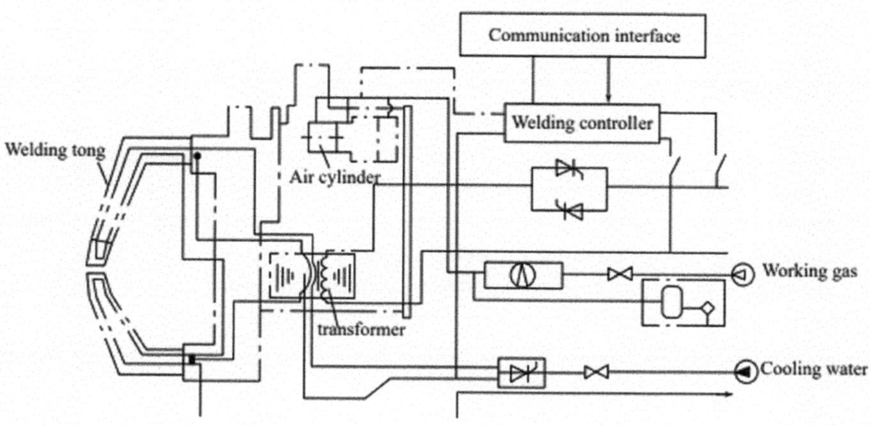

Fig. 5.7 Structure of the integrated welding tong

shown in Fig. 5.9b. The advantages of this type of welding tong include low load at the end of arm of the manipulator, high motion speed and low cost. However, it has disadvantages such as high energy loss, low energy efficiency, restricted working space and welding location, high maintenance cost (connecting cables need to be replaced periodically).

The power supply for spot welding of a built-in welding tong is placed in the robot arm near the welding tong, as shown in Fig. 5.9c. A valuable advantage of built-in tyle welding tong is its compact structure. The secondary cable is short and the transformer capacity is small. However, the design structure of the robot body is complicated.

(2) Welding controller

The function of welding controller is to implement welding parameter input, store welding program, control time sequence of simple or complex spot welding, control current waveform, adjust and control welding pressure, control weld time (including squeeze time, weld time, hold time and off time, etc.), provide fault diagnosis and protection, and communicate with the robot controller and manual teach pendant. The communication mode is generally point-to-point I/O mode.

According to the relationship between the spot welding controller and the robot controller, the spot welding robot system is divided into three types:

(a) Central control type. Management, co-ordination and control are all realized by the host computer. As a module of the whole control system, the welding controller is installed in the cabinet of the robotic controller. The advantage of this control is the high degree of integration of the equipment.
(b) Distributed control type. The welding controller and the robot controller are relatively independent of each other, respectively controlling the welding process and the operation of the welding robot body, and communicate with each other through "answering" mode. When starting welding, the robot controller will send a welding start signal. After receiving this signal, the welding controller will control the welding process and send an end signal to the robot controller upon the completion of welding. After receiving the end signal, the robot controller moves the welding tong to the next welding position. Typical welding cycle is shown in Fig. 5.8.
(c) Group control system. With the group control computer as the center, multiple spot welding robots are connected into a network which is subject to group control. Each spot welding robot is provided with signal ports of "welding requested" and "welding permitted" which are connected to the group control computer to realize the time-staggered welding of the welding machines in the network. This control method can optimize the instantaneous load of the power grid, stabilize the grid voltage, thus improving the weld quality.

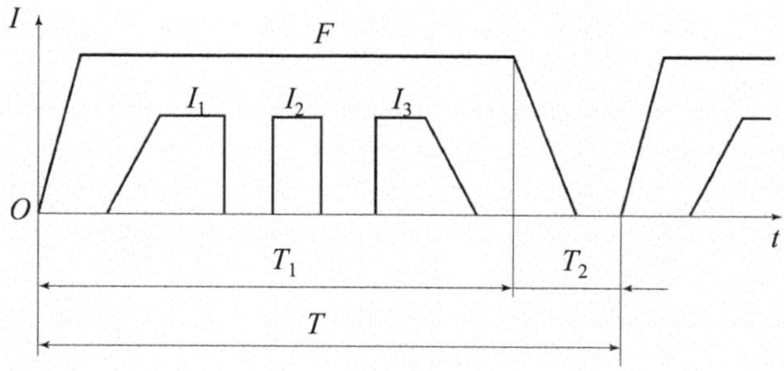

T_1- Control by welding tang controller; T_2-Control by Robot controller; T-Period of welding cycle; F-Electrode pressure; I-Welding current

Fig. 5.8 Typical welding cycle of spot welding robot

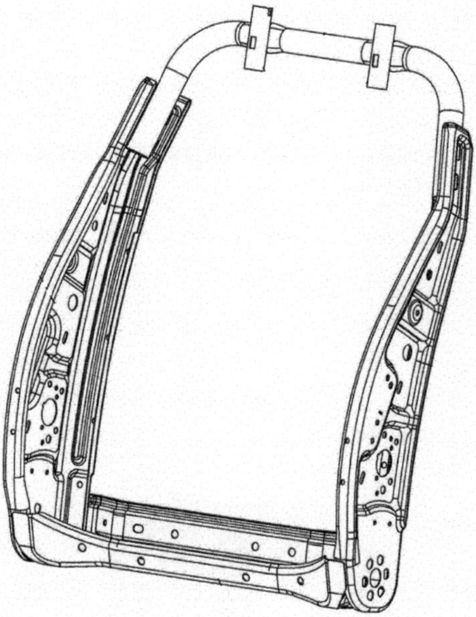

Fig. 5.9 Vehicle seat frame assembly

5.1.4 Application of Spot Welding Robots

(1) Spot welding robot work station for vehicle seat frame assembly

The vehicle seat frame is usually made of tubing and stampingss, as shown in Fig. 5.9. The entire skeleton has a total of 30 weld spots, most of which are located in spatial positions that is difficult to access, with high requirements on production pace. The production time of a single piece shall be no longer than 60 s. Moreover, the product has high requirements on accuracy, and the overall error shall be less than 0.5 mm.

Spot welding robotic workstations are very suitable for welding this type of metal structure.

A typical vehicle seat frame spot welding robot work station mainly consists of a spot welding robot, a spot welding controller, a medium frequency servo welding tong, a workbench table, fixtures and a safety fence (Fig. 5.10). The robot body is of 6 DOF multi-joint type, with a repeat positioning accuracy of ±0. 07 mm, each shaft driven by an AC servo motor with a maximum load of 235 kg. The workbench is selected to be linked with the robot, with its transposition controlled by the robot controller in a unified way. The workpiece clamping is performed by the robotic controller and PLC controller in collaboration. The welding parameters and time sequence control of the welding process shall be preset by the user using a teach pendant. As the materials and thicknesses of workpieces at different parts of the frame of seats are different, parameters such as welding current, weld time and welding pressure shall be set separately for each welding spot.

The defects frequently encountered in the actual production includes nugget deviating, burn-through and over-samll nugget. The main reason for nugget deviating is that the teaching point for the welding spot deviates significantly from the desired position. The main reason for burn-through is that the welding current is too large, the weld time is too long or the gap between the two workpieces is too large. The main reason for the over-small nugget is that the welding current is too low and/or the weld time is too short.

(2) BIW (Body-in-White) Spot Welding Robot Workstation

The BIW spot welding robot work station is usually composed of several spot welding robots which are controlled in group control mode. The group control computer

Fig. 5.10 Seat framework spot welding robot workstation

Fig. 5.11 BIW spot welding robot workstation

communicates with each robot and the robots communicate with one another via network line. With this kind of communication, this computer controls the welding sequence and process of each robot. Under the control of group control computer, all robots perform welding according to predetermined pace. At times, there may be multiple robots simultaneously welding in different locations, while several other robots are ready for welding. Each robot will automatically perform the next predetermined weld after completing the current one. The robots coordinate with each other to make full use of time and space, thus significantly improving the welding productivity. Figure 5.11 shows a typical spot welding robot station for body-in-while (BIW).

5.2 Arc Welding Robot Systems

5.2.1 Structure of Arc Welding Robot System

Arc welding robot system consists of a robot body, a robot controller, a teach pendant, a welding system, an external sensor, a positioner, a safety protection device and a welding torch cleaning station, while welding system mainly consists of a welding torch, a wire feeder and an arc welding power supply. The power supply for arc welding shall be dedicated for the robot and shall have communication interfaces with the robot. Figure 5.12 shows the structure of a typical arc welding robot system. A robot station may incorporate one robot and multiple positioners, where welding and workpiece assembly are performed at the same time; or multiple robots and one positioner, where the welding on different location of the workpiece is performed at the same time. Figure 5.13 shows a robot work station consisting of one robot and

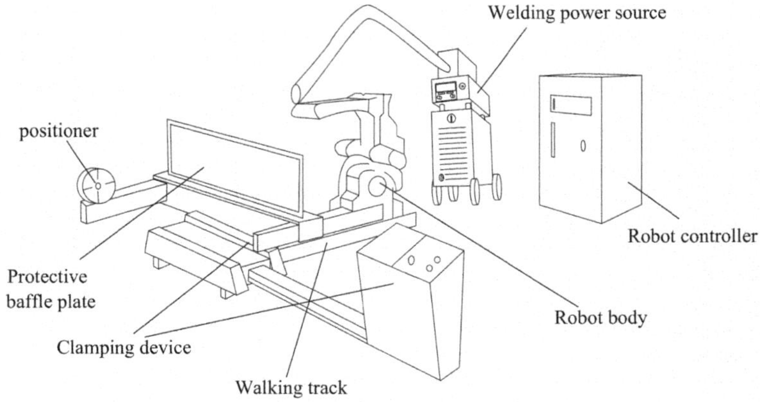

Fig. 5.12 Structure of typical arc welding robot system

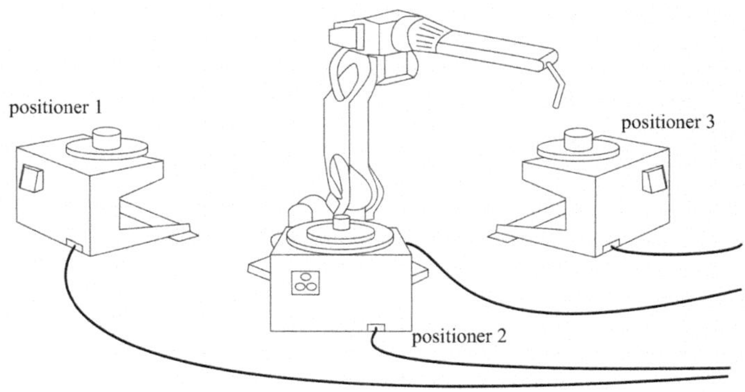

Fig. 5.13 Robotic workstation consisting of one robot and three positioners

three positioners. Figure 5.14 shows a robot work station consisting of two robots and one positioner.

5.2.2 Robot Body and Controller

(1) Robot body

In order to achieve high quality welding, the arc welding robot is required to drive the welding torch precisely along the bisector of the groove (if the torch swings, the weaving center travels along the bisector of the groove) and ensure the orientation of the welding torch. In addition, the control system shoud be able to continuously adjust the welding process parameters (such as welding current, arc voltage, welding

Fig. 5.14 Robot workstation consisting of two robots and one positioner

speed, and position & orientation of welding torch, etc.) according to the groove size or alignment during welding. Generally, the following requirements shall be met:

(a) Degree of Freedom. Six-degree-of-freedom arc welding robot can meet most of the requirements for arc welding. If the workpiece is complicated and high degree of automation is required, the degree of freedom can be extended by means of positioner.
(b) Repeat positioning accuracy. Most arc welding processes require a repeat positioning accuracy of ±(0.05 –0.1) mm, and plasma arc welding process requires a higher positioning accuracy.
(c) Work space. The work space of a robot is generally 1.40 –1.6 m, which can be extended through a gantry, as shown in Fig. 5.15. The work space can also be expanded through a cantilever beam or a traveling track, etc., as shown in Fig. 5.16.

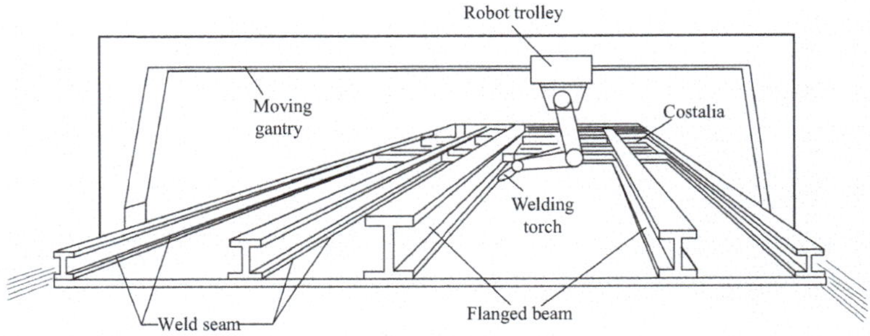

Fig. 5.15 Robot system expanding work space via gantry

(a) Expanding work space via cantilever beam (b) Expanding work space via travelling track

Fig. 5.16 Robot systems expanding work space with different ways

(d) Load capacity. Generally, the weight of the robotic arc welding torch is usually small and the load capacity of a arc welding robot required to be 6 –20 kg.
(e) Easy installation of various sensors for adaptive control.

(2) Robotic controllers and teach pendants

The robot controller is the nerve center of the robot. It stores and executes the work programs created via the teach pendant, processes all the information and controls all the actions of the robot during its work. The arc welding robot controller shall generally meet the following requirements:

(a) Having abundant interface functions for exchanging information with positioners, arc welding power supplies, and sensors, etc. through certain protocols.
(b) Providing with sufficient storage space for storing the programs created in more than 1 hour of teaching. It shall be able to store at least 5000–10000 point positions.

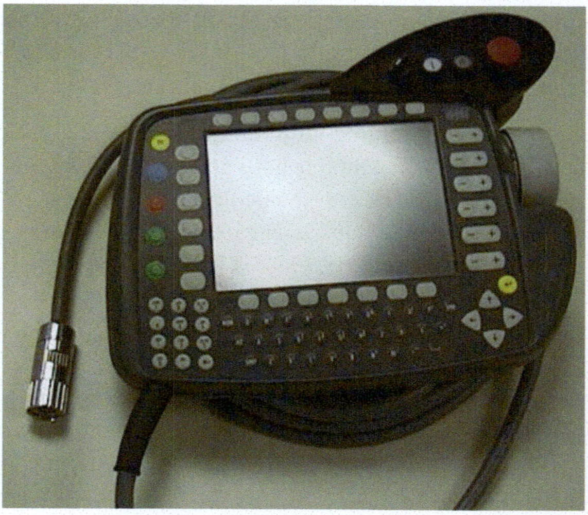

Fig. 5.17 Arc welding robot teach pendant

(c) Having high immunity to interference and high reliability, as well as stable operation in various production environments with failure less than 1 fault per 1000 h.
(d) Having self-testing and self-protection functions, e.g., when welding wires or electrodes are "stuck" to the workpiece, the system is automatically powered off immediately to protect the system; the welding torch shall be able to automatically reset and re-strike the arc when the welding arc is not ignited.

Arc welding teach pendant is used for teaching robot, preparing teaching program, displaying operating status of robot, and executing or test-running the teaching program. It can be connected with the robot controller via interfaces (such as USB interface and CAN bus) and communicate according to certain communication protocols. If a USB interface is used, hot swap shall be available. The teach pendant has rich keyboard functions and touch screens to facilitate robot motion control and programming, as shown in Fig. 5.17. Different robots have different operating systems, most of which are similar to Windows.

5.2.3 Welding Systems for Arc Welding Robots

The welding system for an arc welding robot mainly consists of an arc welding power supply, a welding torch and a welding torch cleaning station.

(1) Arc welding power supply

In many cases, the welding robot needs to continuously adjust the welding parameters during the welding process. Therefore, the power supply for robotic arc welding shall be equipped with a robot interface for communication with the robot control system. In addition, it shall have higher reliability, dynamic performance and adjustment performance. High-performance full digital power supply with expert database or synergic welding program is preferred. In addition, the power supply for robotic arc welding shall have the following functions:

(a) Automatic welding wire re-burning and ball removal function, i.e., preventing the formation of molten balls at the ends of welding wire during the arc extinction process through coordinated control of wire feeding speed and power supply output voltage. This is because the arc welding robot requires 100% success rate of arc initiation. Once a ball is formed at the end of the welding wire, the success rate will be significantly reduced.
(b) Accurately adjusting the the magnitudes, rise rates and decending rates of the arc igniting current and extinguishing current to ensure the quality of weld formation at the igniting and extinguishing points.
(c) Arc welding robot controller connection interface shall be provided to communicate with robot controller with high speed by means of I/O, Device Net, Profibus and Ethernet.
(d) The power supply for gas metal arc welding shall be capable of reliable one-pulse, one-droplet transfer.
(e) The duty cycle shall be 100%.
(f) The power supply for TIG welding shall be also provided with high-frequency shielding function. Otherwise, the high-frequency electric signal will enter the robot control system during the arc igniting process, affecting the robot operation and even damaging the robot control system.

(2) Wire feeder

The wire feeder is usually installed on the shoulder of the robot body. As compared with semi-automatic welding, the arc welding robot requires higher stablity in wire feeding. Therefore, it shall be equipped with four-wheel wire feeding mechanism and driven by servomotor. Especially for CMT-GMAW wire feeders, servo motors with extremely low inertia should be used to achieve fast switching between wire feeding/retraction. The wire feeder for arc welding robots is generally controlled by a microprocessor and equipped with interfaces for communication with the arc welding power source and robot controller. It shall has a jog wire feeding/retraction function, and the welding wire reel should be easy to replace.

(3) Welding torch

The robotic welding torch can be mounted in two ways: Internally mounted and externally mounted, as shown in Fig. 5.18. If the internal mounting method is used, the wrist part of the robot body shall be hollow to facilitate the mounting of welding torch. This mounting method has the advantage of good accessibility of the welding torch, but suffers disadvantages such as complex construction and detrimental effect of rotation of the welding torch on cable life. In case of external installation, a welding torch clamping device needs to be installed on the sixth axis of arc welding robot, which greatly reduces the accessibility of welding torch. However, this installation

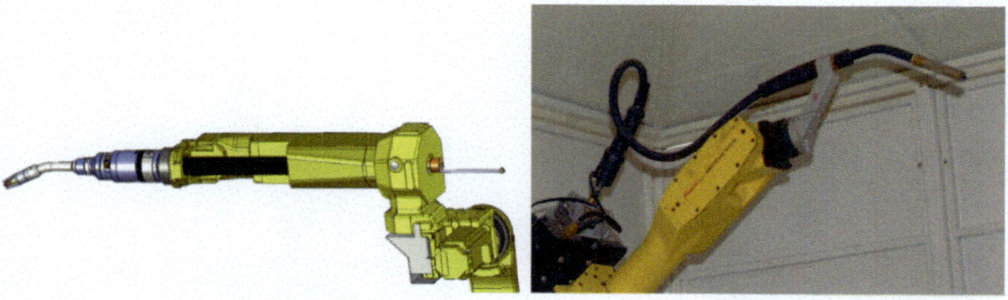

(a) Internally mounted (b) Externally mounted

Fig. 5.18 Installation of a robotic welding torch

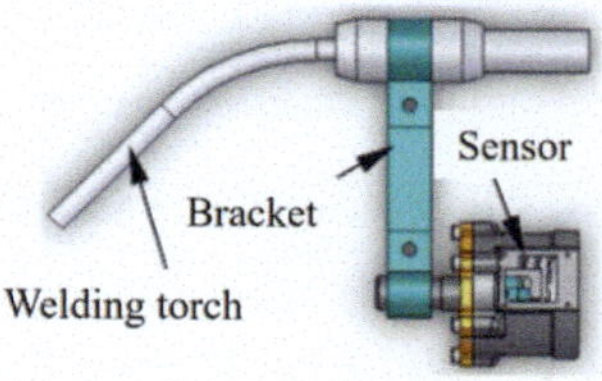

Fig. 5.19 Mounting an anti-collision device to welding torch

method has the advantages of low cost and simple structure. This installation method is usually used on the premise that the accessibility meets the requirements.

It is preferable to install anti-collision sensors on the robotic welding torch, as shown in Fig. 5.19. In this way, the welding torch will stop immediately when encountering any obstacle or person to ensure the safety of personnel and equipment. The commonly used anti-collision sensors are three-dimensional sensors of compression spring type. In case of collision, the collision force will press the spring to retract, which activate the switch to stop the robot immediately. After resetting, the spring will rebound automatically, without the need to re-calibrate the welding torch. This anti-collision device has the advantages of small size, simple structure and high reliability, which can prevent collision in all directions. The degree of protection can also be adjusted by changing the springs with different elasticity coefficients according to actual needs.

(4) Torch-cleaning station

In order to improve production efficiency, arc welding robot systems are usually equipped with a torch-cleaning stations to remove spatter particles and foreign matters from the torches and to apply anti-spatter oil. The torch-cleaning station usually uses a rotating reamer for nozzle cleaning. Insert the reamer into the nozzle and rotate it around the welding wire and contact tube for a few turns to clean up the spatters. After the cleaning, rotate the oil injection nozzle to align with the inner wall of the nozzle to the spray anti-spatter silicon oil, as shown in Fig. 5.20.

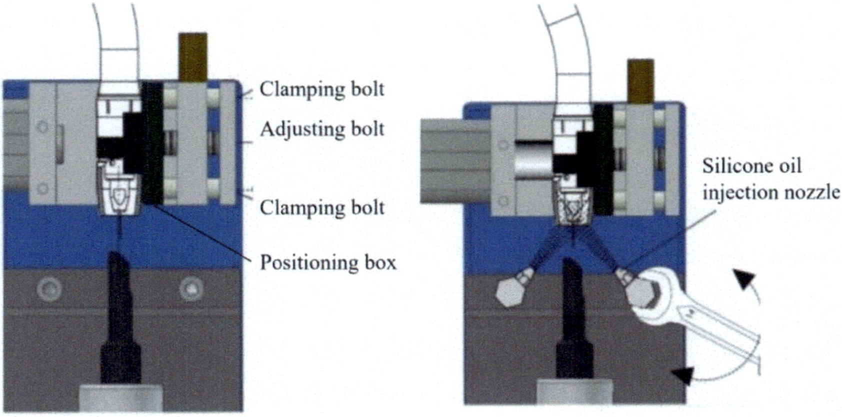

(a) Cleaning spatters with reamer (b) Spraying Silicone oil on the inner wall of nozzle

Fig. 5.20 Schematic diagram for torch-cleaning station

5.3 Welding Robot Positioner

A positioner is an indispensable part of robot system. It is used to position the to-be-welded seam to the most suitable position for robot welding by flipping, rotating and moving the workpiece, so as to improve the weld quality and efficiency.

Positioning can be done before welding or in real time with robot action during welding. If real-time positioning is required during welding, the positioner shall be equipped with communication interfaces for robot and communicate with the robot through certain communication protocols. The controller of the robot shall coordinate and control the movement of the positioner in a unified manner. The motion axis of the positioner directly becomes the externally-extended motion axis of the robot. The control of the positioner by the robotic controller is called linkage. Figure 5.21 shows a typical example of extending the degree of freedom (motion axis) of the arc welding robot system from six to eight lingkage between the positioner and the robot body.

The positioners can be classified into two types according to whether it is linked with robots, namely, linkage positioners and common positioners. Most positioners used in production are linkage positioners. According to the number of motion axis, the positioners can be classified into single-axis, double-axis and triple-axis types.

The most common motions of a positioner are rotary motion and flipping motion (tilt motion). The rotary drive shall realize stepless speed regulation and reverse rotation. The speed fluctuation at the rated maximum load shall not exceed 1% within an adjustable rotating speed range. The flipping motion shall be stable, without any jerk or slip under rated maximum loads. A tilt angle indication scale as well as limit device for controlling the tilt angle shall be provided. The tilt mechanism shall be

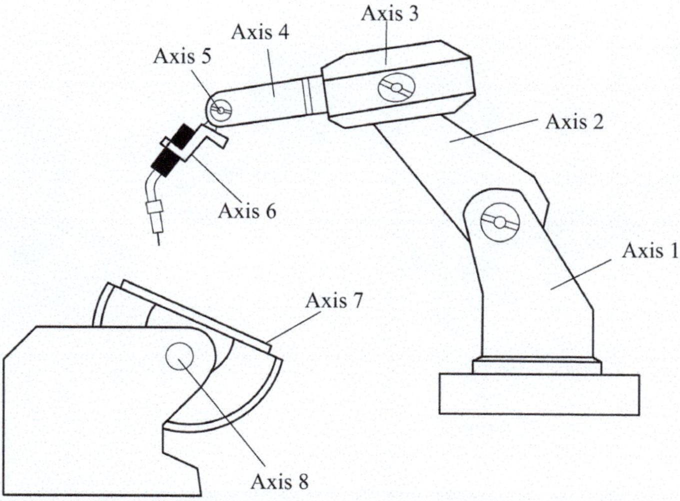

Fig. 5.21 Arc welding robot system with eight axes

equipped with a self-locking feature for safety. Some positioners can also move in certain directions, such as translating or rising.

5.3.1 Single-axis Positioner

The typical single-axis positioner is the headstock-tailstock positioner, with the headstock and tailstock detachable, as shown in Fig. 5.22. This kind of single-axis positioner mainly consists of a drive headstock, a frame, a tailstock and a drive system. The drive headstock is equipped with a servo drive motor and a high precision reducer to provide rotating power. There is no power unit on the tailstock, which is only used to clamp the workpiece. This positioner rotates the workpiece around a horizontal axis by 360°. Its main parameters include load capacity, rotation angle, workpiece diameter, workpiece length, rotation speed, and positioning accuracy, etc.

Another typical single-axis positioner is a double-column single-rotation one, as shown in Fig. 5.23, It consists of a workbench (or a fixture), two columns and a drive unit mounted on one column. The workbench surface is at a certain distance from the ground to to achieve a large-angle flipping. The workbench (or fixture) may also be designed to be lifted and lowered along the column. This type of positioner is suitable for welding large workpieces, such as the rear frame of loader, roller frame, and rectangular structural parts of engineering machinery.

Other single-axis positioners include box-type and T-type ones, as shown in Fig. 5.24. The workbench surface of the box-type positioner is perpendicular to the ground, and its rotating axis is parallel to the horizontal direction and at a certain distance from the ground, as shown in Fig. 5.24a. The workbench surface of T-type

(a) Integrated Type (b) Split Type

Fig. 5.22 Headstock-tailstock positioner

Fig. 5.23 Double-column single-rotation positioner

positioner is parallel to the ground, and the rotating axis is perpendicular to the ground. The workbench is engraved with mounting baselines and mounting holes for mounting various positioning and clamping mechanisms.

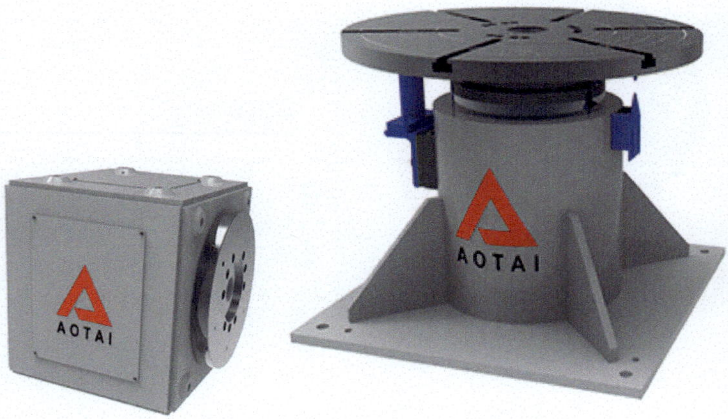

(a) Box-type (b) T-type

Fig. 5.24 Other single-axis positioners

5.3.2 *Double-axis Positioner*

Two types of double-axis positioners are commonly used, namely, A-type, L-type, C-type, U-type and C-type, etc., as shown in Fig. 5.25. Although they have different shapes, their structures are basically the same, consisting of a flipping mechanism, a rotating mechanism, a base, a working platform and a drive system. The flipping and rotating drive mechanism consists of a servo motor and a high-precision reducer, which controls the flipping and rotating with high precision. The workbench is engraved with mounting baselines and mounting holes for mounting various positioning and clamping mechanisms. The workbench has high strength and impact resistance. By linking with a robot, the double-axis positioner can turn the to-be-welded seam at any position to the flat welding position for welding, making it particularly suitable for welding complex structural members with welds on different planes, as shown in Fig. 5.26.

5.3.3 *Three-axis Positioner*

A typical three-axis positioner is an H-type one, which mainly consists of a frame, a rotating support column, a flipping frame, two rotating heads (with each mounting on one end of the flipping frame) as well as drive systems for each axis, as shown in Fig. 5.27. The drive system consists of a servo motor and a RV precision speed reducer. The workpiece can rotate along the vertical axis at the center of the supporting column, flip along the axis of the main beam of the flipping frame, and can also be

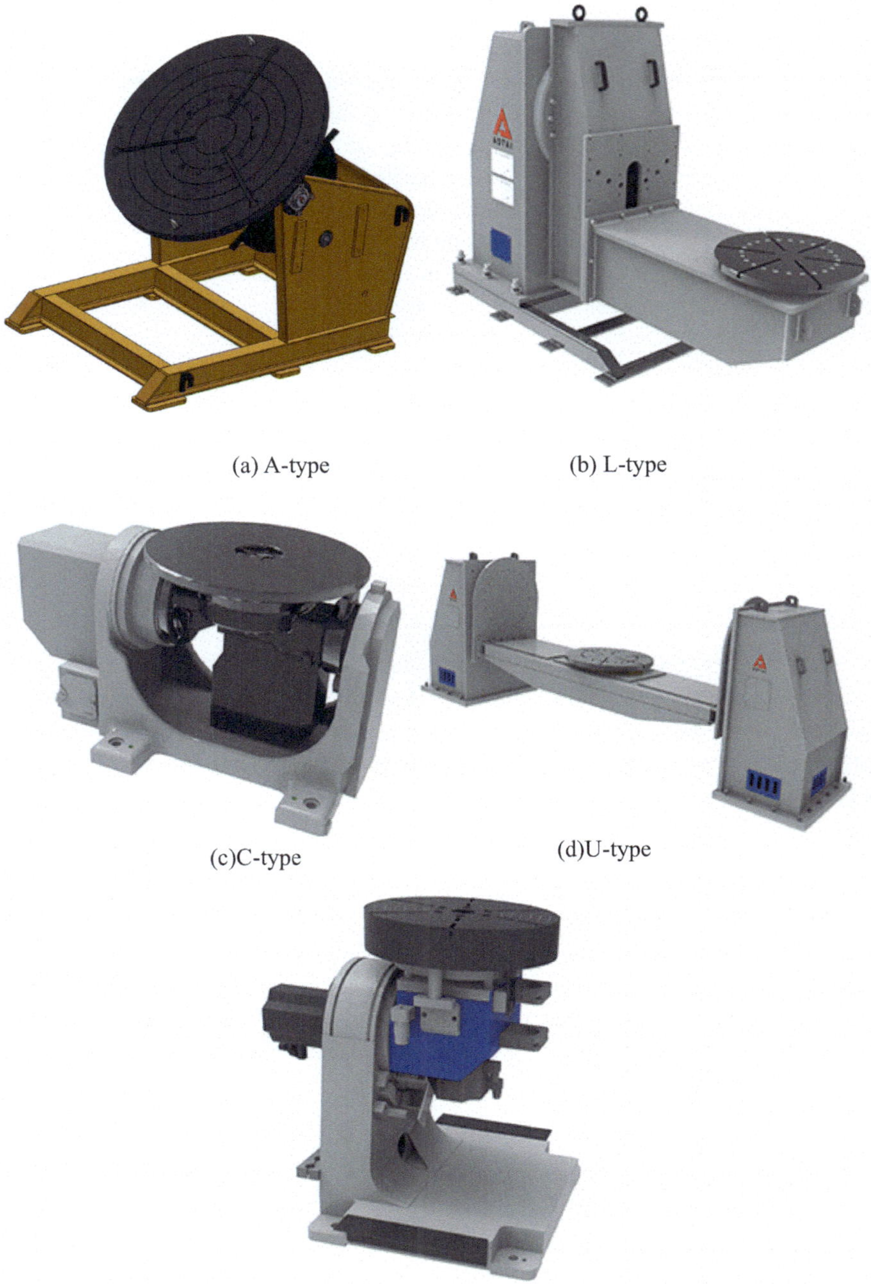

(a) A-type
(b) L-type
(c)C-type
(d)U-type
(e) C-type

Fig. 5.25 Commonly used double-axis positioners

Fig. 5.26 The spatial position is changed to the flat welding position through linkage between the positioner and robot

driven to rotate by a rotating head. The two rotating heads can hold two workpieces to form two workstations. When welding is carried out at one workstation, workpieces can be clamped or removed at the other. Isolation boards are provided between the two workstations to protect the operators.

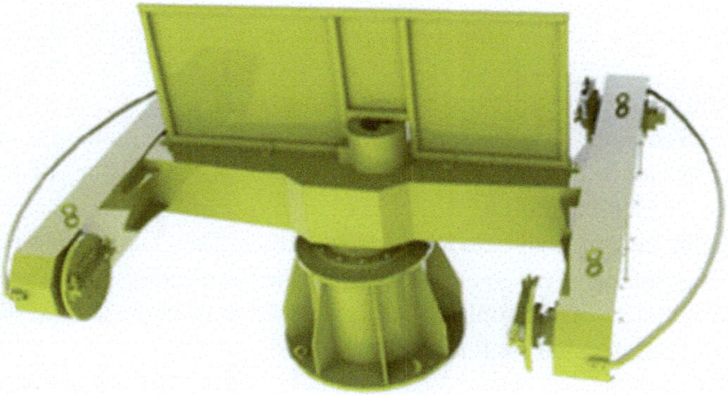

Fig. 5.27 H-type triple-axis positioner

5.3.4 Welding Fixtures

Welding fixtures are used to assemble and clamp workpieces. They are positioning and clamping devices used to improve assembly accuracy and efficiency, ensure dimensional accuracy of weldments and prevent or reduce welding deformation.

(1) Structure of fixture

The fixtures required for different welding workpieces are different. The workpieces vary greatly, so the specific structure of fixtures are also different from with each other. However, the functional structure of the fixtures is basically similar, all consisting of a supporting table, a positioning device, a compression or clamping device, and a measurement device.

(2) Supporting table

In general, the workbench on the welding robot positioner can be used as the support table for the fixtures. This supporting table is provided with holes and grooves for mounting the positioning device, and compression or clamping device.

If the positioner does not have a workbench, sufficient holes or slots shall be provided when designing the surpporting table to facilitate the installation and replacement of measuring devices, positioning devices and fixtures. Frame structure is preferred to save materials and reduce the dead weight, provided that the rigidity and strength meet the requirements.

(3) Positioning device

The positioning device is usually composed of a positioning stop, a positioning pin and positioning template. The following requirements shall be met:

(a) The positioning device shall have sufficient rigidity and hardness, and the working surface shall have sufficient wear resistance to ensure sufficient positioning accuracy during its service life.
(b) In order to enhance the versatility, the positioning device shall be made for easy adjustment and replacement to suit the positioning and fitting of products of different structures or sizes.
(c) Reasonable design shall be adopted to avoid the force from being imposed on the positioning element, so as not to affect the positioning accuracy;
(d) The convenience of assembly and disassembly of workpieces and the accessibility of welding robot shall not be affected.

(4) Clamping or compression device

The installation of clamping or compression devices shall comply with the following requirements:

(a) They shall have sufficient rigidity and hardness, and the working surface shall have sufficient wear resistance to withstand the forces.

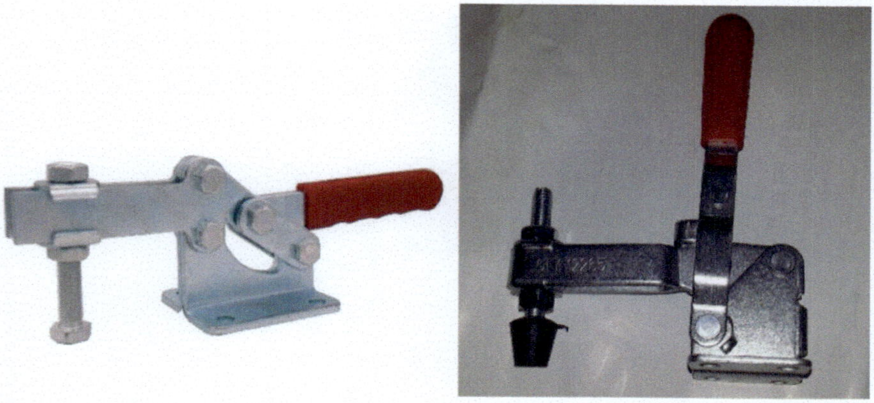

Fig. 5.28 Quick clamping device

(b) They can perform reliable clamping or compression without slipping or generating excessive restraint stress, to avoid damaging the positioning accuracy and affecting the product shape.
(c) They shall ensure the easy assemble and disassemble the workpiece without affecting the accessibility of welding robot.
(d) The the clamping device itself shall be easy to replace.

Currently, the commonly used clamping devices include quick clamping device, and pneumatic clamping device, etc. The quick clamping device is featured with simple structure and quick action. It takes only a few seconds to realize tight clamping, meeting the requirements of mass production. Figure 5.28 shows the typical structure. Quick clamping devices can be used in series or in parallel to achieve secondary clamping or multi-point clamping. Components with lower requirements on positioning accuracy can be clamped and positioned at the same time, eliminating positioning elements. Figure 5.29 shows a pneumatic clamping device, which allows for automatic and more reliable compression.

5.4 Application of Arc Welding Robot Systems

5.4.1 Application in Engineering Machinery Industry

(1) Welding robot workstations for oil pump square tank and donkey head

The oil pumping unit is a kind of machinery for oil extraction, commonly known as "kowtow machine". It mainly consists of a donkey head, a spindle, a link, a crank mechanism, a gearbox, a power equipment and an auxiliary device, as shown in Fig. 5.30. During operation, the motor will drive the up-down movement of the donkey head through the gearbox and crank linkage system. The donkey head will in

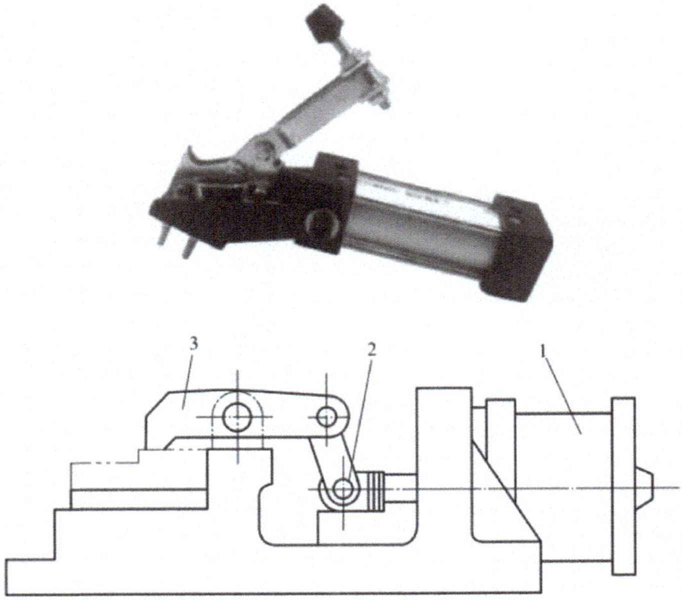

1-Cylinder 2-Lever 3-Press head

Fig. 5.29 Pneumatic clamping device

turn drive the up-down movement of the plunger of the downhole oil pump through the sucker rod, making the crude oil in the well continuously pumped out of the wellhole.

Figure 5.31 shows the square tank structure of oil pump. The welding robot workstation for the oil pump square tank consists of a welding robot body, a MAG welding

Fig. 5.30 Pumps

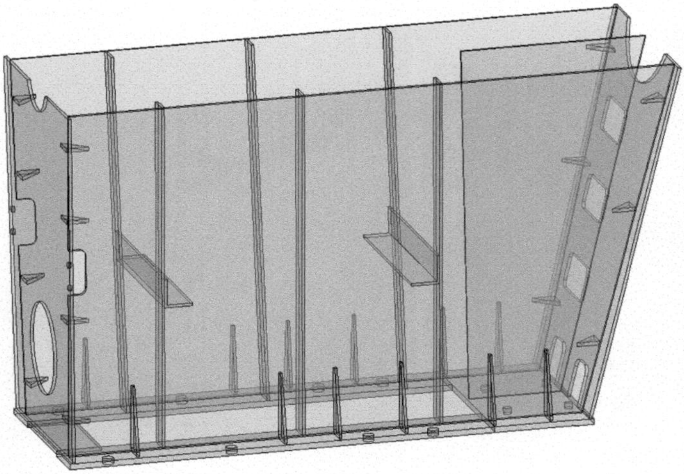

Fig. 5.31 Oil pump square tank

power supply, a tri-axis gantry, a headstock-tailstock positioner, an electrical control cabinet and other peripheral equipment. The system layout is shown in Fig. 5.32. The movement range of the robot gantry is 3000 mm in X direction, 1500 mm in Y direction, and 2000 mm in Z direction. The AC servo drive is usually adopted by the gantry. The effective combination of the welding robot with the gantry expands the effective work space to the maximum extent (Length: 3000 mm + 2 × 1610 mm; width: 1500 mm + 2 × 1610 mm; maximum length of the workpiece to be welded: 2710 mm; maximum width of the workpiece to be welded: 1860 mm), enabling the robot can complete the welding of all the welds of the oil pump square tank.

Figure 5.33 shows the structure of donkey head of oil pump. The welding robot workstation for the donkey head is composed of a welding robot, a welding power supply, a two-axis inverted traveling guide rail, a headstock-tailstock positioner, an electrical control cabinet and other peripheral equipment. The system arrangement is shown in Fig. 5.34. There are a number of internal welds with poor accessibility on the donkey head assembly, and the inverted traveling guide rail is used to turn the robot upside down to facilitate the welding torch to enter the donkey head for reliable welding of these internal welds. The inverted traveling guide rail is driven by AC servo, with the horizontal movement range being 6000 mm and the vertical movement range being 2000 mm, significantly expanding the work space of the robot.

In order to ensure the welding quality and efficiency, both robot workstations are provided with complete self-protection function, original path resumption, fault self-diagnosis, weld seam positioning, multi-layer multi-pass welding function, expert database, weaving function, and welding torch cleaning and wire cutting, etc.

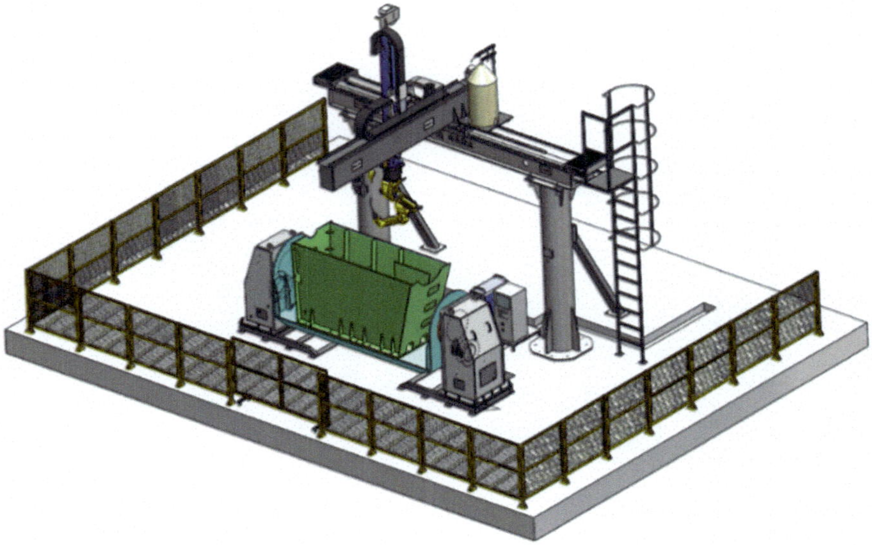

Fig. 5.32 Welding robot workstation for oil pump square tank

Fig. 5.33 Donkey head of pump

(2) Welding process

The surface of the to-be-welded workpiece shall be as clean as possible without oil stains. The assembly and tack weld shall meet the dimensional tolerance requirements of the workpiece drawing. For a fillet weld, if the root gap exceeds 2 mm, back weld

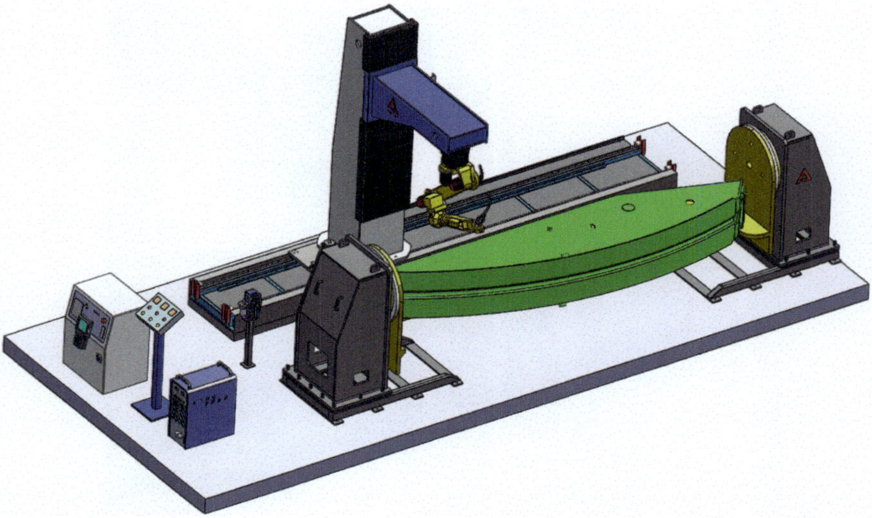

Fig. 5.34 Welding robot workstation for oil pump donkey head

should be first deposited using shielded metal arc welding or semi-automatic CO_2 gas shielded welding. And the tack welding is not allowed to be carried out within 10 cm of the starting point of the weld.

Figure 5.35 shows the flow chart of the welding process.

(a) Assembly: Assmble the workpieces according to the drawing requirements and perform tack welding.
(b) Installation of workpiece: The operator enters the work area of robot, fixes the workpiece to be welded on headstock-tailstock positioner with calmping devices.

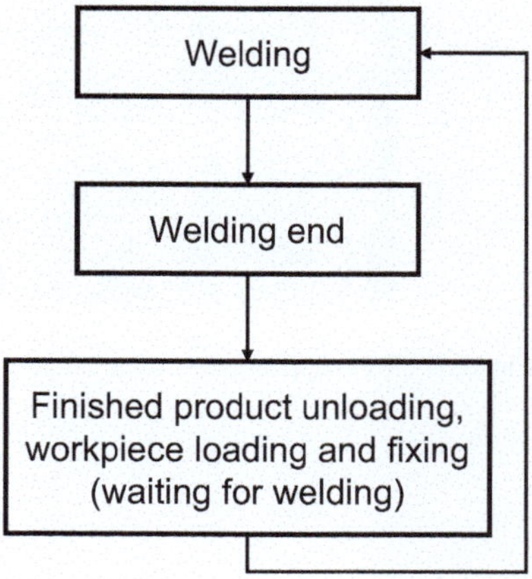

Fig. 5.35 Flow diagram of welding process

Fig. 5.36 Oil pump square tank being welded by the welding robot workstation

(c) Robotic welding: The operator returns to the safe position and presses the startup button, and the robot starts to perform automatic welding from the preset position.
(d) Unloading workpiece: After welding, the operator will enter the work area of robot again to unload the welded workpiece.

Repeat the above steps for the next workpiece.

Figure 5.36 shows an oil pump square tank being welded by the robot workstation. Figure 5.37 shows the typical weld appearance on this oil pump square tank.

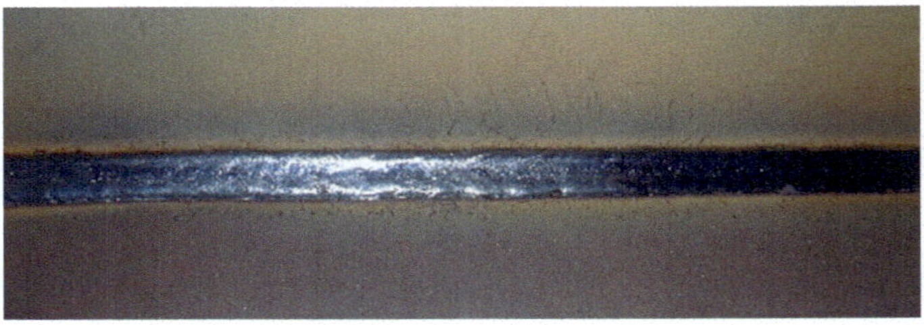

Fig. 5.37 Weld appearance on the oil pump square tank

5.4.2 Application in Building Engineering Industry

(1) Welding robot workstation for construction aluminum formwork

Aluminum formwork is made of aluminum section, aluminum plate, reinforcement bar and fasteners, etc., which are assembled and welded together. During construction, the formwork is used to form a closed cavity required for the construction of concrete structure so as to prevent the concrete leakage and ensure structure formation during concrete pouring, as shown in Fig. 5.38.

A typical construction aluminum formwork welding robot station consists of two welding robot bodies, two MIG welding power supplies and wire feeders, two welding workbenches, a control system and other peripheral equipment. The system arrangement is shown in Fig. 5.39. This welding robot station incorporates intelligent and flexible welding robot, digital pulse inverter welding machine in combination with flexible and variable workbenches and fixtures to achieve the welding of aluminum formwork of various specifications and ensure good welding quality.

(2) Welding workbench

The workbench adopts pneumatic flipping and precise positioning devices, as shown in Fig. 5.40. The pneumatic flipping device is composed of pneumatic drive shaft, positioning plate and compressing beam. The cylinder controls the lifting and lowering of the compression beam by means of a drive shaft to realize the automatic clamping and flipping of workpieces. Each compressing beam can be adjusted at will according to the actual dimensions of different workpieces. The positioning plates are used for fast and precise positioning of workpieces.

Fig. 5.38 Aluminum formwork application

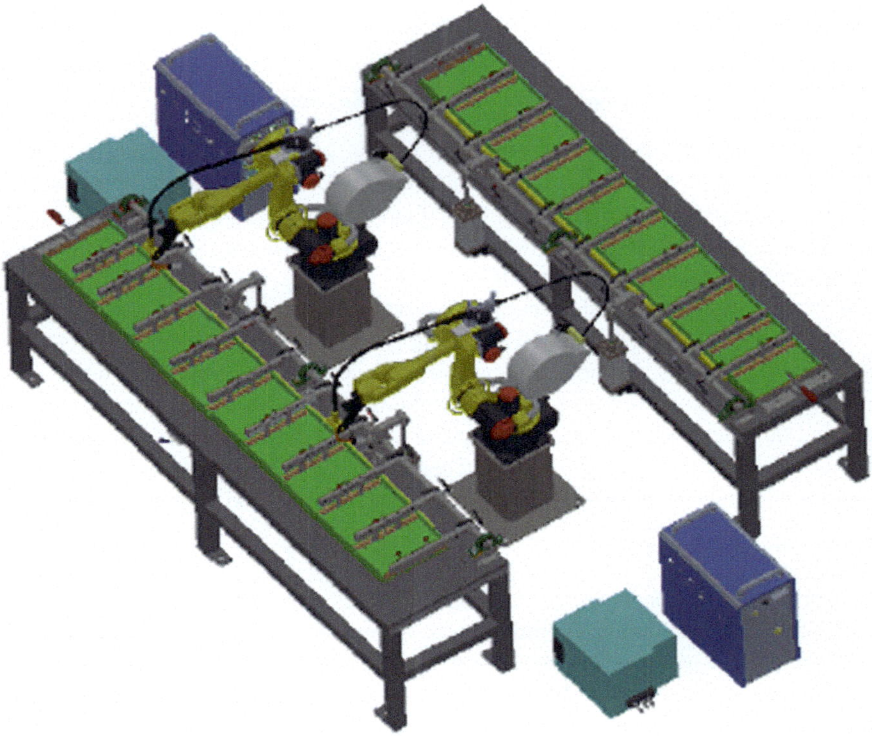

Fig. 5.39 Aluminium formwork welding robot workstation

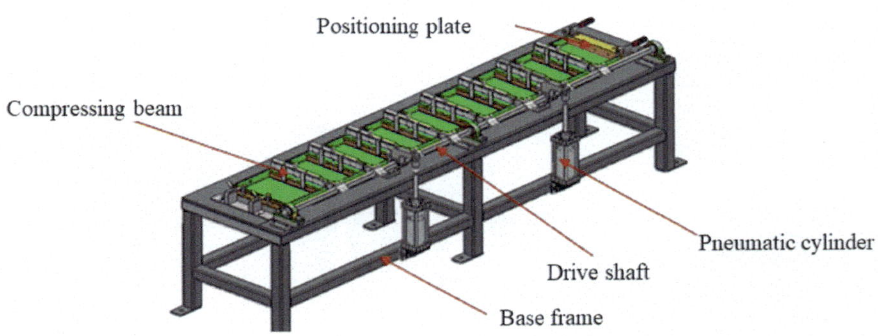

Fig. 5.40 Welding tooling

There are many types of aluminium formworks, and different robotic workstations are used for different aluminum formworks. In addition to the two-robot two-position workstations mentioned above, there are other workstations in various forms, such as single-robot two- workbench and single-robot single- workbench workstations. Figures 5.41, 5.42 and 5.43

show the physical images of the three types of workstations.

(3) Welding process

For a double-robot and double-position aluminum formwork welding robot workstation, two robots weld the same workpiece at the same time. Upon the completion of welding at one position, the workpiece is transferred to another position for welding. The specific process flow is as follows:

(a) Installation of workpiece: The operator enters the work area of robot, places the assembled workpiece on the welding table according to the positioning reference, and presses the button switch of compressing beam; then, the compressing beam will be turned to the horizontal position automatically. The stiffening plate shall be manually attached to the positioning reference on the crossbar before locking the clamp device.
(b) Robotic welding: After installation, the operator returns to the safe position, presses the startup button of robot, and the robot invokes the teaching program to perform automatic welding from the preset position.
(c) Unloading and installation of workpiece: After welding, the robot jumps to another position for welding, and the operator enters the work area of robot

Fig. 5.41 Two-robot and two-workbench system

Fig. 5.42 Single-robot and two-workbench system

Fig. 5.43 Single-robot and single-workbench system

Fig. 5.44 Weld appearance

again, press the button switch to flip the compressing beam, unloads the workpiece that has been welded, and installs the new workpiece on the workbench for robotic welding.

Perform the above cycle for the welding of the next workpiece.

(4) Analysis of welding effect and production efficiency

When using the above-mentioned robot workstation to weld aluminum formwork, the welding electrical parameters, welding speed, and welding gun posture can all be kept stable, thus reducing the influence of human factors on weld quality. The spatter during welding proces is very small and the weld well formed without undercut and other defects, effectively improving the weld quality and stability of welding quality. Figure 5.44 illustrates the appearance of the actual weld.

For a double-station robot workstation, the production time of each aluminium formwork is equal to the weld time of the workpiece, because the loading and unloading can be performed in the other staton while welding in one station. For traditional semi-automatic welding, the production time of a single workpiece equals to the weld time + loading and unloading time + installation time of workpiece + rest time of workers. Table 5.2 shows a comparison of semi-automatic welding and robotic welding production times for two gauges of aluminium formwork. It can be seen that the production efficiency of robotic welding is 2 –2.5 times that of conventional semi-automatic welding.

5.4.3 Application in the Electric Power Construction Industry

(1) Crossbar welding robot workstation

Crossbars are an important part of the power tower. They are used to install insulators and hardware fittings to support the conductors and ground wires and maintain the specified safety distance between them, as shown in Fig. 5.45.

Table 5.2 Comparison of production time between semi-automatic welding and robotic welding

Workpiece No	Dimensions of workpiece (m)	Production time with semi-automatic welding (min/piece)	Production time with robotic welding (min/piece)	Efficiency rise (%)
1	1.2 × 0. 6	6	2.5	140
2	2.7 × 0. 6	11	5	120

Fig. 5.45 Power tower

The welding robot workstation for the crossbar mainly consists of a welding robot body, a MAG welding power supply, two headstock-tailstock positioners, a robot walking mechanism, a control cabinet and other peripheral equipment. The system layout is shown in Fig. 5.46. The box-type crossbar is made of dozens of steel plates and angle steel of various sizes. Assembly errors will inevitably occur after assembly and positioning, and deformation will occur during the welding process. Therefore, this robot workstation should be equipped with a seam tracking device to ensure that the welding torch is always aligned with the weld groove during the welding process. In addition, this robot workstation is usually equipped with a weld positioning device, so that the robot can automatically find the starting position of the weld at the start of welding, thus increasing the welding efficiency.

(2) Welding positioner

The crossbar welding robot worksation adopts a linear walking mechanism, which effectively expands the work space of the robot and can accommodate crossbar welding of different lengths. The headstock-tailstock positioner can turn the workpiece to make the welds reach the optimum position and orientation, and allow for flat position welding of butt joint and fillet joint. A compressing device consiting of a pair of lead screws and compressing plates is adopted for clamping and positioning the workpiece. This compressing device can ensure the positioning accuracy and positioning repeatability of the workpiece.

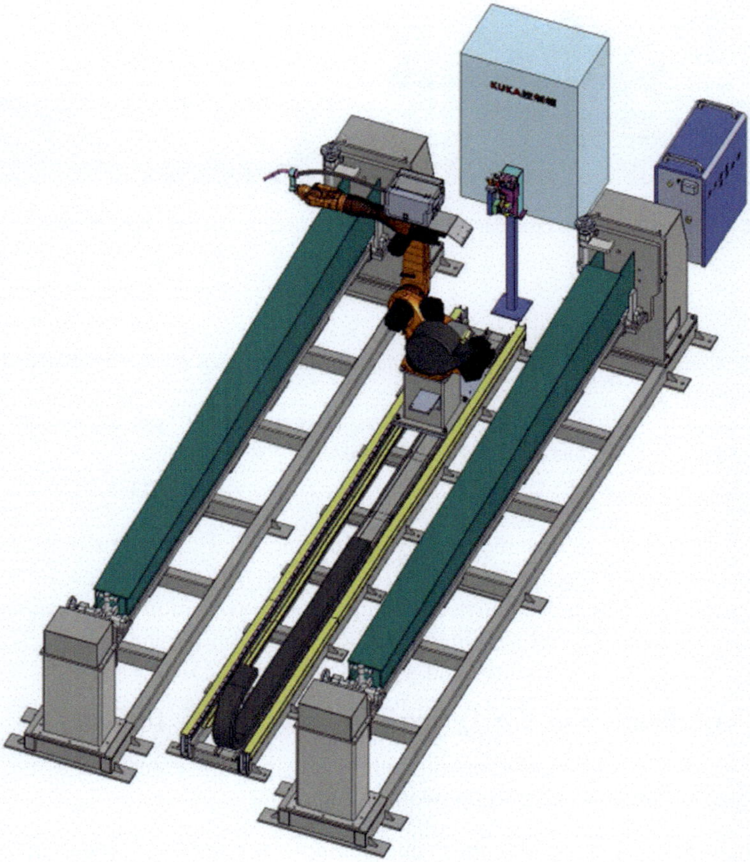

Fig. 5.46 Two-station crossbar welding robot workstation

(3) Welding productivity

Figure 5.47 shows an industrial welding site of crossbar welding robot station and the weld appearance made by this station. Welds are well formed with small residual deformation and no spatter around it. Table 5.3 compares the single crossbar production time between robot welding and semi-automatic welding.

Fig. 5.47 Physical images of crossbar welding robot workstation and weld appearance

Table 5.3 Comparison of single crossbar production time between robotic welding and semi-automatic welding

Dimensions of workpiece/m	Semi-automatic welding time	Robotic welding time (min/piece)	Efficiency rise (%)
1.0	16	12	33
2.0	40	22	80
2.5	60	30	100
3.0	72	38	111
3.5	84	46	117

5.4.4 Application in the Agricultural Machinery Industry

(1) Maize harvester welding robot workstation

The maize harvester is a kind of mechanical equipment capable of corn stem cutting, picking, peeling, stripping, straw treatment and post-harvesting tillage operations, as shown in Fig. 5.48.

Welding robot system can be used for welding of various parts of the harvester, such as straw-pulling wheel, stem-pulling wheel, ears-picking rack and front bridge, as shown in Fig. 5.49.

The welding robot workstation for maize harvester is mainly composed of a welding robot body, a MAG power supply, two headstock-tailstock positioners, a control cabinet and other peripheral equipment, as shown in Fig. 5.50. The digital pulse inverter welding power supply is used as the MAG welding power supply which operates with one-pulse, one-droplet spray transfer, so as to reduce the heat input and deformation of the workpiece. The headstock-tailstock positioner can turn the workpiece over to make the welds reach the optimum welding position, and allow for flat position welding of butt joint and fillet joint. In addition, the the compressing plate mounting on the positioner features easy operation, which enables quick loading/unloading of various components.

Fig. 5.48 Maize harvester

(a) Straw-pulling wheels

(b) Stem-pulling wheels

(c) Ears-picking rack

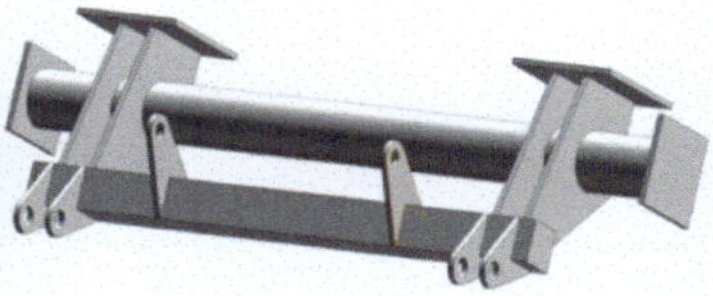

(d) Front bridge

Fig. 5.49 Parts of maize harvester

Fig. 5.50 Welding robot workstation for maize harvester

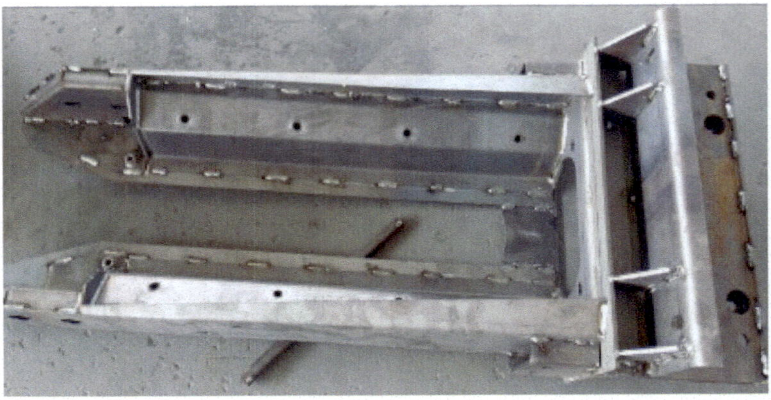

(a) Cutting table made by the welding robot station

(b) Weld appearance (robotic welding)

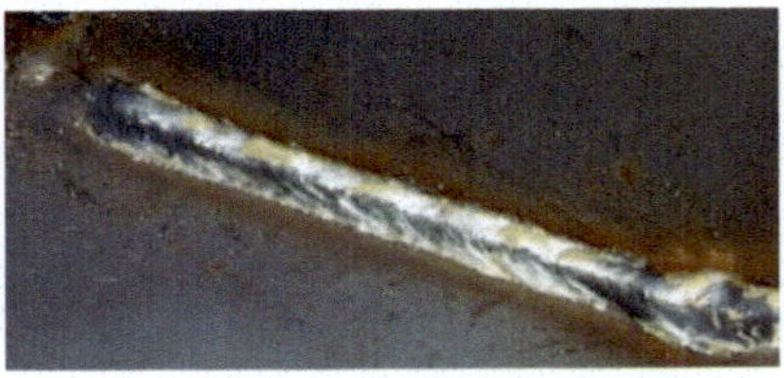

(c) Weld appearance (semi-automatic welding)

Fig. 5.51 Cutting table and appearance of welds made by welding robot station and semi-automatic welding

(2) Welded component

Figure 5.51 shows the cutting table, a component of maize harvester welded by the welding robot workstation as well as a comparison of the weld appearance between robotic welding and semi-automatic welding. It can be seen that the weld made by the welding robot station is more aesthetically pleasing than that made by semi-automatic welding.

5.4.5 Application in Steel Structure Industry

(1) Welding robot workstation for corbel

Steel structure buildings have obvious advantages over brick-concrete ones in terms of environmental protection, energy saving, high efficiency and industrialized production. Large buildings such as the 325 meter high Diwang Tower in Shenzhen, the 421 meter high Jinmao Tower in Pudong, Shanghai, Beijing's Jingguang Center, Bird's Nest, CCTV New Building, Water Cube, and Guangzhou Tiger Leaping Gate Bridge all use steel structures. Figure 5.52 shows the steel-structured Hutiaomen bridge.

The corbel in the building steel structure is used to attach the cantilever beams to the suspension beams and transmit the loads from the suspension beams. Fig. 5.53 shows several typical corbel. Their welding quality plays a crucial role in ensuring the safety and stability of the entire building.

The corbel has a maximum weight of 1000 kg, the length of the workpiece is 300–1500 mm, and the dimensional range of H-shaped section is 250 * 150 * 10 * 10 -1250 * 600 * 50 * 50. All welds are fillet welds with three types of grooves: square groove (when the plate thickness is less than 10 mm), one-side bevel groove (when the plate thickness is 10–30 mm) and K groove (when the plate thickness is 30 –50 mm). After the pre-welding assembly the workpieces shall be fixed through tack welding. The leg length of the tack weld shall be less than 3 mm. The root gap shall

Fig. 5.52 Guangzhou Hutiaomen bridge

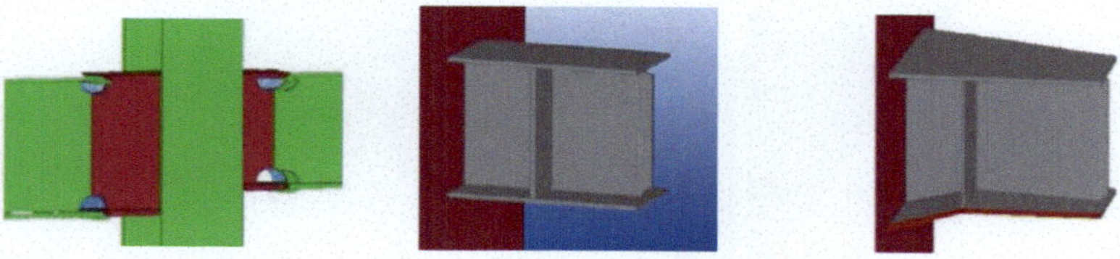

(a) Standard corbel (b) Stiffened corbel (c) H-shaped variable-cross-section corbel

Fig. 5.53 Corbel components

be consistent along the whole seam and be less than 2 mm. All welds are required to be full-penetration welds and conform to Class I non destructive testing standard.

The corbel welding robot workstation mainly consists of a welding robot body, a MAG welding power supply, 2 L-type positioners, a control cabinet, 2 welding fixtures and other peripheral equipment. Figure 5.54 shows the arrangement of single-robot two-station corbel component welding robot workstation system. After the assembly and tack welding are completed, The corbel workpiece is lifted to the robot workstation with a crane and fixed on the positioner through the fixtures. The positioner rotates the workpiece and turn the weld to the flat welding position, then the welding robot automatically seeks to return to the initial welding position to restart welding. After welding one section of the weld, the positioner rotates the workpiece again and turns the next section of the weld to the flat welding position for welding. Repeat the above process to complete all the welding for the corbel.

(2) Welding positioner

The corbel welding robot workstation employs two L-type double-axis positioners for double-station welding. The welding workbench is designed to be arranged in a double-layer structure, as shown in Fig. 5.55. Pneumatic compression mechanism and positioning blocks are installed on the upper layer; the lower layer is equipped with drive motors, screw transmission mechanism and guide rails, etc., which are used to drive the positioning blocks installed on the upper layer.

The workbench is usually in a horizontal position. After the workpiece has been placed on the centre of the workbench by means of a walking crane, it is positioned by means of electrically operated positioning blocks in the X- and Y-directions, and then tightened by means of pneumatic compression mechanisms in the Y-direction. When the robot is started, the positioner flip the workpiece to the flat welding position for welding. After the welding is completed, the positioner turns back to the initial

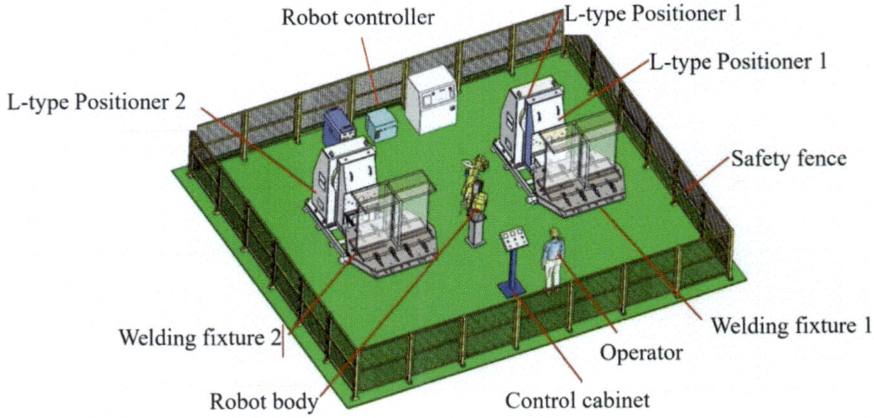

Fig. 5.54 Corbel welding robot workstation

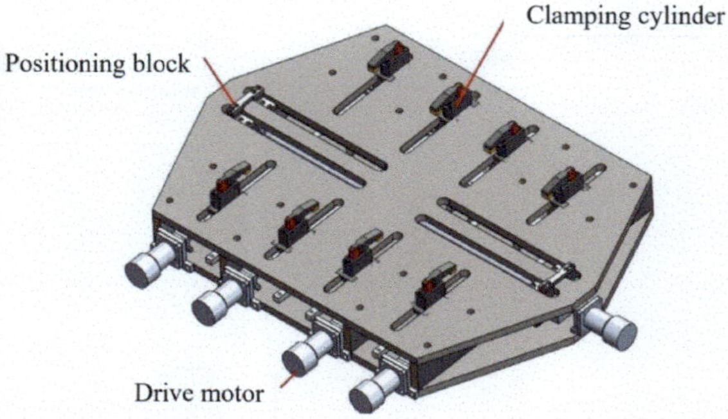

Fig. 5.55 Welding positioner

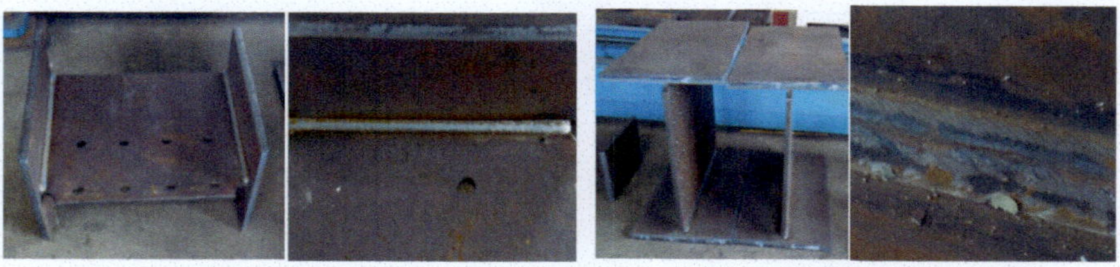

(a) 10 mm thick fillet weld (b) 20 mm thick fillet weld

Fig. 5.56 Corbels and weld appearance

position, the pneumatic compression mechanism is automatically released, and the positioning mechanism automatically returns to the initial position. Afterwards, the finished workpiece can be moved to the finished product area by the walking crane.

In order to ensure the welding quality and efficiency, the corbel welding robot workstation is generally provided with the following functions: Original path resumption, fault self-diagnosis, weld searching, multi-layer and multi-pass welding, automatic seam locating, expert database, weaving, and welding torch cleaning and cutting, etc.

Figure 5.56 shows corbels of several sizes and appearance of welds made at the corbel welding robot workstation.

5.5 Other Welding Robot Systems

5.5.1 Laser Beam Welding Robot System

The laser beam welding robot system consists of a robot body, a controller, a laser welding system and a positioner, as shown in Fig. 5.57. The welding system consists of a laser and a laser beam welding head or a scanning laser head. The robot body is

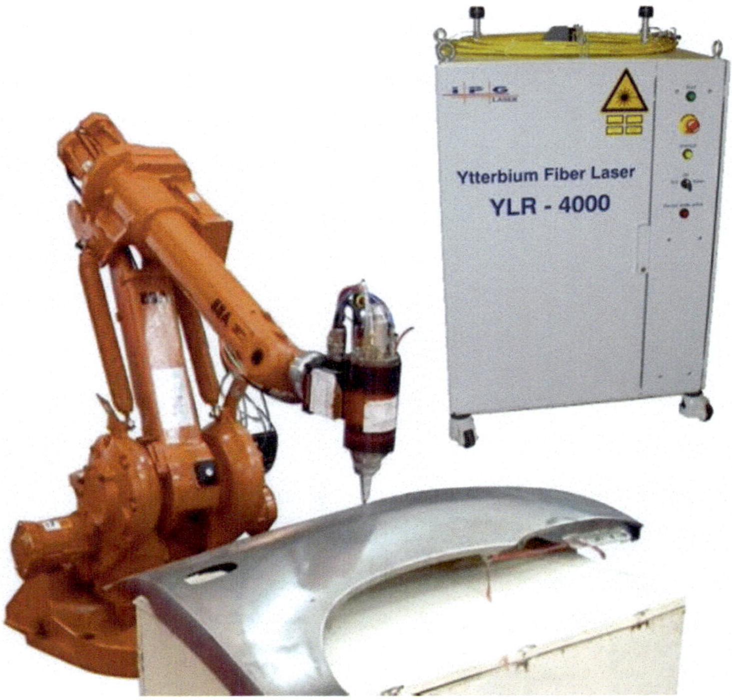

Fig. 5.57 Laser beam welding robot system

required to have high positioning accuracy and repetition accuracy (repetition accuracy: ±0.05 mm) and a load capacity of no less than 30–100 kg. The welding system of laser robot is usually equipped with CCD cameras to detect the welding process and the actual position of the seam, thereby realizing automatic alignment of the laser beam with the seam to compensate the machining and assembly errors of the workpiece. Typical positioner consists of rotating-flipping axes, which are linked with the robot to form the 7th and 8th axes of robot. These axes are used to turn the complicated workpiece to the position where it is easier to weld and increase the accessibility of the welding torch. As the laser beam welding speed is high, the welding productivity mainly depends on the the efficiency of loading and unloading the workpiece. Therefore, to improve the welding efficiency, the laser beam welding robot system usually incorporates two or more workbenches. When welding is performed on one workbench, the uploading or downloading is carried out on the other workbenches. The system is usually equipped with CAD/CAM off-line programming software. After the edited three-dimensional digital/analog files of workpieces are imported into the robot control system, the welding program is automatically generated, thus eliminating the need for teaching process, and significantly saving time and improving production efficiency. The system is usually equipped with a laser safety cabin which includes laser safety glass and automatic doors.

There are two types of high-power lasers commonly used in laser welding robots: solid-state lasers and gas laser. Solid lasers are generally Nd: YAG lasers with a wavelength of 1.06 mm. They can be transmitted via optical fiber, which simplifies the optical path system and allows long-distance transmission, facilitating remote welding. Currently, this kind of laser has a power capacity of up to 40 kW. The gas laser uses CO_2 gas as its working medium and its wavelength is 10.6 mm. It is characterized with good safety and high power (up to 50 kW at present). Laser beam welding robot systems have been widely used in butt welding of vehicle body and ship casings.

5.5.2 *Friction-Stir Robot System*

The friction-stir welding robot system is an advanced automation equipment that integrates heavy-duty industrial robot and friction-stir welding main shaft system. It consists of a robot body, a controller and a friction welding main shaft system, as shown in Fig. 5.58 (robot controller is not shown in the Fig. 5.58). During welding, the welding probe applies a large force and moment on the workpiece, which results in a large force and moment on each bearing of the robot. Therefore, the robot body is required to have a large load capacity (generally greater than 500 kg) and maintain high stability, repeat positioning accuracy and position & orientation accuracy under large load. To achieve reliable robotic control, the friction-stir welding robot system shall be equipped with sophisticated sensing systems such as a pressure sensor, a temperature sensor, and a seam tracking sensor, etc. The welding head shall be made as small as possible to improve accessibility.

The friction-stir welding robot has the following advantages.

(a) The flexibility of friction stir welding operation is significantly improved.
(b) Complex motion path can be realized, which is suitable for welding products with complex structure.
(c) Workspace and degree of freedom can be easily extended by linking external axes.
(d) It is easy to realize multi-mode process control, such as pressure control, torque control, etc. to improve the quality of welded joints.
(e) Green, energy saving, high efficiency, low welding cost. According to statistics, the single piece welding cost with friction stir welding robot is 20% lower than that with TIG welding robot.

Fig. 5.58 Friction-stir welding robot system

Chapter 6
Robotic Welding Processes

The commonly used robotic welding processes include resistance spot welding, gas metal arc welding (GMAW), tungsten inert gas arc welding (TIG), laser beam welding and friction-stir welding.

6.1 Resistance Spot Welding Process

6.1.1 Operation Principles and Characteristics

(1) Principle of operation

Resistance welding is a joining process in which the coalescence of metals is achieved with the heat generated by the welding current flowing through the workpieces under a certain pressure. Resistance welding process includes resistance spot welding, resistance seam welding, resistance projection welding, upset welding and flash welding. Figure 6.1 illustrates the mechanisms of these 5 resistance welding processes. Resistance welding robots generally adopt resistance spot welding process, which is widely used in automobile, motorcycle and agricultural machinery manufacturing industries, etc. This section mainly presents the resistance spot welding process.

For resistance spot welding, two water-cooled cylinder copper alloy electrodes are used to conduct current and apply pressure, as shown in Fig. 6.2. The coordinated application of mechanical pressure is necessary for spot welding operation. Usually, the mechanical pressure is applied to the workpieces prior to the conduction of welding current. The purpose of the pre-pressurization is to bring the surfaces of the two parts to be welded into reliable contact via plastic deformation. The plastic deformation zone is surrounded by a plastic ring. Outside the plastic ring, the surfaces of the workpieces are out of contact, so that the current path between the two workpieces is restricted to the inside of the plastic ring. What's more, the plastic ring

M. Chen et al., *Technologies of Robotic Welding*, Advanced and Intelligent Manufacturing in China, https://doi.org/10.1007/978-981-97-3149-7_6

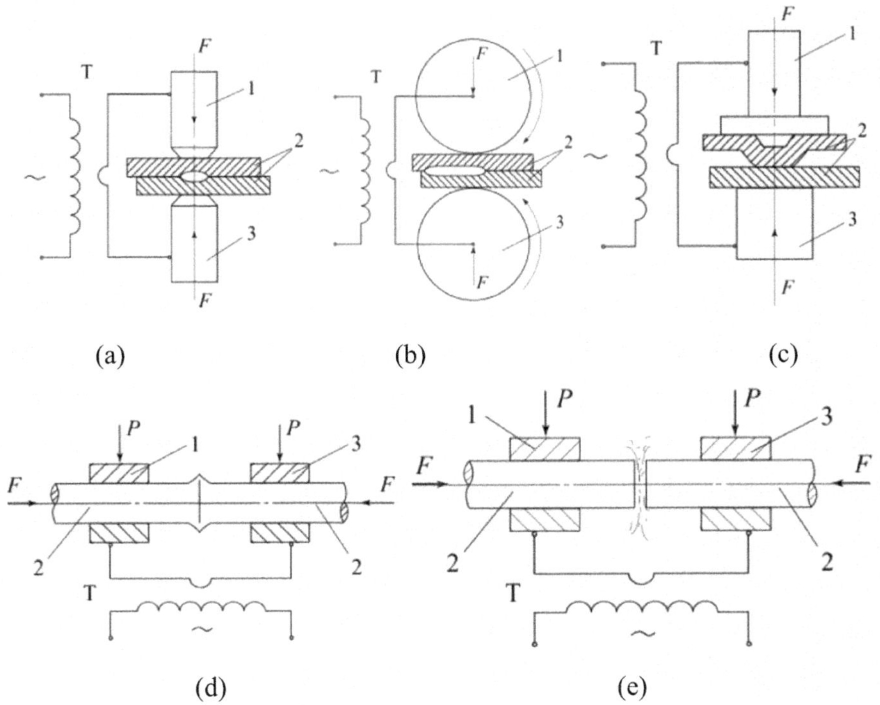

Fig. 6.1 Schematic diagram of various resistance welding processes. **a** Spot welding. **b** Seam welding. **c** Projection welding. **d** Resistance butt welding. **e** Flash butt welding. 1, 3-Electrodes 2-Workpiece F-Electrode Pressure (Upsetting force) P-Clamping Force T-Power Supply (Transformer)

prevents the ingress of air during welding. The heat generated by the resistance of workpiece (including contact resistance) to the welding current can be calculated by the following formula:

$$Q = I^2 Rt \tag{6.1}$$

where: Q-Heat generated (J); I-welding current (A); R-resistance between the two electrodes (Ω); t-duration of current (s).

The resistance R between the two electrodes is composed of the resistance R_W of the two workpieces, the contact resistance R_c between them and the contact resistance R_{cw} between the electrode and workpiece, as shown in Fig. 6.2. That is

$$R = 2R_W + R_C + 2R_{ew} \tag{6.2}$$

As the contact resistance R_c between workpieces is very much higher than the others (see Fig. 6.2), the heat released from the contact surface is the maximum and concentrated. The metal in this location is rapidly heated above the melting point

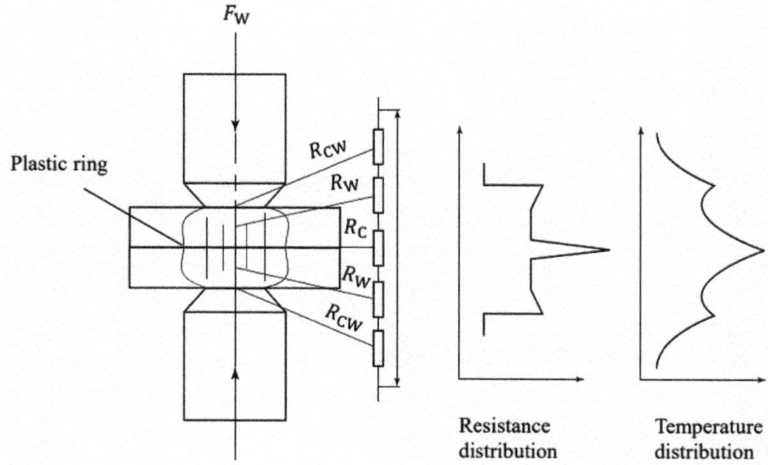

Fig. 6.2 Schematic diagram of resistance spot welding

T_m of the base material to form a nugget. The liquid metal in the nugget is rapidly homogenized in terms of temperature and composition by strong electromagnetic stirring. Generally, the temperature of nugget is about 300–500 K higher than the metal melting point T_m. Due to heat dissipation of electrode, the nugget grows at a higher rate along the surface of workpiece than that along the direction perpendicular to the surface, so it is ellipsoidal. Though the contact surface between the workpiece and the electrode has a high resistance and heat generation, the temperature usually does not exceed (0.4–0.6) T_m due to the heat dissipation caused by water cooling of the electrode. In spot welding, a large fraction of the resistive heat is lost by conduction away into the cooling water flowing through the electrodes and the base metal around the nugget, for this reason, the heat utilization rate is only 20–30%.

(2) Welding cycle

For resistance spot welding, the evolution of all the parameters (pressure and current) involved in completing a weld spot are called a welding cycle. The welding cycle for spot welding consists of four basic stages: squeeze time, weld time, hold time and off time, as shown in Fig. 6.3.

1) Squeeze time t_1

It is the time interval from the timer initiation to the delivery of welding current. This time is to ensure that the electrode is pressed against the workpiece before welding current delivery and appropriate pressure is applied between the workpieces, so that a plastic ring is formed to confine the welding current flow path.

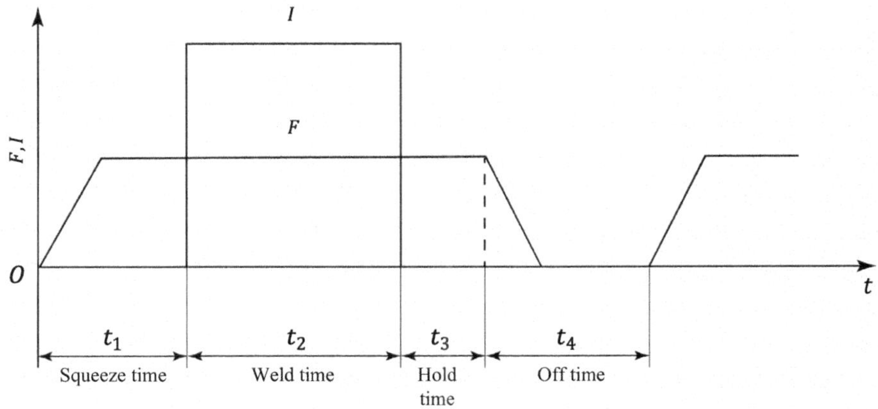

Fig. 6.3 Welding cycle for resistance spot welding

2) Weld time t_2

It refers to the time interval during which the welding current flows through the workpieces to produce a nugget.

3) Hold time t_3

It is the time interval in which the electrode pressure is maintained after the welding current is cut off. The nugget cools and solidifies under pressure. The purpose of the remained pressure in this duration is to obtain a dense spot weld without defects such as shrinkage cavity and cracking, as well as to develop strength.

4) Off time t_4

It is the time during which the electrodes are lifted off the workpieces and the workpieces are moved to the next spot for repetitive welding. This duration is applicable only if the welding cycle is repeated.

(3) Characteristics of resistance spot welding

The advantages of resistance spot welding are as follows:

(a) No or very less metallurgical reaction occurs during welding because that the molten metal is isolated from the air by the plastic ring.
(b) High welding quality. The heat-affected zone is narrow and deformation and residual stress are low. No distortion straightening or heat treatment is required after welding.
(c) No need for filler metal and shielding gas.
(d) Easiness to be mechanized and automated, and easiness to be included in the production line including various manufacturing processes.
(e) High productivity and low production cost.

Resistance spot welding has the following limitations:

(a) Lack of reliable NDT methods for spot weld.
(b) Lap joint is the main joint type for spot welding, which increases the size and weight of the product and reduces the tensile strength and fatigue strength of the joint.
(c) High current and power requirement.

6.1.2 Dimension Parameters of Nugget

The strength of a spot-welded joint is determined by the geometric dimensions in case that there is no any defects in the solidified nugget. The geometric dimensions of nugget are shown in Fig. 6.4. The nugget diameter shall increase with the increase of plate thickness, and the following relation is generally applied:

$$d_n \geq 5\sqrt{\delta} \tag{6.3}$$

where: d_{n}—Nugget diameter, mm; δ—Minimum workpiece thickness, mm.

The Nugget height h_{n} can be in a large range which is expressed as

$$h_{\mathrm{n}} = (0.2 - 0.8)\,(\delta - \Delta) \tag{6.4}$$

where: h_{n}—Nugget height, mm; Δ—Indentation depth on the surface of workpiece, generally = (0.1–0.15) δ, (mm).

The percentage of the nugget height h_n in one workpiece with respect to the workpiece thickness δ is called penetration rate A. It can be given by

$$A = (h_n/\,\delta) \times 100\% \tag{6.5}$$

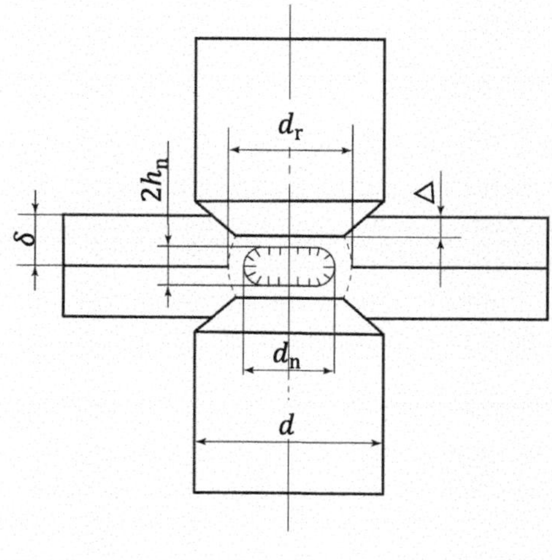

Fig. 6.4 Geometric dimensions of the nugget, δ—Minimum Workpiece thickness; d—Electrode diameter; d_n—Nugget diameter; d_r—Outer Diameter of Plastic Ring; h_n—Nugget height; Δ—indentation depth

The generally recommended value for A is 20–80%. A penetration rate A greater than 80% will make the nugget come too close to the surface of the workpiece, causing overheating of the surface metal, grain coarsening, spatter, shrinkage cavity, crack, and other defects, thus reducing the bearing capacity of the joints. On the other hand, if the weld penetration rate A is lower than 20%, the nugget is too small to get the required bearing capacity.

The indentation caused by electrode pressure on the surface of workpiece is a sign indicating that the nugget has been forged. However, it shall not be too deep; otherwise, the surface quality of the product will be affected and the section area at this location will be reduced, resulting in excessive stress concentration and a decrease of the bearing capacity of the nugget. The higher the electrode pressure, the longer the weld time; or the higher the welding current, the deeper is the indentation. The indentation depth can be reduced by using larger flat-faced electrode or a hard welding procedure with increased current and decreased welding time.

6.1.3 Welding Parameters

The welding parameters for spot welding mainly include welding current I_w, weld time t_w, electrode pressure F_w and electrode working surface dimension d_e.

(1) Welding current

An increase of welding current will increase the nugget size and strength. There is a reasonable upper limit and lower limit of the current density for a desirable nugget. When it is lower than the lower limit, the heat is too low to form a desirable nugget. Excessive current density, on the other hand, tends to result in molten metal splash, internal void and cracking, and reduce the bearing capacity. Generally, the upper limit of the welding current at which the splash occurs increases with increasing electrode pressure. At a given electrode pressure, the welding current shall be adjusted to a value slightly lower than the splash current to achieve the maximum spot weld strength.

(2) Weld time

The influence of weld time on the size of nugget is basically similar to that of welding current. The nugget size will increase with the prolonging of weld time. However, excessively long weld time tend to cause defects such as overheating of weld zone, molten metal splash, etc..

Figure 6.5 shows the recommended welding current (I_w) and weld time (t_w) as a function of workpiece thickness for spot welding of different materials

(3) Electrode pressure

Electrode pressure affects the heat production, heat dissipation through electrode, formation of plastic ring and compactness of welding spot. When other parameters remain unchanged, increasing the electrode pressure will result in reduction in the

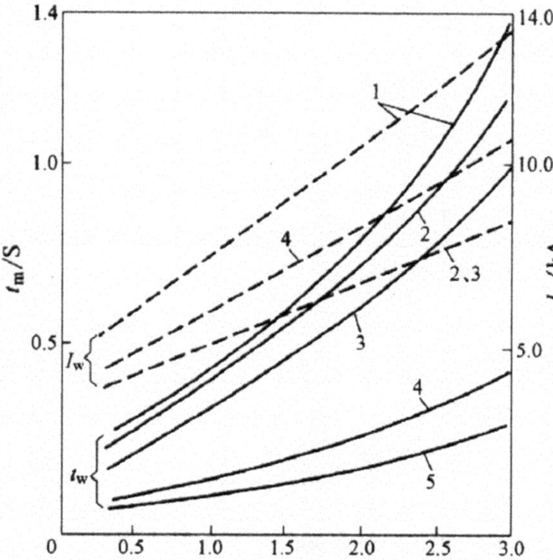

Fig. 6.5 Welding current and weld time as a function workpiece thickness. 1-Low and medium alloy steel; 2-Special high temperature alloys; 3-High temperature alloys; 4-Stainless steel; 5-Copper alloys

contact resistance and hence the resistance heat generation, but an increase in the heat dissipation. As a result, the size of nugget and the penetration rate will be reduced and the indentation depth will be over increased. On the other hand, and over decreasing in the electrode pressure will lead to an over increase in resistance heat generation, and destroy the enclosure of molten metal by the plastic ring, causing weld defect such as molten splash, internal cavity, burning-through.

If the capacity of welding machine is large enough, welding procedure with high electrode pressure and high welding current is recommended to be used to improve stability of welding quality.

For the welding of metallic structural members with high rigidity, high elevated-temperature strength, large coefficient of linear expansion or high crack sensitivity, stepped or saddle-shaped electrode pressure shall be used to avoid molten metal splash and internal void due to shrinkage, as shown in Fig. 6.6b, c.

(4) Shape and size of electrode face

The geometric dimensions and strength of nugget are subject to the influence of structural shape, size and cooling conditions of the electrode body and electrode flat-face contacting the work. For commonly used conical electrodes, the larger the taper angle of the electrode face, the better is the heat dissipation. However, when the cone angle is too large, the diameter of the working face of the electrode will become over-sized rapidly due to repetitive heating and abrasion. If the cone angle is too small, the heat dissipation condition is poor and the surface temperature of electrode

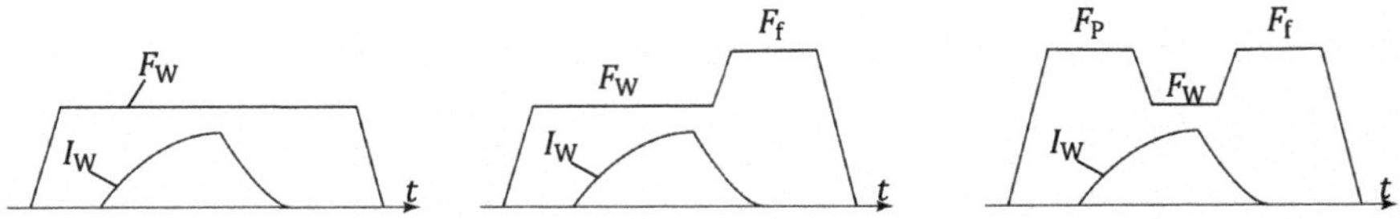

Fig. 6.6 Various types of electrode pressure cycling. **a** Constant electrode pressure; **b** stepped electrode pressure; **c** saddle-shaped electrode pressure. I_w-welding current; F_w-welding pressure; F_f-upsetting force; F_p-pre-welding pressure

is high, making it more susceptible to deformation and wear. In order to improve the stability of spot welding quality, the change in diameter d_e of electrode working face shall be minimized during welding. Therefore, the cone angle is generally selected in the range of 90°–140°. The spherical electrode has a large head size, the contact area between the electrode face and the workpiece is enlarged, thus the welding current density is reduced and the heat dissipation capacity is enhanced. As a result, the weld penetration and nugget diameter will be reduced. However, the indentations on the surface of workpiece will be shallow without causing large stress concentration. In addition, the current density and electrode pressure in the weld zone are evenly distributed, and it is easy to maintain stable quality of the nugget. What's more, the alignment requirement for upper and lower electrodes are lower and the deviation has little impact on the quality of nugget. Therefore, for welding metals with low thermal conductivity, such as stainless steel, spherical electrodes with large working face should be recommended.

(5) Welding parameter coordination

The above-mentioned welding parameters are interrelated in term of effect on welding quality. The welding current I_w, weld time t_w, electrode pressure F_w and electrode working surface diameter d_e all affect the heat generation of the welding zone, among which, F_w and d_e directly affect the heat dissipation, while t_w and F_w determine the size of plastic zone of the nugget. If I_w and t_w are increased and F_w is decreased, the resistive heat will increase, resulting in an increased nugget size. In this case, if the heat dissipation is not enough (e.g. small d_e), molten metal splash and overheating may occur.

To achieve a certain nugget dimensions and penetration rate, either welding procedure with high welding current and short weld time or procedure with low welding current and long weld time can be adopted.

The procedure with high welding current and short weld time is referred to as hard procedure, which is characterized by high heating rate, steep temperature gradient in the weld zone, narrow heating area, less overheated microstructure, good surface quality of weldment, high mechanical properties, and high productivity. Therefore, this kind of procedure shall be used as long as the power of the welding machine permits and the welding parameters can be controlled accurately. However, due to the fast heating rate, increased electrode pressure and improved heat dissipation conditions are required. Otherwise, defects such as molten metal splash may be result.

The process procedure with low welding current while long weld time is called soft procedure. It is characterized by low heating rate, gentle temperature gradient in the weld zone, wide plastic zone and easy plastic deformation. The soft procedure is usually adopted in situations when the power capacity of the spot welding machine is low, workpiece thick is high, or the parent material has low plasticity or high quenching-sensitivity.

6.2 Gas Metal Arc Welding (GMAW)

6.2.1 Operation Principles and Characteristics

(1) Principles of operation

GMAW is a welding process in which the arc between the continuously-filled welding wire and the workpiece is used as the heat source, and a externally-supplied pure gas or mixed gas is used to shield the welding wire, arc and molten pool. See Fig. 6.7 for its schematic diagram. Welding wire is used as both electrode and filler material, and there are two types of wire, i.e. solid and flux-cored wires. The commonly used gas includes CO_2, Ar, He and their mixture.

(2) Classification

According to the type of shielding gas and welding wire used, GMAW is classified as follows:

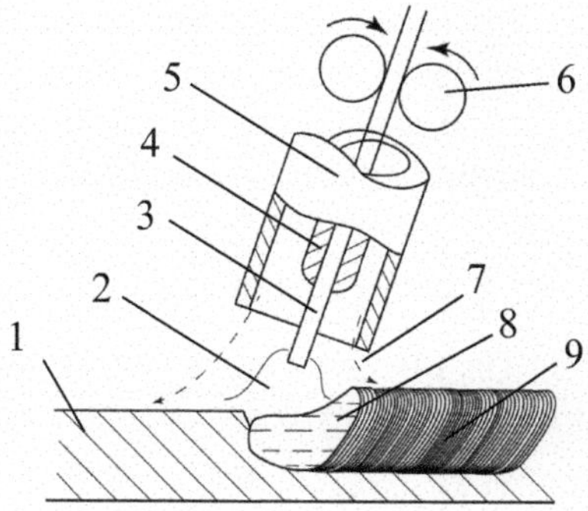

Fig. 6.7 Schematic diagram of gas metal arc welding. 1-Base metal; 2-Arc; 3-Welding wire; 4-Contact tip; 5-Nozzle; 6-Feeding roll; 7-Shielding gas; 8-Molten pool; 9-Weld metal

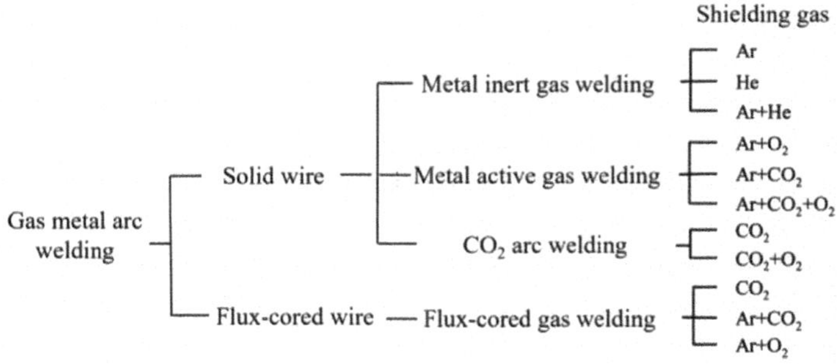

(3) Characteristics

The process characteristics of gas metal arc welding are as follows:

(a) Wide scope of application. MIG is especially suitable for the welding of aluminum and aluminum alloys, titanium and titanium alloys, copper and copper alloys and stainless steel. MAG welding and CO_2 gas shielded welding are suitable for the welding of ferrous metals. It is suitable for sheet of thin section as well as plates of medium and thick sections, and is suitable for welding in all positions.
(b) High productivity and low welding deformation. The use of welding wires as electrodes results in higher allowable current density, which leads to greater penetration capacity and faster deposition rate. For metals with higher thickness such as aluminum, copper and their alloys, the productivity is higher than that of TIG welding, and the deformation of weldments is lower than that of TIG welding.
(c) Easy automation. The arc of argon arc welding is open arc. The parameters of the welding process are stable, and easy to monitor and control. Therefore, it is easy to realize automation and robotic operation.
(d) Insensitive to oxidation film. DC reverse welding is generally adopted for argon arc welding. When welding aluminum or aluminum alloy, it will result in strong cathodic clean-up action, and therefore, there is very low requirements on removal of oxide film before welding. CO_2-shielded welding is also insensitive to oil stains and rust.

GMAW has the following disadvantages:

(a) The MIG welding of aluminum and its alloys is susceptible to air pores.
(b) Weld quality is not as good as TIG weld.

(4) Application

(a) Weldable Metals

MIG may be used for welding of aluminum, copper, titanium and their alloys, stainless steel, heat resistant steel, etc. MAG welding and CO_2 gas-shielded welding are mainly used for carbon steel, low-alloy high-strength steel; MAG welding is

Table 6.1 Thickness range applicable for GMAW

Thickness/mm	0.13	0.4	1.6	3.2	4.8	6.4	10	12.7	19	25	51	102
Single-pass weld with I groove		⟸	═	⟹								
Single-pass weld with V groove		⟸	═	═	═	⟹						
Multi-layer weld with V groove			⟸	═	═	═	═	⟹---	---	---	---	

mainly used for important metal structures, and CO_2 gas-shielded welding is used for ordinary metal structures.

(b) Welding position

Gas metal arc welding has good adaptability and can be performed at all positions. The highest production efficiency is achieved at flat position. At every welding position, its production efficiency is higher than that of shielded metal arc welding.

(c) Metal thickness

Table 6.1 shows the range of thickness applicable for GMAW. In principle, there is no restriction on the thickness of multi-layer groove welding, it is only limited by economic factors.

6.2.2 Metal Transfer

The metal transfer modes in GMAW include short circuit transfer, globular transfer, fine droplet transfer, spray transfer and pulsed spray transfer. The globular usually occurs under the condition of high arc voltage and low welding current, which is very unstable and easily leads to defects such as poor fusion, incomplete penetration and excessive reinforcement, and therefore is not applicable for application in the industrial welding production.

(1) Short-circuit transfer

Short-circuit transfer occurs when both the welding current and arc voltage are low. As the arc length is short due to low voltage, the molten droplet at the welding wire touches the molten pool before the molten droplets becomes large enough to detach from the wire. A liquid metal bridge forms, which shorts the welding circuit. As the short-circuit current increases rapidly, the short-circuit bridge shrinks under the effect of increasing electromagnetic pinch force. The local resistance and current density at the neck will increase sharply, resulting in sharp increase of resistive heat. The sharp increase of resistance leads to rapid evaporation of the liquid metal and electrical explosion will occur when the evaporation rate reaches a certain value, transferring the droplets into the molten pool, as shown in Fig. 6.8. This transfer is called short-circuit transfer. It is characterized by a small and fast-solidifying molten pool, thus very suitable for the joining of thin-section materials, for all-position

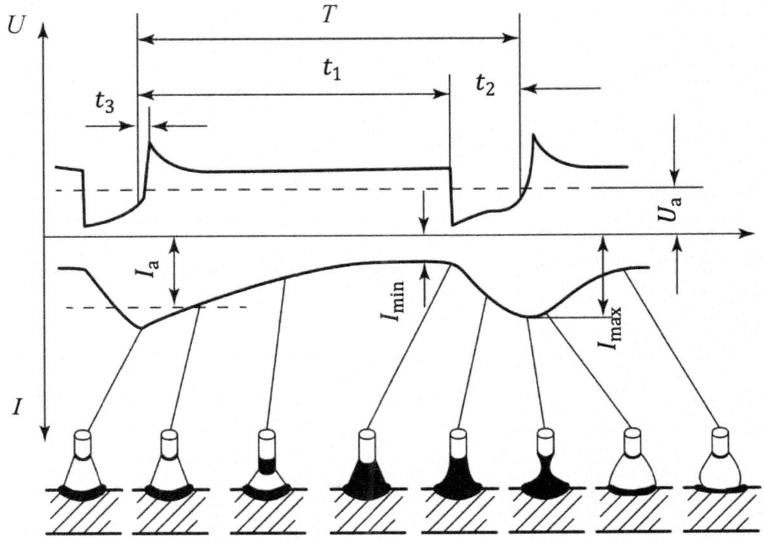

Fig. 6.8 Schematic illustration for Short-Circuit transfer process. T-short-circuit transfer period; T_2-short-circuit duration; T_l-arcing time; T_l-open-circuit voltage recovery time; U_a-arc voltage; I_{min}-minimum current; I_{max}-maximum current; I_a-average current

welding, and for bridging large root-openings. This mode of transfer is commonly used with CO_2 shielding gas, small-diameter wires (the welding wire diameters is generally no more than 1.6 mm), rarely used in MAG and MIG welding. Low current MIG/MAG welding usually adopts pulsed spray transfer.

(2) Fine droplet transfer

The fine droplet transfer occurs in CO_2 welding with large-diameter wire (the welding wire diameter is generally no less than 1.6 mm), high current and high voltage. In this mode of transfer, droplets detached from the welding wire goes in the molten pool through the arc at a high speed with spatters. It is characterized by a deep molten pool, a partially buried or fully-buried arc (depending on the magnitude of the current), as shown in Fig. 6.9. The fine droplet transfer is commonly used in high current CO_2 welding of thick section mild steel.

(3) Spray transfer

The spray transfer is a commonly used form of transfer in conventional MIG/MAG welding. For a given wire diameter, there is a transition current I_T at which the transfer frequency increases sharply, the droplet size decreases sharply, as shown in Fig. 6.10. This indicates a transition from globular transfer to spray transfer. For welding of aluminum and aluminum alloys with a current higher than I_T, the molten wire metal is transferred drop by drop, thus the transfer is referred to as spray transfer. For welding of steel with a current greater than the transition current, the metal transfer

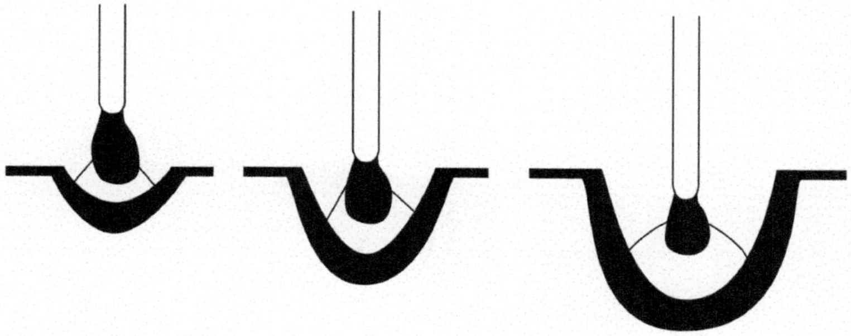

Fig. 6.9 Schematic illustration for fine droplet transfer

proceeds in the form of liquid stream, thus it is called streaming transfer. The mode of transfer charactered by a stable arc, little spatter and a large and slow-solidifying molten pool due to high current, as a result, conventional MIG/MAG welding can only be used for flat welding position and thick section materials.

(4) Pulsed spray transfer

Pulsed spray transfer occurs only in pulsed current MIG/MAG welds. As long as the pulse current is greater than the transition current, a stable spray transfer can be achieved in the range as high as hundreds of amperes and as low as tens of amperes, either on thick plates or on thin plates.

Pulse spray transfer has three types: One-pulse one-droplet transfer, one-pulse multiple-droplet transfer, and multiple-pulse one-droplet transfer. The form of the transfer is mainly determined by pulse current and pulse duration, as shown in Fig. 6.11. Among the three transfer modes, the one-pulse on-droplet mode is the best, and the multi-pulse one-drop process has worst performance.

Fig. 6.10 Volume of droplets and transfer frequency as a function of welding current

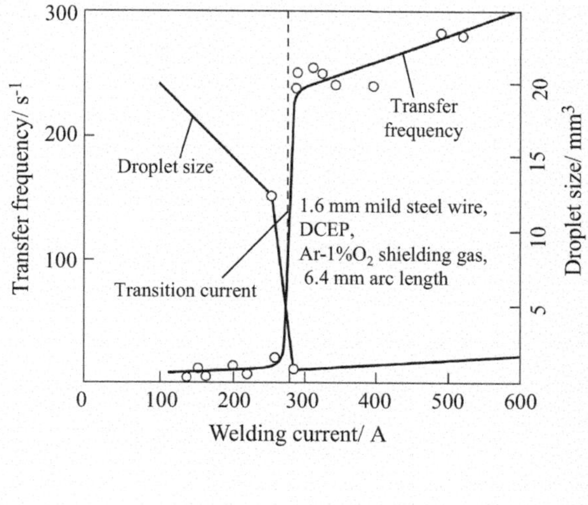

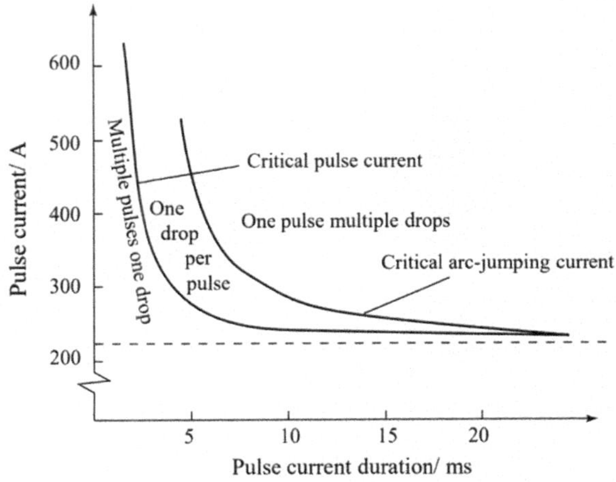

Fig. 6.11 Dependence of pulsed spray transfer on pulse current and pulse duration

6.2.3 Welding Parameters

For gas metal arc welding, the process parameters include the type of shielding gas, the welding wire diameter, the welding current, the arc voltage, the welding speed, the flow rate of shielding gas and the nozzle to work distance.

(1) Shielding gas

The composition of shielding gas is usually selected according to the type of base metal, usually mixed gas. For aluminum and aluminum alloy, Ar or Ar + He is generally used; For mild steel and low alloy steel, CO_2, Ar + O_2, Ar + CO_2 or Ar + CO_2 + O_2 are used; for stainless steel welding, Ar + O_2 or Ar + CO_2 is used; for copper and copper alloy, Ar + N_2 or Ar + He is generally used.

(2) Welding wire diameter

The welding wire diameter shall be determined based on the workpiece thickness and welding position. Generally, small-diameter welding wires (diameter $\leq$1.6 mm) are used for the welding of thin-section plate and out of position welding, large-diameter wire for welding of medium and large thickness plates in flat position. Table 6.2 provides the applicable scope of welding wires with a diameter of 0.8–2.0 mm. When large-thickness plates are welded in the flat position, welding wires with a diameter of 3.2–5.6 mm are preferred, and the welding current used can be up to 500–1000 A. The advantages of this type of large-diameter wire high current welding are high penetration capacity, less number of weld passes, high welding productivity and low welding deformation.

(3) Welding current

The welding current is the most important welding process parameter. During practical welding, the welding current shall be determined based on the workpiece thickness, welding wire diameter, droplet transfer method, and welding position. In the

Table 6.2 Selection of wire diameter

Wire diameter (mm)	Workpiece thickness (mm)	Welding position	Metal transfer mode
0.8	1–3	All-position	Short-circuit transfer
1.0	1–6	All-position, one-side welding with back formation	Short-circuit transfer
1.2	2–12		
	Medium and large thickness	Backing welding	
1.6	6–25	Flat, horizontal or vertical	Spray transfer
	Medium and large thickness		
2.0	Medium and large thickness		

case of welding by means of an constant-speed wire feeder plus constant-voltage power source syetem, the welding current is adjusted by the wire-feed speed. It shall be noted that for welding wires with a certain diameter, there is a certain range of allowable current. The arc will be unstable when the current is below or above this range. Table 6.3 shows the typical range of welding currents available for MAG welding of low-carbon steel with various wire diameters.

For pulsed argon arc welding with consumable electrode, the current parameters shall be: Background current I_b, pulse current I_p, pulse duration t_p, background duration time t_b, pulse period $T = (t_p + t_b)$, pulse frequency $f = (1/T)$, pulse amplitude ratio $F = (I_p/I_b)$, pulse width ratio $K = [t_p/(t_b + t_p)]$. At present, almost all robots use pulse MIG/MAG welding power supplies which allow one-pulse, one-droplet transfer. When welding, only the average welding current (or directly set plate thickness) needs to be set via the power supply, and other power supplies will match automatically. Thus, the adjustment of the welding parameters does not differ in any way from the normal MIG/MAG.

(4) Arc voltage

The arc voltage mainly affects the weld width, but has little effect on penetration. The arc voltage shall be selected according to the current, the composition of the shielding gas, the type of the materials to be welded and the droplet transfer method. Table 6.4 shows the arc voltages for different shileding gases.

Table 6.3 Typical welding current range for MAG welding of low-carbon steel

Wire diameter (mm)	Welding current/A	Metal transfer mode	Wire diameter (mm)	Welding current/A	Metal transfer mode
1.0	40–150	Short-circuit transfer	1.6	270–500	Spray transfer
1.2	80–180		1.2	80–220	Pulsed spray transfer
1.2	220–350	Spray transfer	1.6	100–270	

Table 6.4 Arc voltage for welding with different shielding gases/V

Metal	Spray transfer					Short-circuit transfer			
	Ar	He	Ar + 75% He	Ar + (1–5%) O_2 or Ar + 20% CO_2	CO_2	Ar	Ar + (1–5%) O_2	Ar + 25% O_2	CO_2
Aluminium	25	30	29	–	–	19	–	–	–
Magnesium	26	–	28	–	–	16	–	–	–
Carbon steel	–	–	–	28	30	17	18	19	20
Low alloy steel	–	–	–	28	30	17	18	19	20
Stainless steel	24	–	–	26	–	18	19	21	–
Nickel	26	30	28	–	–	22	–	–	–
Nickel-Copper Alloy	26	30	28	–	–	22	–	–	–
Nickel-chromium-ferroalloy	26	30	28	–	–	22	–	–	–
Copper	30	36	33	–	–	24	22	–	–
Copper-nickel alloy	28	32	30	–	–	23	–	–	–
Silicone bronze	28	32	30	28	–	23	–	–	–
Aluminum bronze	28	32	30	–	–	23	–	–	–
Phosphorus bronze	28	32	30	23	–	23	–	–	–

Note The welding wire diameter is 1.6 mm

(5) Gas flow

The flow rate of shielding gas is usually determined based on the current, the nozzle diameter and the joint type. There is an optimum flow range for a nozzle with a certain diameter. Excessive flow will cause turbulence; over-small flow causes poor stiffness of shileding gas flow and worse protection. The nozzle diameter is usually 20 mm and the flow rate of shielding gas is 10–20 L/min. The optimum range of gas flow is usually determined by experiment.

(6) Nozzle to work distance

The nozzle to work diatance shall be selected according to the current, as shown in Table 6.5. If the distance is too large, the protection effect will be weakened, the wire extension will increase and the welding current will decrease, which will easily lead to defects such as incomplete penetration and non-fusion. However, if the wire extension is too small, the nozzle may be clogged by spatters.

Table 6.5 Recommendations for nozzle to work distance

Welding current/A	<200	200–250	350–500
Nozzle to work distance/mm	10–15	15–20	20–25

6.2.4 High-Efficiency GMAW

In recent years, GMAW has achieved great development in efficiency and spatter suppression. Some of these new processes are briefly described below.

6.2.4.1 CMT (Cold Metal Transfer) GMAW

(1) Principles of operation

CMT is an improved variant of spatter-free short-circuit transfer. It is a new "cold" welding technology based on advanced digital power supply and wire feed. Through coordinating the welding current waveform with welding wire motion (feeding/retracting) according to the arc status, stable short-circuit transfer can be achieved by retracting the welding wire at a very low current, thus avoiding spatters completely.

Figure 6.12 shows the waveform of welding current and the waveform of motion speed of welding wire during a CMT GMAW. When the droplet and molten pool are short-circuited, the current in the welding circuit is immediately switched to a low value (around 20 A), and the short-circuit bridge quickly becomes cold. At the same time, the motion of welding wire changes from feeding to retraction. After a while of wire retraction, the short-circuit molten metal bridge is broken, the droplet in cold state is transferred into the molten pool and the arc is re-ignited. Once the arc is ignited, the welding current is rapidly increased to a pulsed current, which melts the wire to form a droplet. At the same time, the wire movement changes from retraction to feeding. After the droplet grows to a certain size, the welding current is decreased to the background current to prevent the droplet size from becoming too large. As welding wires are fed forward, the droplet contacts the molten pool and then the next CMT cycle begins. The welding current in the short-circuit phase is switched to such a low value (around 20 A) that the short-circuit molten metal bridge will never electrically explode. The molten metal bridge is necked off by the mechanical force caused by wire retraction, as a result spatter is completely avoided. The workpiece undergoes a "hot/cold" alternating process during welding, significantly reducing the heat input. This welding process mainly adopts robotic operation and less frequently semi-automatic operation.

(2) Characteristics and application

1) Advantages

CMT GMAW offers the following advantages:

(a) Low noise of the arc, even droplet size and transfer period, easiness to achieve spatter-free short-circuit transfer welding and brazing.
(b) Precise control of arc length. The arc length is adjusted through mechanical monitoring and adjustment. The arc length is basically immune from the unevenness of workpiece surface and welding speed. This makes the CMT arc more stable without arc interruption even at very high welding speeds.

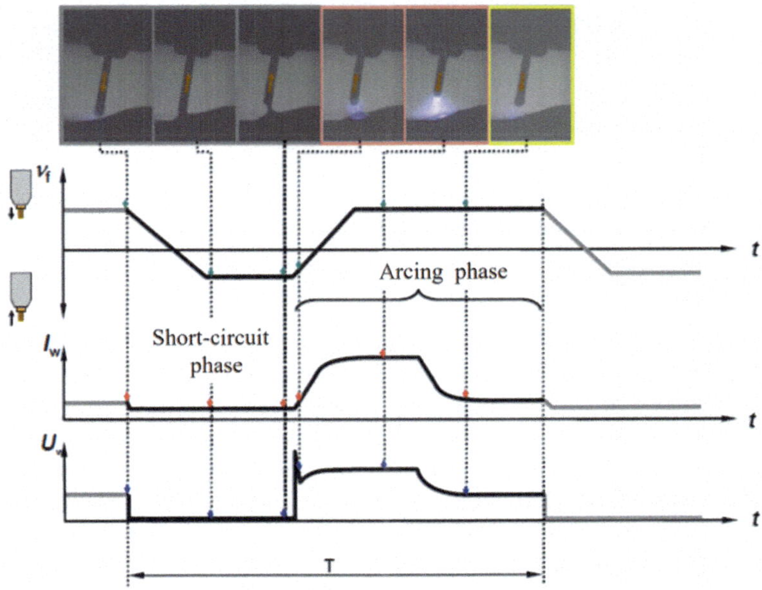

Fig. 6.12 CMT transfer process

(c) High arc igniting speed. The arc igniting rate is twice as fast as conventional gas metal arc welding (30 ms for CMT GMAW and 60 ms for MIG), which allows the base metal to be melted in a very short time.
(d) High quality and repeatability of welds. Combining CMT and pulse control techniques can effectively adjust heat input and improve weld formation, as shown in Fig. 6.13.
(e) Low heat input, small weld heat affected zone and weld deformation. Figure 6.14 compares the application range of welding parameters for different droplet transfer modes of gas metal arc welding. It can be seen that the minimum welding current and arc voltage and minimum heat input are used for CMT welding. The low heat input significantly reduces weld heat effects and weld residual deformation.

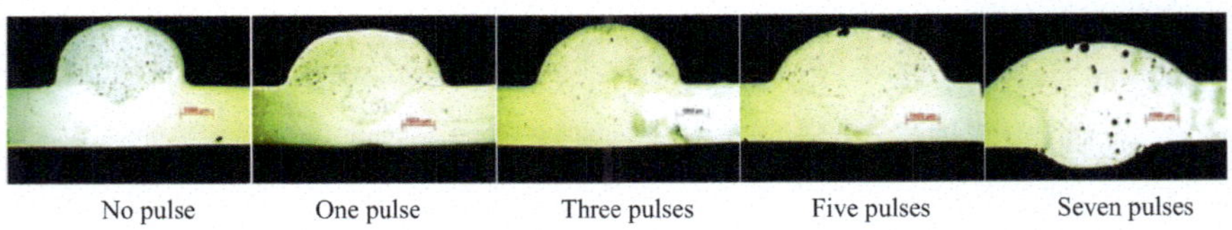

Fig. 6.13 Effect of welding current pulse on weld formation

Fig. 6.14 Comparison of welding parameters between CMT GMAW and conventional GMAW

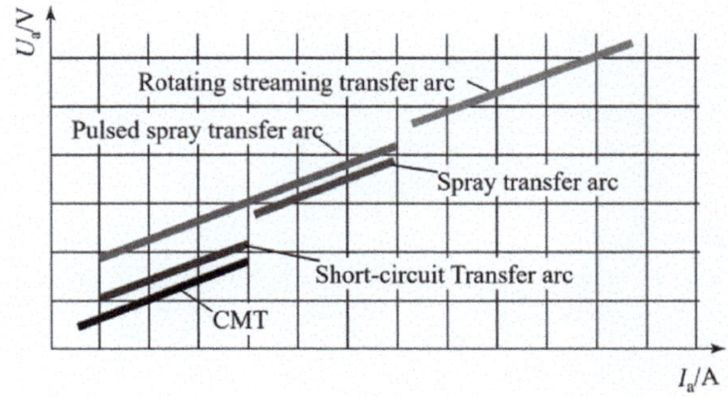

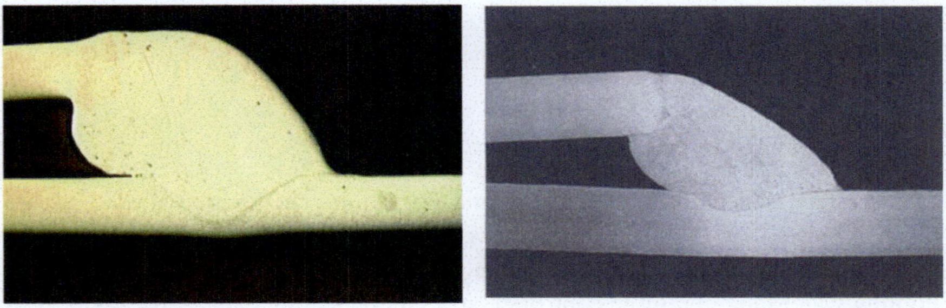

(a) CMT, 1.0 mm thickness, 1.3 mm gap

(b)MIG, 1.2 mm thickness, 1.2 mm gap

Fig. 6.15 Comparison of gap bridging capacity of CMT GMAW with MIG welding

(f) Improved gap bridging capability. Due to the reduced heat input and the small arc impact force, the bridging capacity of CMT GMAW is significantly higher than that of ordinary MIG welding, as shown in Fig. 6.15.

2) Application

(a) Materials suitable for CMT GMAW are as follows:

A. Welding of aluminium, steel and stainless steel sheets or ultra-thin plates (0.3–3 mm) without fear of collapse and burn-through;
B. Spatter-free CMT brazing for galvanized or hot-galvanized plates.
C. Used for dissimilar metal joining between galvanized steel sheet and aluminum sheet. High quality weld can be acheived, in term of both weld appearance and mechanical properties of welded joint.

(b) The joints suitable for CMT GMAW include lap joint, butt joint, fillet joint and crimping joint.
(c) CMT GMAW can be used at various welding positions, such as flat welding, horizontal welding, overhead welding and vertical welding.

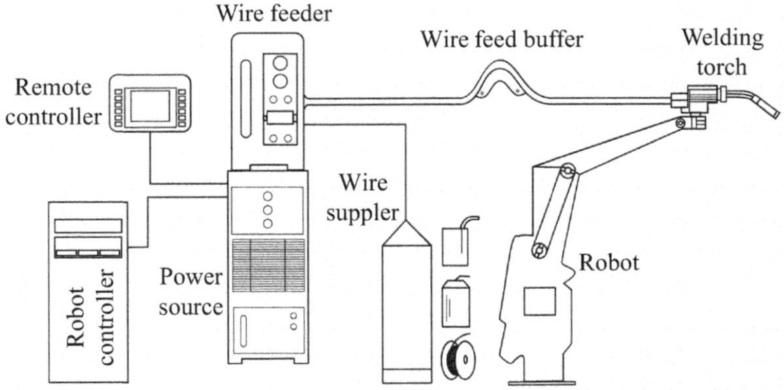

Fig. 6.16 Structure of CMT welding robotic system

(3) Robotic CMT GMAW system

CMT GMAW is usually used with robotic operation. The robotic CMT GMAW system consists of a digital welding power supply, a special CMT wire feeder, a CMT welding torch with wire withdrawing mechanism, a robot, a robot controller, a cooling water tank, a remote controller, special connecting cables and a wire feed buffer, as shown in Fig. 6.16.

6.2.4.2 STT (Surface Tension Transfer) GMAW

(1) Principles of operation

STT GMAW is an improved variant of short-circuit transfer gas metal arc welding process that suppress spatters by current waveform control approach. Spatters during short-circuit transfer mainly occur at two moments: One is initial stage of short-circuit and the other is the moment of electrical explosion of the necking liquid metal bridge. At the initial stage of short-circuit when the molten droplet begins to contact the molten pool, the contact area is very small, and the current in the surface of the droplet is opposite to that in the surface of the molten pool, thus producing mutually repulsive electromagnetic forces. If the short-circuit current grows too fast, the rapidly increased electromagnetic repulsion force will expel the droplets out of the molten pool, resulting in large-drop spatter. At the end of short-circuit, the necking part of the liquid metal bridge explodes, and the explosion force inevitably causes spatters. The level of spatter is related to the explosion energy. The higher the explosion energy, the higher the spatter level is. Hence one can see that spatters can

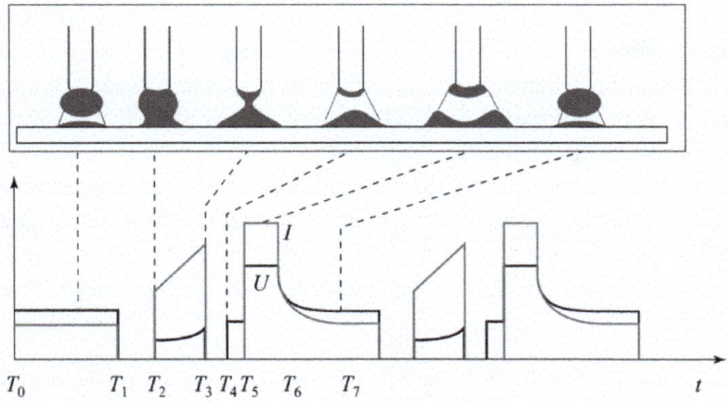

Fig. 6.17 Schematic diagram of STT GMAW process

be effectively suppressed by reducing the current at these two moments. The STT technologies adopts this strategy, as shown in Fig. 6.17. The STT process consists of six stage

(a) Background current stage T_0–T_1: In this stage, the current is set to an appropriate value (50–100 A) to allow for wire melting and formation of droplets, meanwhile to prevent the droplet from growing too rapidly.
(b) Bridge-forming stage T_1–T_2: Once the droplets come into contact with the molten pool, the current is rapidly switched to a value about 20–50 A to avoid expulsion between the droplet and weld pool. Under the effects of gravity and surface tension, the droplet flows divergently into the molten pool, forming a stable liquid metal bridge.
(c) Necking-down stage T_2–T_3: In this stage, the welding current increase at a certain rate, causing the molten metal bridge to shrink rapidly.
(d) Necking-off stage T_3–T_4 is. When the bridge necks down to a certain critical degree at which the neck is about to explode, the current is reduced to about 50 A again within a few microseconds to prevent the bridge from exploding, and then the bridge will be mechanically broken under the effects of gravity and surface tension with no or very little spatter.
(e) Boost current stage T_4–T_6: In this stage, the arc reignites, and the current is increased to a higher value to allows for rapid wire melting and droplet formation. On the other hand, high current is required to provide the necessary weld penetration and width. Then, the current then decreases to a background value.

(2) Characteristics and application

1) Advantages

(a) The spatter rate is reduced to about 0.2%. There is no need to clean the workpiece and nozzle after welding, which saves labor time and improves efficiency.

(b) The weld surface is neat and the root of weld seam tend to have reliable fusion, making particularly suitable for thin plate welding in various positions as well as back-welding of thick plates or thick-walled pipes. When used for back-welding for pipings as an alternative for TIG welding, the welding can be performed at higher speeds and lower costs.
(c) At the same penetration depth, the heat input is 20% lower than that of ordinary CO_2 gas shielded arc welding, resulting in less weld deformation and less heat-affected zone.
(d) Good bridging capacity, e.g. 3 mm thick plates, with clearance allowed up to 2 mm.

2) Disadvantages

(a) It is can only be used for welding thin plates, not thick plates.
(b) The range of welding parameters for stable welding process and quality is narrow. For example, for 1.2 mm wire, the range of welding current is only 100–180 A.

3) Application

STT GMAW is suited not only for welding of non-alloyed steels with CO_2, but also for the welding of stainless steels with pure Ar. In addition, it can also be used for the welding of high-alloy steels, cast steels, heat-resistant steels, and zinc-plated steels, etc. It is widely used for the welding of thin plates and back-welding of oil & gas pipelines.

6.2.4.3 T.I.M.E GMAW

(1) Principles of operation

T.I.M.E (Tranferred Ionized Molten Energy) GMAW utilizes a large wire extension, high wire feed speed and a special quadruple-element gas mixture to achieve extremely high deposition rate and welding speed. The T.I.M.E process has strict requirement for the welding equipment, demanding the use of a high performance inverter power supply, a high performance wire feeder and a dual-circuit cooled welding torch.

The gases used for TIME GAMW are 0.5% O_2 + 8% CO_2 + 26.5% He + 65% Ar. The following gases may also be used:

$$30\%\ \text{He} + 10\%\ CO_2 + 60\%\ \text{Ar};$$

$$8\%\ CO_2 + 92\%\ \text{Ar} + 300 \times 10^{-6}\text{NO};$$

$$2\%\ O_2 + 25\%\ CO_2 + 26.5\%\ \text{He} + 46.5\%\ \text{Ar}.$$

In a quadruple-element gas mixture, Ar is mainly used to shield and stabilize the arc. He has a high thermal conductivity and ionization voltage, which increases the arc temperature and power, thereby increasing the penetration capability; also, it increases the fluidity of molten metal and improves weld formation. CO_2 decomposes easily under the high temperature of the arc, which can cool and compress the arc and improve the concentration of energy and the straightness of the arc. O_2 can reduce the size of droplets, which tends to result in spray transfer, reduces the surface tension of the molten pool, improves the wetting property, and prevents defects such as undercuts and humps.

There are three modes of droplet transfer in T.I.M.E GMAW: Short circuit transfer when the welding current (low wire speed) is low. In case of moderate welding current, spray transfer is used; in case of high welding current, rotating streaming transfer is used. For traditional GMAW welding, during the rotating streaming transfer, the liquid metal in the shape of a rotating cone continuously throws out a large number of small spatters in all directions, which makes the welding process unstable and deteriorates the the weld formaiton. However, the quadruple-element gas mixture reduces the rotating speed of the rotary molten metal cone, stabilizing the transfer process. In addition, the droplets transferred to the molten pool are no longer concentrated on the center line of the molten pool, the impact force of droplets on the molten pool is reduced, the concave depth of the molten pool is decreased, and the velocity of liquid metal in the molten pool is lowered, which helps to avoid humps and undercuts that may occur during high-speed and high-current welding. In addition, the quadruple-element gas reduces the surface tension of the molten metal and improves its mobility, which also helps to avoid undercuts and humps.

(2) Equipment

The TIMW gas metal arc welding machine consists of an inverter power supply, a wire feeder, an relaying wire feeder, a special welding torch, a cooling water tank with refrigeration compressor and a special gas mixing device. Due to the large welding current and wire extension, the T.I.M.E GMAW process places strict requirements on the cooling of the torch nozzles and contact tips, requiring a dual-circuit cooling system, as shown in Fig. 6.18. The dedicated gas mixer provides an accurate mix of multiple gases required for the TIME process and provides 200 L of spare gas per minute for at least 15 welding machines. If a certain kind of gas is exhausted, the gas mixing unit will automatically terminate the gas mixing and trigger alarm. This special mixing device saves up to 70% of gas as compared with conventional gas cylinder plus proportioner syetem.

Due to the high welding speed and deposition speed, this welding process usually adopts automatic welding or robotic welding, and semi-automatic welding is rarely adopted.

(3) Features and applications

1) Advantages

(a) High deposition rate. For the same wire diameter, T.I.M.E GMAW high-speed welding may be performed with a higher current and a stable rotating streaming transfer, resulting in higher wire feed and deposit rate. The deposition rate is up to 10 kg/h for flat welding and 5 kg/h for non-flat position.

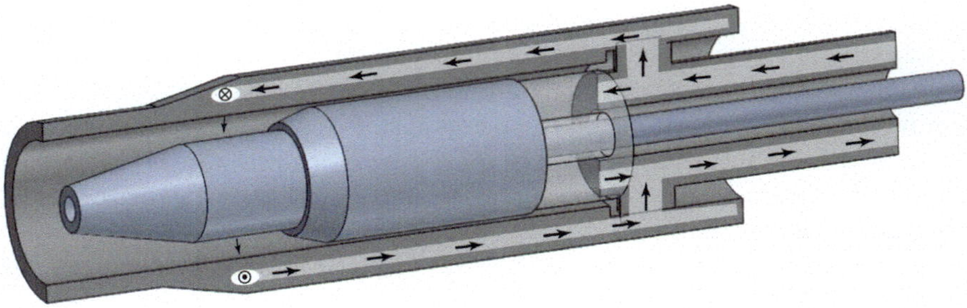

Fig. 6.18 Dual-circuit cooling welding torch

(b) High penetration and high welding speed.
(c) High adaptability. T.I.M.E GMAW has a wide range of welding processes, including short-circuit transfer, spray transfer and rotating streaming transfer. It is suitable for workpieces of various thicknesses and various welding positions.
(d) High weld quality. Stable rotating streaming transfer helps to ensure the fusion of the side walls; the addition of helium improves the fluidity and wettability of the molten metal, resulting in a neat weld seam. The special shielding gases for T.I.M.E GMAW reduces the hydrogen, sulphur and phosphorus content of the weld metal and improves the mechanical properties of the weld, especially the cryogenic toughness.
(e) Low production costs. Due to the greater penetration capacity, smaller groove sizes can be used, which reduces welding wire consumption. The high depositing and welding speed will save man-hours, resulting in a significant reduction in production costs. 25% cost reduction can be achieved compared to conventional GMAW.

(2) Application

T.I.M.E GMAW is suitable for welding of carbon steel, low alloy steel, fine grain high-strength steel, low-temperature steel, high-temperature heat-resistant steel, high yield strength steel, and special steel. Its application fields include ship, steel structure, automobile, pressure vessel, boiler manufacturing and military industry.

6.2.4.4 Tandem GMAW

(1) Principles of operation

Tandem GMAW (Phase controlled dual-wire pulsed GMAW) is a highly efficient GMAW variant of welding using two synergistically controlled pulsed arcs. This method uses two independent digital power sources and a double wire torch. The welding torch is of compact design. The two contact tubes are installed inside the nozzle at a certain angle and distance, as shown in Fig. 6.19. The two welding wires are respectively powered by an independent digital power source, forming two pulsed arcs which can independently adjust all electrical parameters. The two pulsed arcs are controlled by a synchronizer SYNC, which maintains a certain phase relationship to ensure a more stable welding process, as shown in Fig. 6.20. Two arcs form a molten pool as shown in Fig. 6.21.

As compared with single-wire welding, the number of parameters that affect the weld fomation increases, including the welding current, arc voltage, welding speed, shielding gas, angle of welding torch, extension length, wire diameter, angle between

Fig. 6.19 Typical double-wire torch

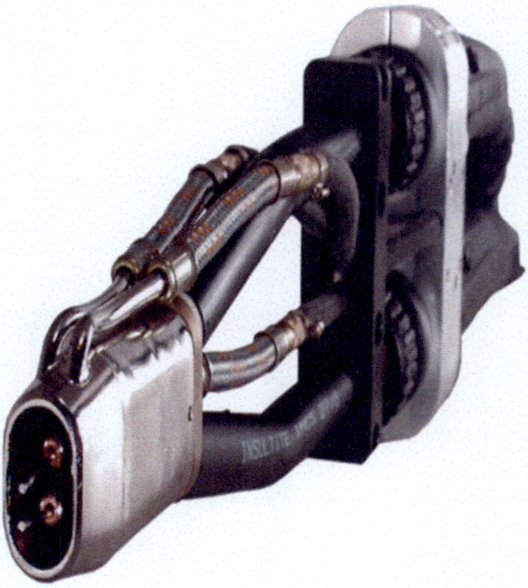

Fig. 6.20 Power supply configuration for Tandem GMAW

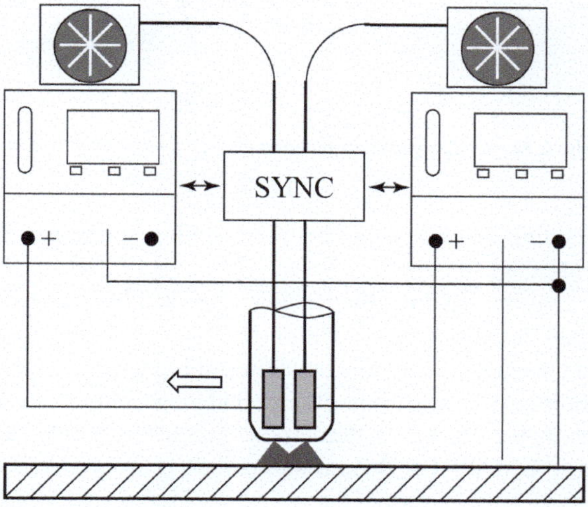

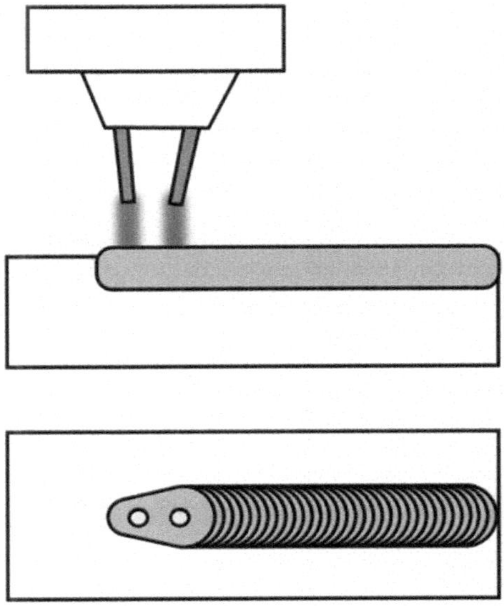

Fig. 6.21 Schematic diagram of the Tandem GMAW

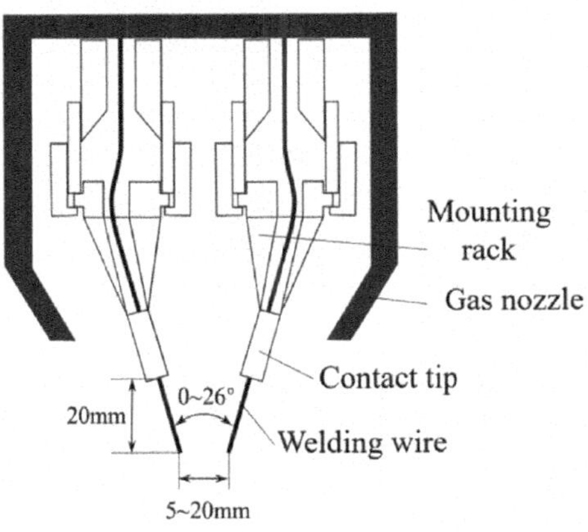

Fig. 6.22 Welding wire arrangement

welding wires, spacing between welding wires and phase difference between two pulsed arcs.

During welding, the leading wire shall be tilted backwards and the trailing wire shall be tilted forward. The current in the trailing wire shall be slightly less than that in the leading wire. The rapid flow of liquid metal is prevented by the arc force of the trailing wire to prevent humps, undercuts and incomplete fusion defects. For a give workpiece thickness, the current of each wire is only half that of a conventional single wire GMAW. For this reason, the arc pressure is reduced significantly, and the arc cavity at the surface of the molten pool is reduced, which reduces the rate at which the pool metal moves backward, thereby preventing humps, undercuts and incomplete

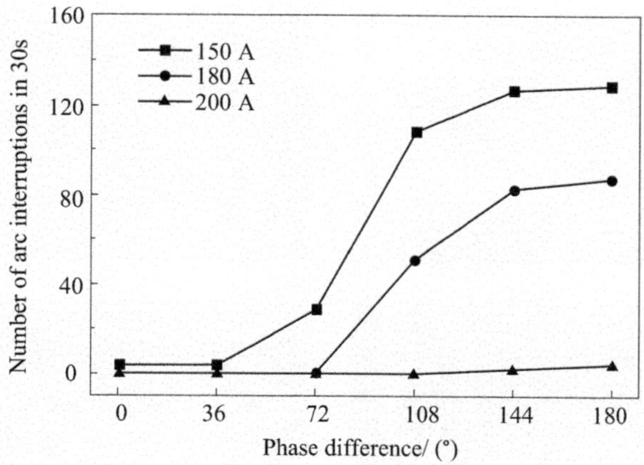

Fig. 6.23 Effect of welding current and phase difference on arc interruption times in 30 s

fusion defects. The angle between two welding wires shall generally be 0–26°, and the spacing between welding wires shall be 5–20° (the most common spacing is 8–12°), as shown in Fig. 6.22. Proper matching of the phase difference between two pulsed arcs according to the current can effectively control the stability of arcs and molten pools, achieve good quality of weld formation, and significantly increase the deposition and welding speed. Figure 6.23 shows the effect of welding current magnitude and phase difference on arc stability. When the welding current is small, the phase difference between two pulsed arcs has a significant impact on the number of arc interruptions. This is because the greater the phase difference, the longer the time in which the background current arc is drawn and elongated by the peak current arc, and the background current arc will be extinguished when being elongated to a certain point, as shown in Fig. 6.24. Therefore, at low currents, the phase difference between the two pulsed arcs should be maintained at 0°. When the welding current is large, the number of arc interruptions is independent of the phase difference, and the phase difference between two pulsed arcs shall be maintained at 180°; When one arc acts on the pulse current, the other arc is under the background current, and the electromagnetic force between the two arcs is small, which effectively reduces the electromagnetic interference between the two arcs and the mutual interference between the two transferred droplets.

Due to the large weight of welding torches, this process can only be used for robotic and automatic welding, rather than semi-automatic welding.

(2) Characteristics and application

1) Advantages

With two independent and distance-adjustable arcs, Tandem GMAW has a highly controllable process, which has the following advantages:

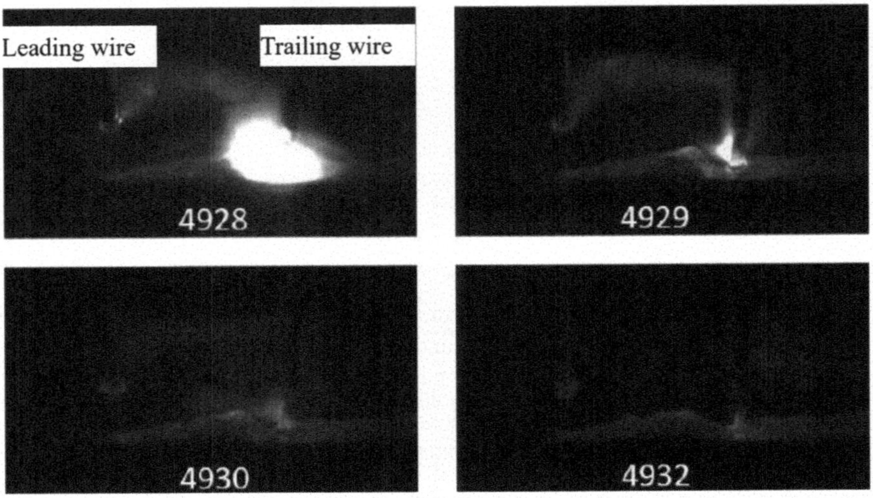

Fig. 6.24 Background current arc (leading arc) extinguished after being over-drawn by the peak current arc (trailing arc) (welding current 100 A + 100 A, phase difference 180°)

(a) Significant increase in welding speed and deposition rate. Total welding current for two arcs is up to 900 A, deposition rate up to 30 kg/h. For the welding of thin plates, welding speed can be significantly increased, up to 1–4 times more than traditional single-wire GMAW welding.
(b) When welding workpieces with a certain plate thickness, the heat input required is lower than that required by single-wire GMAW welding, reducing the width of HAZ and the level of residual deformation.
(c) Extremely stable arc, stable droplet transfer and low spatter rate.
(d) The large nozzle of the double-wire welding torch increase the shielding gas coverage, resulting in higher shielding effect and reduced porosity rate.
(e) High adaptability. In case of multi-layer welding, the main wire and auxiliary wire may be defined at any discretion, and there is no necessity to change the direction of welding torch when changing the welding direction.
(f) Easy to adjust energy distribution. By adjusting the energy parameters of the two arcs, the energy can be reasonably distributed on both sides of the joint, suitable for welding the plates of different thicknesses and materials.

2) Application

Two-wire GMAW can weld carbon steels, low alloy high strength steels, Cr-Ni alloys, and aluminium and aluminium alloys. There are significant economic benefits in the manufacturing of automobiles and automobile parts, ships, boilers and pressure vessels, steel structures, railway locomotives and rolling stock.

6.2.4.5 Plasma-Metal Inert Gas (Plasma-MIG) Hybrid Welding

(1) Basics of plasma-MIG hybrid welding

This welding process uses two power supplies, one for plasma arc and the other for MIG welding, and a special PA-MIG torch, as shown in Fig. 6.25. The welding torch is provided with three gas channels used for delivering the central gas, plasma gas and shielding gas. Generally, these three gases are argon gas. During welding, two arcs are generated between the welding torch and the workpiece, namely, plasma arc between tungsten electrode and workpiece and MIG arc between welding wire and workpiece, as shown in Fig. 6.26. The welding wires, MIG arcs and molten pools are all surrounded by plasma. Generally, this welding process adopts robotic welding or automatic welding rather than semi-automatic welding.

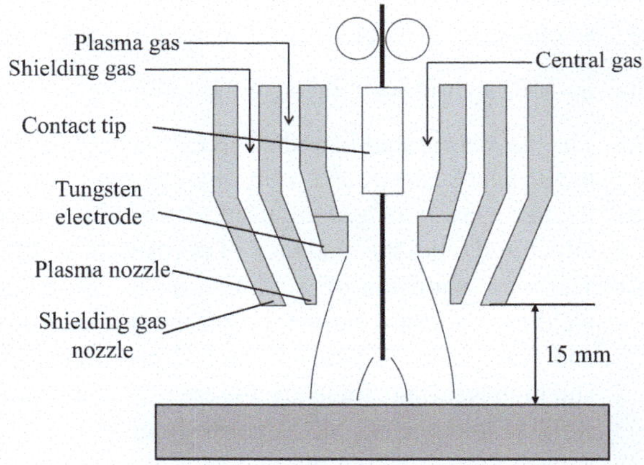

Fig. 6.25 Plasma-MIG hybrid welding torch

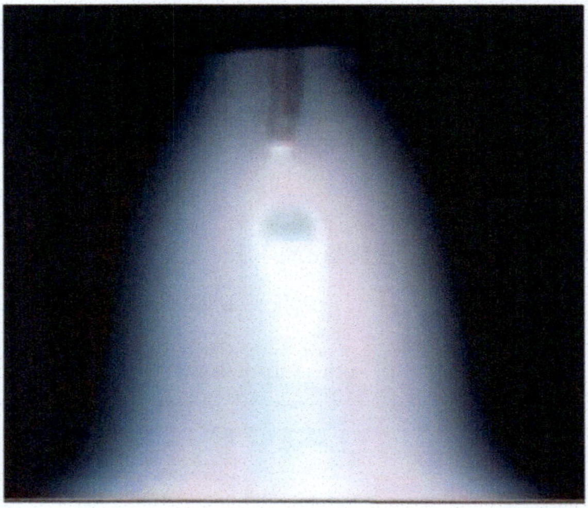

Fig. 6.26 Plasma-MIG hybrid welding arc

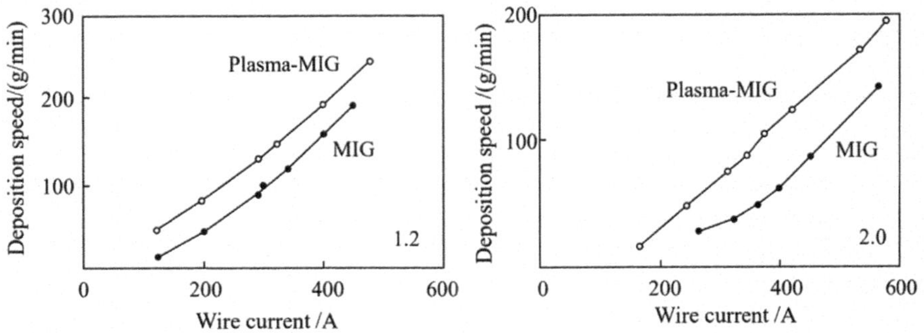

Fig. 6.27 Comparison of deposition speeds between MIG and plasma-MIG hybrid welds

(2) Characteristics of plasma-MIG hybrid welding

1) Advantages

Plasma-MIG hybrid welding have the following advantages:

(a) The plasma arc stabilizes the MIG arc and the droplets at the wire ends, improves the droplet transfer, and overcomes the arc drift.
(b) As three layers of gas are used, the welding torches provide higher shileding effect, reducing the tendency of porosity in welds.
(c) The wire extension of welding wire is larger than that of conventional MIG welding, and the compressed plasma arc additionally heats the welding wire and workpiece, resulting in high deposition speed, high penetration capacity and high welding efficiency. Figure 6.27 compares the deposition speed of MIG welding and plasma-MIG hybrid welding.
(d) By properly selecting the diameter of tungsten electrode used for plasma arc, the temperature of the molten pool can be increased and the wetting property of the molten metal can be improved.
(e) Because the plasma arc stabilizes the cathode spots of MIG arc, stable arc and good weld formation can also be achieved when pure argon gas is used to weld high-strength steel, which significantly reduces the oxygen content in weld and improves its mechanical property.

2) Disadvantages

The disadvantages of plasma-MIG are:

(a) Complex welding torches, involving many welding parameters with narrow matching range;
(b) Not suitable for semi-automatic welding, only for automatic welding or robotic welding.

6.3 Tungsten Inert Gas Welding (TIG Welding)

6.3.1 Operation Principles, Features and Applications

(1) Principles of operation

Tungsten Inert Gas (TIG) Welding is a welding process in which the base metal and filler wire are heated and melted by the arc between the tungsten electrode and the workpiece under the protection of inert gas. Figure 6.28 shows the schematic diagram of TIG welding.

Common inert gases include argon (Ar), helium (He) and argon-helium mixture. In some cases (e.g. stainless steel or nickel-base alloy welding) argon plus a small amount of hydrogen may be used. The protective effect of these gases is basically the same, but there are significant differences in the arc process characteristics. Helium gas significantly increases arc power, which in turn increases welding speed and penetration. Helium gas cost is much higher than argon gas. The most commonly used gas for TIG is pure argon. Therefore, the TIG welding is also known as tungsten argon arc welding.

TIG welding can be used with DC (direct current) and AC (alternating current), and AC TIG welding can be operated with sine wave AC and rectangular (square) wave AC modes. The AC TIG welding is used for welding active metals such as aluminum and aluminum alloy, magnesium and magnesium alloy. DC TIG welding is used for other metals than aluminum and magnesium, where usually DC straight polarity (DCSP) is adopted. DCSP TIG welding has the advantages of arc stability, large penetration capability and good weld formation. However, the oxide film on the surface of the workpiece cannot be removed due to the anode connection of the workpiece.

Sheet welding is generally performed by means of pulsed TIG welding. Pulsed TIG welding can be further classified into low frequency (0.1–10 Hz), medium frequency (1 Hz–1 kHz) and high frequency (20–40 kHz) pulsed welding.

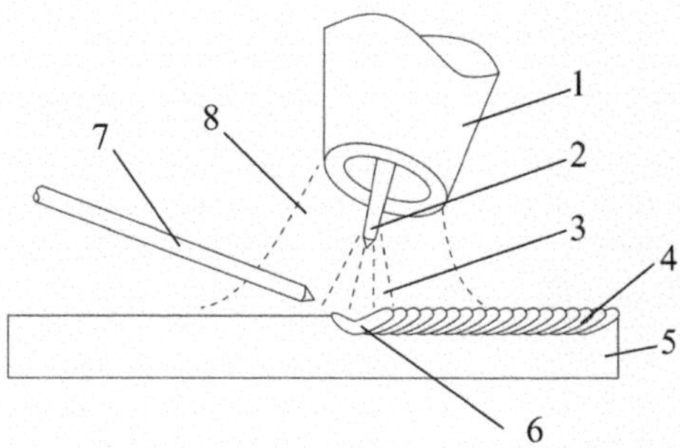

Fig. 6.28 Basic principle of TIG welding. 1-Nozzle; 2-Tungsten electrode; 3-Arc; 4-Weld; 5-workpiece; 6-Molten pool; 7-Welding wire; 8-Shielding gas

(2) Characteristics

1) Advantages

(a) Almost all metals can be welded. It is especially suitable for welding aluminum, magnesium and their alloys which have strong chemical activity and suffer from rapid formation of tough high-melting-point oxides.
(b) During welding, the tungsten rod is not melted, there are relatively few interference factors affecting the change of arc length. In addition, the arc electric field strength are low, so the welding process is very stable.
(c) Welds are well formed and of high quality.
(d) Even with a low current of a few amperes, the TIG arc can run stably and the generated heat is relatively concentrated, so it can weld thin section sheet low up to 0.3 mm thick. When used with pulsed current, this process can weld even thinner material and is especially suitable for all-position welding, heat sensitive material welding, and one-side welding with back formation without backing.
(e) The TIG welding uses an open arc with stable parameters which can be easily monitored and controlled. It is an ideal automatic and robotic welding process.

2) Disadvantages

(a) The tungsten electrode has limited current carrying capacity and low arc thermal efficiency coefficient, resulting in shallow penetration, low deposition rate and low welding productivity.
(b) The shielding gas has poor resistance to lateral wind. Windproof measures shall be taken for welding in case of lateral wind.
(c) High requirements on workpiece cleaning. Since there is no metallurgical deoxidation and hydrogen removal action due to the use of inert gas for protection, the oil stains and rust on the workpiece must be removal before welding in order to avoid porosities, cracks and other defects.

(3) Applications

1) Weldable metals

TIG welding can be used for welding almost all metals and alloys, but because of its high cost, it is mainly used for welding stainless steel, heat resistant steel and non-ferrous metals (aluminum, magnesium, titanium, and copper, etc.) and their alloys in production. For ordinary ferrous metals, it is mainly used for backing welding of important welds.

Table 6.6 Applicable scope of workpiece thickness for TIG

Thickness/mm	0.13	0.4	1.6	3.2	4.8	6.4	10	12.7	19	25	51	102
Single-pass weld with I groove	←	—	—	→								
Single-pass weld with V groove			←	—	→							
Multi-layer weld with V groove				←	—	—	- - -	- - -	- - -	- - -	- - -	- - →

2) Joints and positions

TIG welding is mainly used for butt joints, lap joints, T-joints and fillet joints. Flanged edge joint may be used for butt joining of thin plates (≤2 mm). It is suitable for all welding positions.

3) Workpiece thickness and product structure

Table 6.6 shows the range of workpiece thickness applicable to TIG welding. For productivity and cost considerations, TIG welding is generally not used for workpieces with thickness greater than 10 mm.

Thin-section products such as boxes, compartments, diaphragms, casings, skinning, jet blades, cooling fins, fins, pipe joints, encapsulation of electronic components may be manufactured by TIG welding.

In order to ensure the welding quality, TIG welding is generally adopted for the fully penetration root weld of butt joint of important thick-walled members such as pressure vessel, piping and turbine rotor, etc., and the backing weld of narrow gap welds of other structures.

6.3.2 *Welding Parameters*

(1) Current type and polarity

Aluminum, magnesium and their alloys are usually welded with AC, while other metals are preferably welded by direct current straight polarity (DCSP). Pulse current shall be used for plate welding as far as possible. Square-wave AC pulses shall be used for aluminum, magnesium and their alloys, while DC pulses shall be used for other metal sheets.

(2) Electrode diameter and end shape configuration

The selection of the diameter of tungsten electrode shall be such that the allowable current of tungsten electrode is larger than the welding current used. The allowable current of tungsten electrode depends on the diameter, type and polarity of tungsten electrode. The larger the diameter of the tungsten electrode, the larger the allowable current. The allowable current of tungsten electrode is the maximum when DCSP is used and the minimum when DCRP (direct current reverse polarity) is used. The allowable current of tungsten electrode operating with AC is between DCSP and

DCRP. The current waveform also has an important effect on the allowable current of tungsten electrode. During the pulsed TIG welding, since the tungsten electrode is cooled in background duration, the allowable current of the tungsten electrode of a certain diameter will increase significantly.

The shape of the tungsten end has an important effect on the stability of the arc. For TIG welding with DCSP, the tungsten electrode is usually gound to a conical shape for low current or frustum shape for high current, as shown in Fig. 6.29. Table 6.7 shows the recommended end shape and operating current range of tungsten electrodes. For AC TIG welding, the end of the tungsten electrode is usually ground to a spherical shape. As the current increases, the diameter of the ball increases, and the maximum is equal to the radius of the tungsten electrode (i.e., without cone angle).

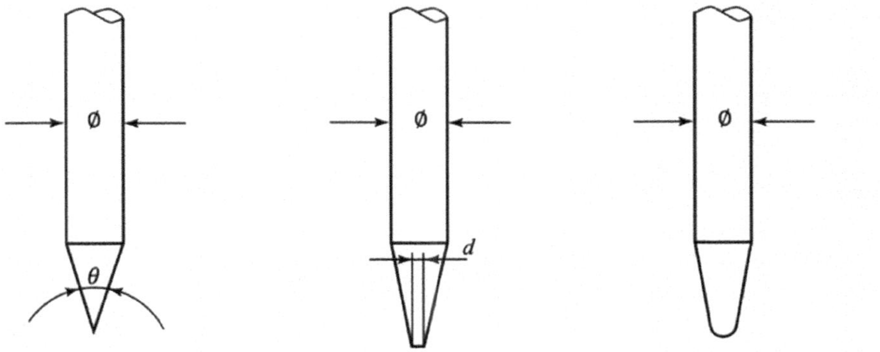

a) Conical shape b) Frustum shape c) Conical shape with round head

Fig. 6.29 Recommended electrode end Shape

Table 6.7 Recommended end shape and size of electrode for different DCSP TIG welding current

Electrode diameter ϕ/mm	Tip diameter d/mm	Cone angle θ/(°)	DCSP	
			Constant current range/A	Pulse current range/A
1	0.125	12	2–15	2–25
	0.25	20	5–30	5–60
1.6	0.5	25	8–50	8–100
	0.8	30	10–70	10–140
2.4	0.8	35	12–90	12–180
	1.1	45	15–150	15–250
3.2	1.1	60	20–200	20–300
	1.5	90	25–250	2–300

(3) Welding current

The welding current is the main parameter for determining the weld penetration. It is generally selected based on the factors such as base metal type, workpiece thickness, joint type, and welding position, etc.

For pulsed TIG welding, the welding current parameters includes: background current I_b, pulse current I_p, pulse duration t_p, background duration time t_b, pulse period $T(=t_p + t_b)$, pulse frequency $f(=1/T)$, pulse amplitude ratio $F(= I_p/I_b)$, and pulse width ratio $K(= t_p/t_b + t_p)$. Four of these parameters are independent.

1) Pulse current I_p and pulse duration t_p

The product of pulse current and pulse duration $I_p t_p$ is called the electric flux, which determines the shape and size of the weld, especially the penetration. Therefore, the pulse current and pulse duration shall be selected according to the base metal type and plate thickness first. Pulse current and duration can be selected according to Fig. 6.30 for workpieces of different materials and plate thicknesses.

For workpieces with thickness less than 0.25 mm, the pulse current shall be reduced appropriately and the pulse duration shall be extended accordingly. For plates with thickness greater than 4 mm, the pulse current value shall be increased appropriately and the pulse duration shall be shortened accordingly.

2) Background current I_b

The background current mainly serves to maintain the arc, and, therefore, as long as the arc stability is ensured, the background current shall be selected as low as possible to highlight the characteristics of pulse TIG welding. However, in the case of welding materials with a high tendency of cold cracking, the background current shall be set to be a little higher so as to prevent crater crack. The background current is generally 10–20% of the pulse current.

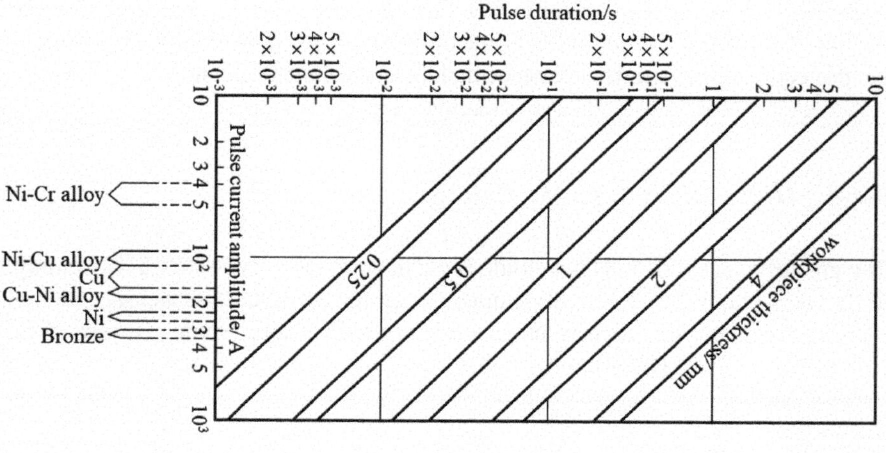

Fig. 6.30 Recommended pulse current and duration for plates of different thicknesses and materials

3) Background duration t_b

The background duration has little effect on the shape and size of the weld. Excessive backgroun duration, however, significantly reduces the heat input and results in a discontinuous weld pass.

(4) Shielding gas flow

Under any given condition, there is an optimum matching window between the gas flow and the nozzle diameter, in which the shielding gas coverage and the effectiveness of the shielding are maximum. The flow rate selection shall ensure that the aspiration of the surrounding air is completely prevented. For the common nozzle diameter range of 5–20 mm, the typical flow rate range is 5–25 L/min. If the gas flow is too low, the stiffness of the gas flow will be insufficient, causing aspiration of the surrounding air. On the other hand, excessive flow rate cause turbulence, also reducing the protection effect.

(5) Extension length of tungsten electrode

The extension length of the tungsten electrode is defined as the length of the tungsten electrode exposed outside the nozzle. If the extension is too large, the tungsten electrode will overheat and the effectiveness of the shielding is reduced, while if the extension is too small, the nozzle will overheat. Therefore, the extension must be maintained at a suitable value. In butt welding, the extension length of tungsten electrode shall generally be kept at 5–6 mm; when welding T-shaped welds, the extension length should preferably be 7–8 mm.

(6) Nozzle-to-work distance

The distance between the nozzle and the workpiece shall be commensurate with the extension length of the tungsten electrode. Generally, it shall be be kept at 8-14 mm. If the distance is too small i.e. the arc length is too short, the tungsten electrode will be prone to come into contact with the molten pool, causing contamination of weld metal by tungsten and reduced life of the electrode; while if the distance is too large, the protection effect will be weakened and the arc will be unstable.

6.3.3 High Efficiency TIG Welding

The greatest weakness of TIG welding is the low productivity. Active TIG welding, TOP-TIG welding, K-TIG welding, double tungsten electrode TIG welding and hot wire TIG welding are all new processes proposed to raise welding productivity. Among them, TOP-TIG welding, K-TIG welding, double-tungsten electrode TIG welding and hot-wire TIG welding mainly used with automatic and robotic operation modes.

6.3.3.1 Hot-Wire TIG Welding

(1) Principle of operation

The operation principle of hot wire TIG welding is shown in Fig. 6.31. The filler wires are heated by a special power supply called hot-wire supply or wire-heating power supply. The welding wire extending beyond the contact tip is heated to a temperature close to the melting point by the resistance heat generated by the hot wire current, and melts rapidly after contacting the molten pool. Therefore, the deposition speed of this method is significantly improved.

The heating effect of welding wires depends on the hot wire current, extension length and wire feeding speed. The wire extension length is generally kept at 15–50 mm. For a given wire extension length, a delicate balance exists between the wire feed speed and hot-wire current. If the hot wire current is too high, the wire will fused in large length, making the solid wire out of contact with the workpiece. As a result, the hot wire will be interrupted periodically, resulting in discontinuous weld bead. If the wire current is too low, the wire will be inserted into the molten pool, resulting in a solid state short-circuit.

The hot wire current flowing through the welding wire generates a magnetic field, which may cause the arc to deflect. To avoid this kind of magnetic deflection, the following measures shall be taken:

(a) The angle between welding wire and tungsten electrode shall be reduced. The angle between cold wire and tungsten electrode is usually close to 90° during cold wire TIG welding, and the angle between hot wire and tungsten electrode shall be controlled at 40–60°, as shown in Fig. 6.32.
(b) Both the hot wire current and welding current are pulsed, and the phase difference between them is controlled to be 180°, as shown in Fig. 6.33. At the peak welding current, the hot wire current is zero and no magnetic blow occurs. The

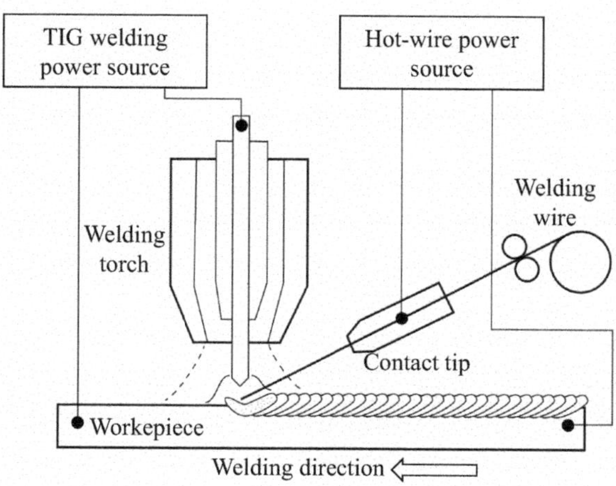

Fig. 6.31 Schematic diagram of hot-wire TIG welding

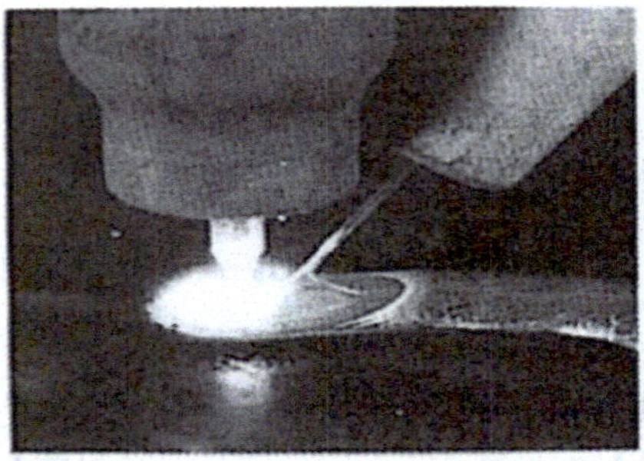

(a) Hot-wire TIG

(b) Cold-wire TIG

Fig. 6.32 Angles of hot and cold filler wires during filling

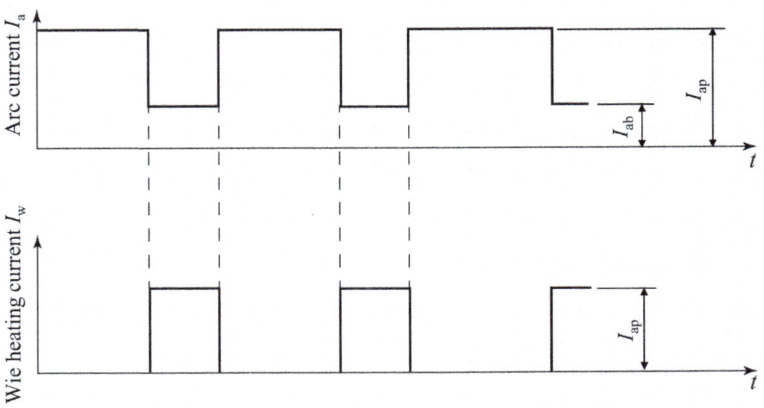

Fig. 6.33 Phase matching of hot wire current and welding current

arc heat is used to heat the workpiece to form a molten pool. When the welding current is the background current, the hot wire current is the peak current, and the arc will deflect towards the welding wire as it is attracted by its magnetic field. The magnetic blow occuring in this time interval has little effect on the penetration depth and the behavior of the molten pool because the background current is low and the arc mainly acts as a pilot arc.

(2) Features

Hot-wire TIG welding has the following advantages over traditional TIG welding:

(a) High deposition rate. Under the same current, the deposition rate can be increased by up to 60%, as shown in Fig. 6.34.
(b) High welding speed. Under the same current, the welding speed can be increased by more than 100%.

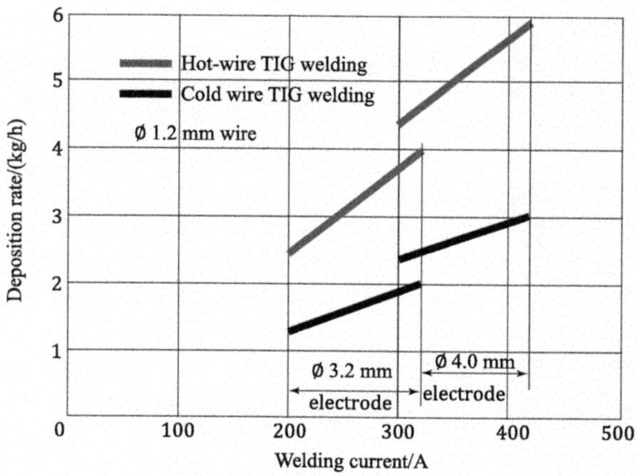

Fig. 6.34 Comparison of deposition rates between hot-wire TIG and cold-wire TIG welding

(c) Low dilution. It can be reduced by up to 60% compared with the cold wire TIG welding.
(d) Minimal weld deformation. As the welding wires are preheated with hot wire current, the welding current required for the same penetration is small, which can help to reduce heat input and welding deformation.
(e) Low porosity sensitivity. The heating of hot wire current causes the welding wire to reach a very high temperature before it is filled into the molten pool. Contaminants such as organic matters volatilize in advance, causing the hydrogen content in the welding area to decrease.
(f) Less alloy oxidation. Lower heat input is required for the same penetration, which reduces the temperature of the molten pool and the loss of alloys.

(3) Applications

Hot wire TIG welding is used for thin and medium-thick plate welding of carbon steel, alloy steel, stainless steel, nickel-based alloys, duplex or multi-phase steel and titanium alloy, etc. It is especially suitable for surface overlay welding of tungsten-chromium-cobalt alloys. But it is not suited for aluminium alloy and copper alloy due their low resistivity.

6.3.3.2 TOP-TIG Welding

Ordinary filler wire TIG welding is called cold-wire TIG welding. The welding wire is almost 90° with the electrode during welding, which is not suitable for robotic welding. Its disadvantages are as follows.

(a) Increased size of the welding torch and poor accessibility.
(b) The wire-feed tip limits the robot flexibility.
(c) For the robotic welding of complex components an extra linkage axis (7th axis) has to be added to coordinate the filling position of the wire.

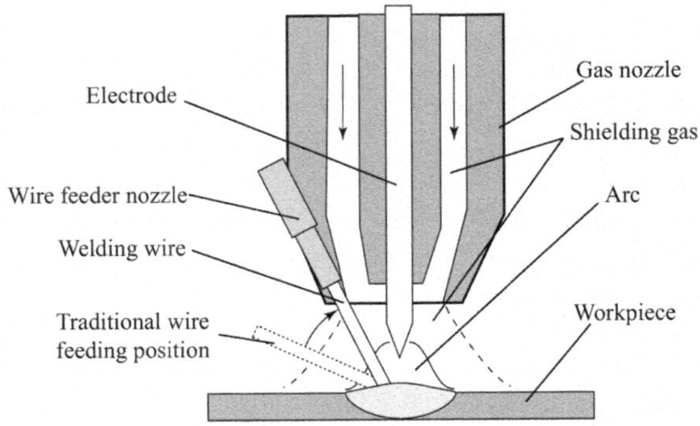

Fig. 6.35 Schematic diagram of TOP-TIG welding

(d) The arc heat of the filler wire TIG welding is used to melt the workpiece and the welding wire respectively, as about 30% of the arc heat is used to melt the welding wire, further limiting the increase of welding speed.

Therefore, the welding robots currently used for TIG welding generally do not use cold wire TIG welding for the sake of flexibility. TOP-TIG welding is a new method developed for robotic filler wire TIG welding.

(1) Principles of operation

TOP-TIG welding is a TIG welding process in which the welding wire is fed into the molten pool through a wire-guide tube integrated in the side wall of the gas nozzle, as shown in Fig. 6.35. The welding wire is fed into the arc directly from the side wall of the nozzle through which it enters the molten pool. The angle between the welding wire and the axis of tungsten electrode is usually maintained at about 20°. In order to prevent the welding wire from contacting the tungsten electrode, the cone surface of the tungsten electrode tip shall be parallel to the welding wire, and for this reason, the end of the tungsten electrode needs to be machined into a 40°cone. The welding wire is first preheated by the high-temperature nozzle when passing through the feeding nozzle, and by the arc column when entering the arc, thus allowing for increased wire feeding speed and resulting in arc energy utilization, as shown in Figs. 6.36 and 6.37.

TOP-TIG welding exhibits three metal transfer modes in different welding condition: globular transfer, intermittent bridging transfer and continuous bridging transfer, as shown in Fig. 6.38. The main influencing factors are the wire feeding speed, welding current and wire feeding direction, which also affect the droplet transfer mode. The droplet transfer is globular transfer when the wire feeding speed is low. As the speed increases, globular transfer will be transformed to intermittent bridging transfer and then to continuous bridging transfer. Figure 6.39 shows the effect of wire feeding speed on the droplet transfer mode under certain welding current and arc voltage.

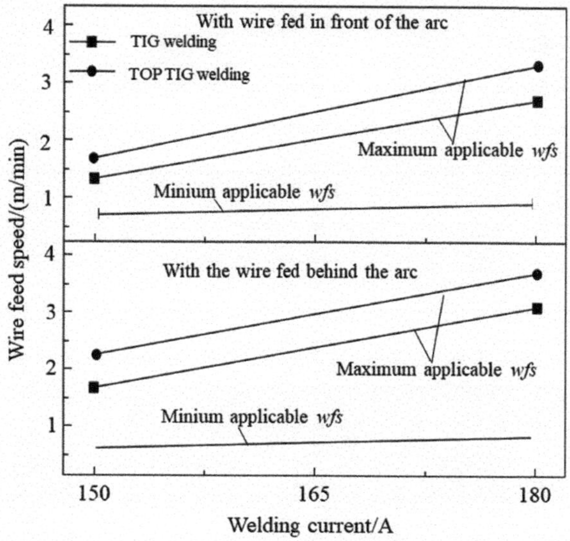

Fig. 6.36 Comparison of applicable ranges of wire feed speed between TOP-TIG and conventional filler TIG welding

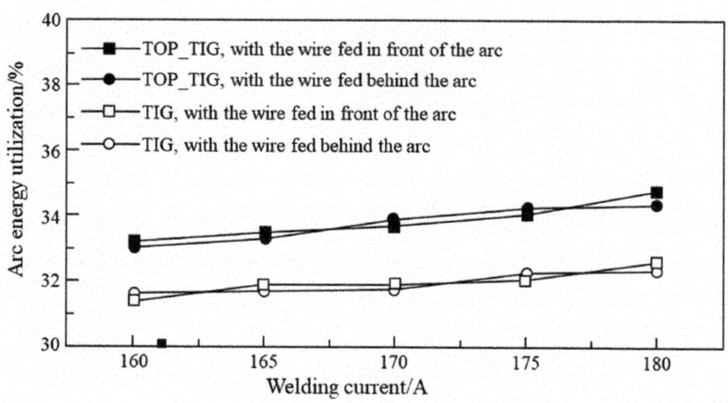

Fig. 6.37 Arc energy utilization rate of TOP-TIG welding and conventional filler wire TIG welding

(2) Characteristics and Applications

1) Advantages of TOP-TIG welding

(a) As compared with ordinary filler wire TIG welding, the operation is easy and flexible, and there is no need the change the wire feeding direction when the weld direction changes.

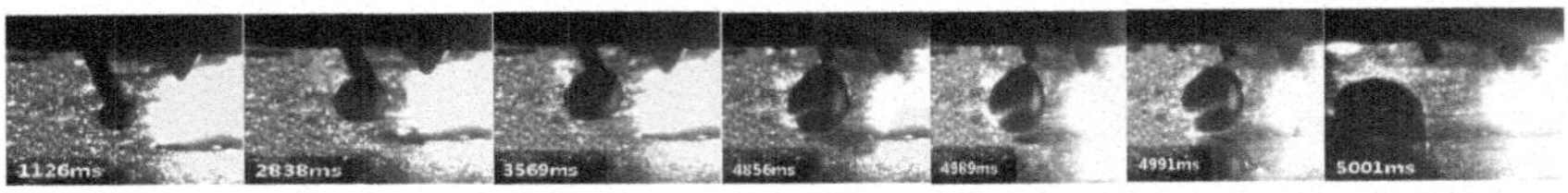

(a) Globular transfer (welding current 180 A, wire speed 0.3 m/min, welding speed 0.5 m/min)

(b) Intermittent bridging transfer (welding current 180 A, wire speed 1.3 m/min, welding speed 0.5 m/min)

(c) Continuous bridging transfer (welding current 150 A, wire speed 1.7 m/min, welding speed 0.5 m/min)

Fig. 6.38 Droplet transfer of TOP-TIG welding with the wire fed in front of the arc

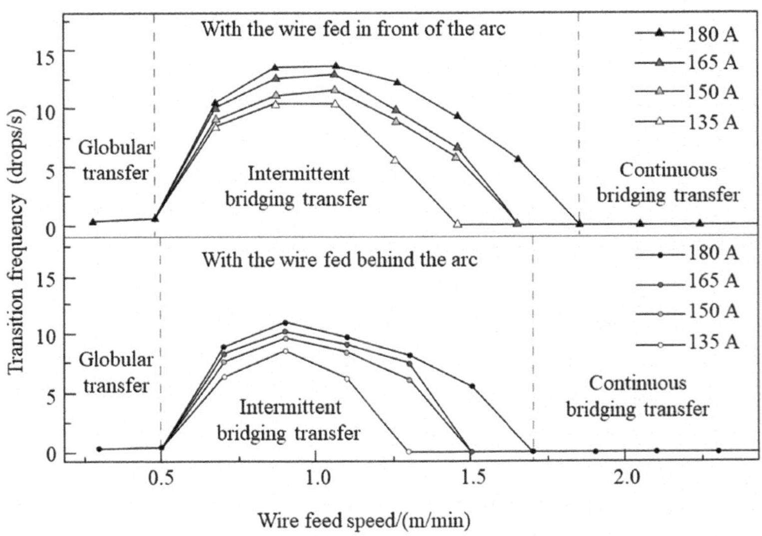

Fig. 6.39 Effect of wire feeding speed on droplet transfer mode

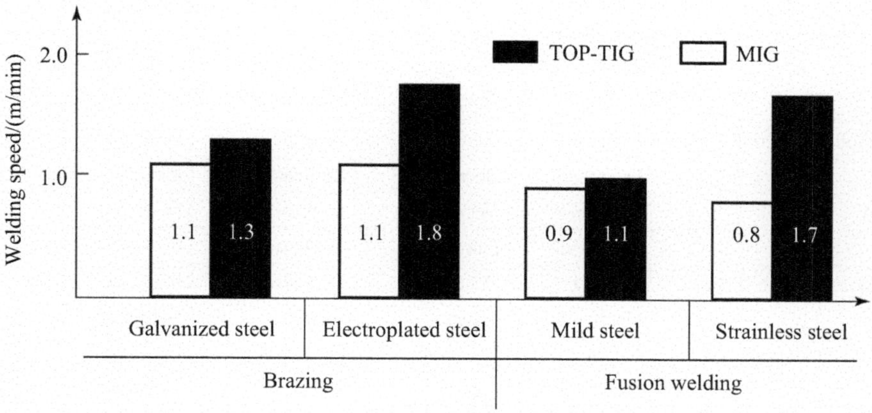

Fig. 6.40 Comparison of the welding speed of TOP-TIG and MIG welding

(b) Fast welding speed and high energy utilization rate. The intensive preheating of the welding wires in the high temperature nozzle and in the high temperature arc column area near the tungsten electrode significantly increases the rate of arc heat utilization and increases the deposition rate and welding speed. When plates less than 3 mm thick are welded, the welding speed of TOP-TIG will be equal to or even higher than MIG, see Fig. 6.40.

(c) Compared with MIG/MAG welding, the weld quality is good with no spattering and low noise.

(d) The distance from tungsten electrode to workpiece does not affect the welding quality as much as TIG welding, thus widening the process window.

2) Weaknesses of TOP-TIG welding

TOP-TIG welding has very strict requirements on the tungsten electrode end, therefore, it can only be performed with DCSP welding and not be performed with AC arc.

3) Application

TOP-TIG welding can be used for galvanized steel, stainless steel, titanium alloys, nickel metal alloys, etc., being more efficient than MIG/MAG welding. As AC arc cannot be used, this method is generally not used for welding active metals such as aluminum, magnesium and their alloys.

(3) TOP-TIG welding process

The main process parameters of TOP-TIG welding include wire-to-electrode distance (distance from tungsten electrode to wire end), diameter of tungsten electrode, welding wire diameter, welding current, wire feeding speed and welding speed. The wire-to-electrode distance is generally 1–1.5 times the welding wire diameter. The diameter of the commonly used tungsten electrodes is 2.4 and 3.2 mm, with the upper current limits being 230 and 300 A, respectively. The commonly used welding

Table 6.8 Rule of influence of main welding parameters on weld formation

Parameter	Trend	Change trend of weld formation		
		Penetraton	Weld width	Reinforcement
Welding current	Increase	Increase	Increase	Decrease
	Decrease	Decrease	Decrease	Increase
Arc voltage	Increase	Decrease	Increase	Decrease
	Decrease	Increase	Decrease	Increase
Wire feeding speed	Increase	Decrease	Decrease	Increase
	Decrease	Increase	Increase	Decrease
Welding speed	Increase	Decrease	Decrease	Decrease
	Decrease	Increase	Increase	Increase

wires are 0.8, 1.0 and 1.2 mm in diameter. Table 6.8 shows the effect of main welding parameters on weld formation of TOP-TIG welding.

6.3.3.3 Keyhole TIG Welding

(1) Principles of operation

Keyhole TIG welding (K-TIG welding) is a kind of improve TIG welding process in which a enhanced water-cooling torch with a large-diameter tungsten electrode (usually more than 6 mm) and a high current of more than 300 A are used to increase the welding current density. During welding, the high welding current density significantly increases the electromagnetic pinch force and plasma force and the arc straightness. Once the workpiece is melted through, a small-diameter keyhole through the thickness of the workpiece is produced in the molten pool directly below the arc, as shown in Fig. 6.41. The keyhole achieves stable condition when the evaporative reaction force of metal vapor, arc pressure and surface tension of molten metal offset. The shape of keyhole is mainly affected by parameters such as the welding current and the density, surface tension and thermal conductivity of parent materials. The ideal shape of keyhole is shown in Fig. 6.42.

(2) Welding equipment

The conventional TIG welding power source is not suitable for use in keyhole TIG welding due to the high welding current. A specially designed TIG power source with high rated current and high-power high frequency oscillator shall be used. If there is no such a power source, parallel combination of two conventional TIG power sources or a submerged arc welding power source can be substitute. The submerged arc welding power source shall be used with high-frequency oscillator to allow for arc ignition and stabilization.

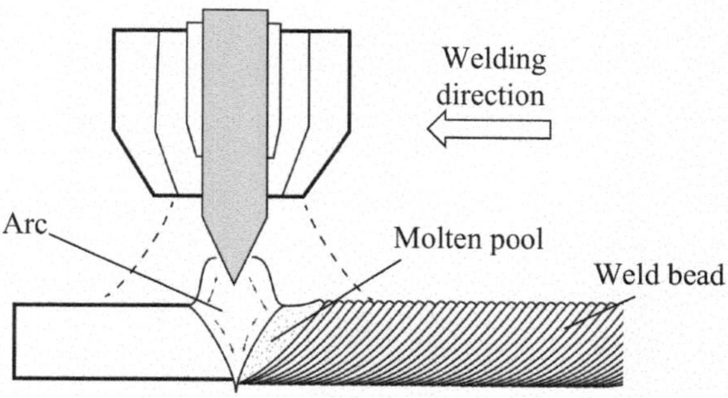

Fig. 6.41 Schematic diagram of K-TIG welding principle

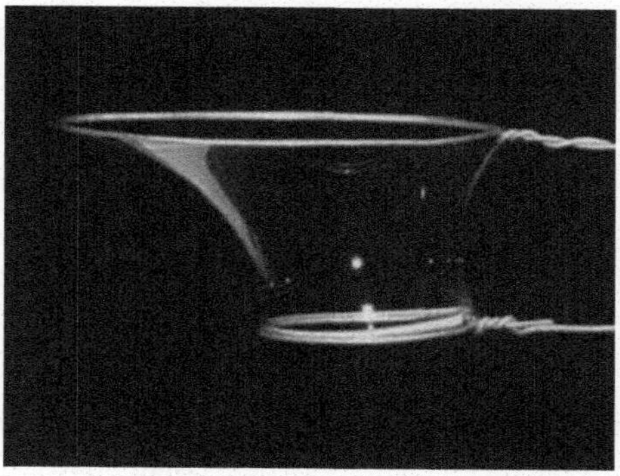

Fig. 6.42 Ideal keyhole shape

Keyhole TIG torches are much larger in size and weight than conventional torches. Figure 6.43 shows a schematic of keyhole TIG welding torch construction. The nozzles are generally made of copper and have a large diameter. In order to ensure good gas protection effect, the shielding gas flow used in welding is generally greater than 20 L/min. The tungsten electrode diameter is greater than 6 mm and requires strong water cooling. The cooling water shall flow directly surround most of the tungsten electrodes to ensure that the cooling is sufficient to constrict the electrode electron emission to a 1 mm-radius area. Only in this way, can the arc pressure penetration capability and the stable formation of the keyhole be ensured.

(3) Advantages

Keyhole TIG welding provides the following advantages:

(a) High penetration capability. At normal welding speed (0.2–0.3 m/min), it is possible to weld through 12 mm thick stainless steel or titanium alloy of at one time without bevelling or wire addition.
(b) High welding speed, up to 1 m/min for stainless steel with a thickness of no more than 3 mm.
(c) Welds are well formed with small weld deformation.

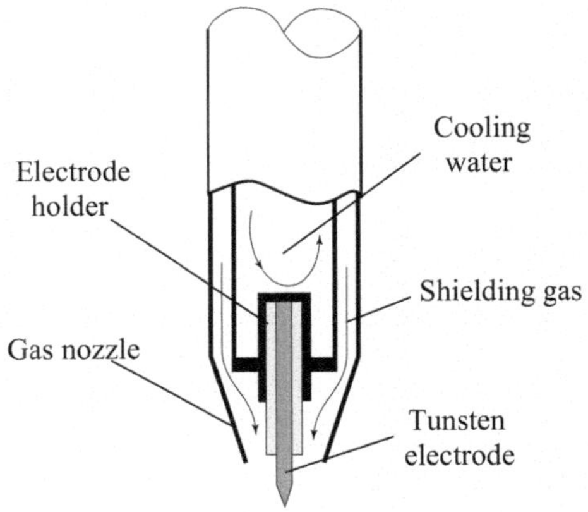

Fig. 6.43 Schematic of keyhole TIG welding torch

(d) High arc thermal efficiency coefficient and energy utilization. Due to the large current, the tungsten electrode has a high thermionic emission capability and its conduction mechanism is close to that of the ideal thermionic emission cathode. The cathode fall voltage is extremely low and the arc heat dissipated on the tungsten electrode is very small. As a result, the thermal efficiency coefficient of the arc is increased.

The disadvantage of keyhole TIG welding is that only the metals with low heat conductivity such as stainless steel, titanium alloy and zirconium alloy can be welded. For the metals with high heat conductivity such as aluminum alloy and copper alloy, the root width of keyhole is large and the stability of molten pool is very low, making it difficult to ensure the weld formation.

(4) Welding parameters

The current used for key-hole TIG welding is usually in the range of 300–1000 A, the arc voltage in the range of 16–20 V, the diameter of tungsten electrode in the range of 6.0–8.0 mm with a tip cone angle of 60°.

6.4 Laser Beam Welding

6.4.1 Operation Principle, Characteristics and Application

(1) Principles of operation

Laser beam Welding is a high-energy-density fusion welding process using a focused laser beam as the heat source. The workpiece heating is essentially a process in which the laser interacts with an opaque substance.

Once striking on the surface, a portion of the laser beam energy is absorbed by the workpiece, resulting in a complex process involving temperature rise, surface melting, vaporization, plasma formation and key-holing, etc., as shown in Fig. 6.44.

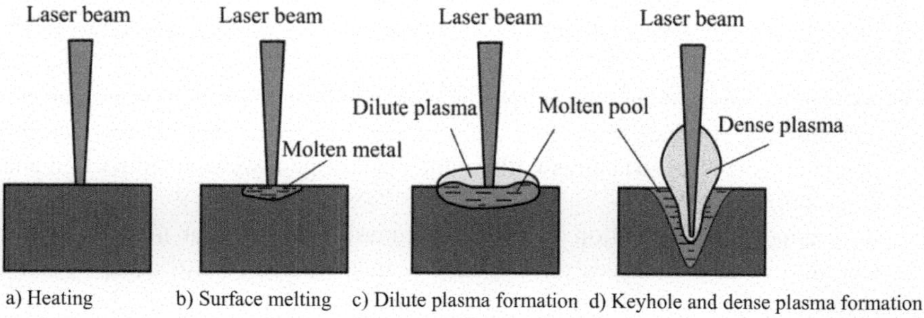

Fig. 6.44 Physical state evolution of the workpiece when a laser beam impinges on the surface

If the laser has a low power density (<10^4 W/cm^2) and a short radiation time, the absorbed energy by the metal from the laser can only cause a temperature rise in the workpiece surface. This kind of laser beam is applicable to surface heat treatment.

When the laser power density reaches 10^4–10^6 W/cm^2 and the radiation time is long, the surface of the material melts and the liquid-solid phase interface gradually moves deep into the material. This kind of laser beam applies to metal surface remelting, alloying, cladding and melt-in mode welding.

When the power density of laser exceeds 10^6 W/cm^2, the material surface not only melts, but also evaporates, and the metal vapor accumulates near the material surface and causes weak ionization. The ionized metal vapor is called plasma, which is beneficial to the absorption of laser by workpiece. In addition, the reaction force of the metal vapour causes depression on the metal surface of the molten pool. This kind of laser beam applies to melt-in mode welding.

When the power density is higher than 10^7 W/cm^2, the surface of material evaporates intensively to form strong plasma. The dense plasma has shielding effect on laser and significantly reduces the absorption rate of laser by the workpiece. In addition, when the metal evaporates, an evaporation reaction force is applied to the surface of the molten pool, under which a small hole (also known as keyhole) is formed in the front of the molten pool through the workpiece. As long as the laser beam is applied, the keyhole remains open by the evaporation reaction force. As the laser beam moves forward, molten metal in the front of molten pool flows backward around the hole to merge into a continuous melt in the rear of the pool and solidify here to form a sound weld. The keyhole significantly increases the beam energy absorption as it confines the reflected beam. This kind of laser beam applies to keyhole mode welding, material cutting, and punching, etc.

(2) Features

The laser beam welding has several advantages over the arc welding process:

(a) After focusing, the laser spot diameter can be concentrated to as small as 0.01 mm, with a high power density (up to 10^{13} W/m^2), and welding is mostly performed in keyhole mode.

(b) Laser heating area is small (<1 mm), and the welding speed can be up to 10 m/min.
(c) As the heat input of welding is low, the weld and heat affected zone are narrow, and welding residual stress and deformation are small, suitable for welding precision parts and structures without the need of post-welding correction and machining.
(d) By changing the direction of laser transmission via optical fiber or prism, welding at long distances or at locations that are difficult to access can be performed. Due to its ability to penetrate transparent bodies such as glass, lasers are suitable for welding highly toxic materials such as beryllium alloys in sealed glass containers.
(e) It can be used for welding materials that are difficult to be welded by arc welding process, such as high melting point metal, ceramic, and organic glass, etc.
(f) In contrast to electron beam welding, laser beam welding does not require vacuum chambers, does not produce X-rays and the beam is not subjected to electromagnetic fields. But the weldable thickness is lower than that of the electron beam.
(g) Laser is very low in electrical-optical conversion and overall energy utilization. In addition, the laser light is partially reflected or refracted by the surface of the smooth metal, affecting the energy transfer to the workpiece. Therefore, it is still difficult to weld some metals with high reflectivity.
(h) High investment in equipment, especially high power continuous lasers. In addition, it has high requirements on processing and assembly accuracy of workpieces, and also relatively high requirements on the precision of fixtures. The economy can only be shown at high productivity.

(3) Applications

Solid laser beam welding or pulse gas laser beam welding may be used to weld metals such as copper, iron, zirconium, tantalum, aluminum, titanium, niobium and their alloys. It can also be used for welding non-metallic materials such as quartz, glass, ceramic or plastic. Continuous CO_2 gas laser beam welding can weld most metals and alloys, but it is difficult to weld copper, aluminum and their alloys because of their high reflectance and low absorption rate.

Laser beam welding has been widely used in aerospace, aviation, electronic instruments, precision instruments, automobile manufacturing, yachting, and medical equipment, etc. It can be used either for small precision parts consisting of wire or foil or for metal structural parts with large thickness.

Laser beam welding can also be combined with electric arc, resistive heat, friction heat and other heat sources to conduct hybrid welding, such as laser-MIG hybrid welding, which significantly improves welding quality and efficiency and reduces manufacturing cost.

6.4.2 *Welding System*

The laser beam welding robotic system mainly consists of a laser beam welding system (including a laser, a beam transmission system, a focus system and a welding torch), a robot, a positioner, a power supply, a control device, an air supply, a water supply, an operation panel and a numerical control device, as shown in Fig. 6.45.

6.4.2.1 Laser

A laser is a device that produces a coherent beam of light by causing the electrons of an excited atom or molecule to jump from a higher energy level to a lower energy level. The lasers may be classified into solid lasers and gas lasers, depending on the type of working medium.

(1) Solid lasers

The working medium of a solid laser is red gems, YAG or neodymium glass rod, etc. The laser mainly consists of working medium, a concentrator, a resonant chamber (full and partial reflectors), a pump lamp, a power supply and a control device, as shown in Fig. 6.46. The power supply charges the energy storage capacitor and discharges the pump lamp (xenon lamp) under the control of the trigger circuit. Then the pump lamp gives off a bright light which is concentrated on the working medium. The working medium is excited to produce lasers. The laser is output through the partially reflecting mirror after oscillation amplification in the resonant chamber. By adjusting the charge voltage on the energy storage capacitor, the laser can output lasers with different energies. Solid lasers can be transmitted via optical fibres.

The wavelength of the solid laser is related to the working medium, is 0.69 μm for red gem and 1.06 μm for YAG.

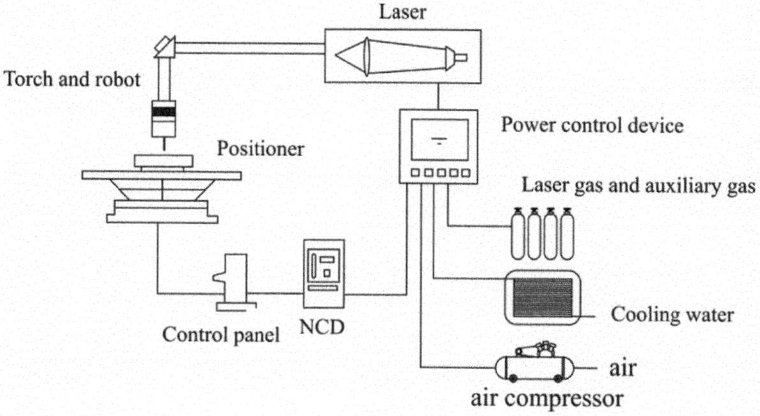

Fig. 6.45 Laser beam welding equipment

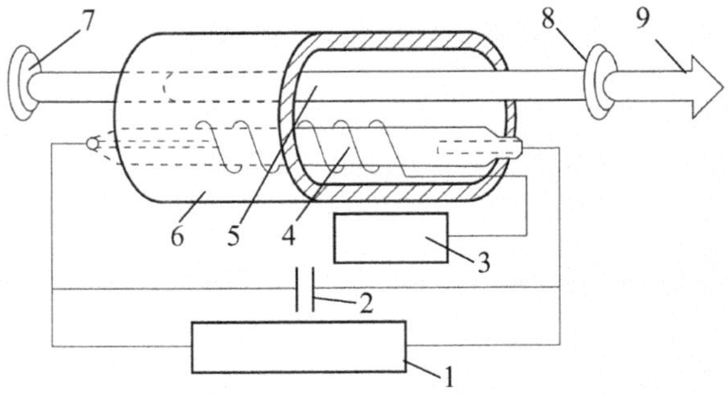

Fig. 6.46 Schematic diagram of structure of solid laser. 1-High-voltage power supply 2-Energy storage capacitor 3-Trigger circuit 4-Pump lamp 5-Laser operating medium 6-Concentrator 7-Full reflector 8-Partial reflector 9-Laser

At present, continuous Nd: YAG lasers with an output power of more than 10 kW is available. The available industrial pulsed YAG laser is much lower in the average output power, but its peak power can reach more than 15 times the average power.

Solid lasers that use xenon lamps as excitation devices are called lamp pump lasers. When laser diodes are used as excitation devices, they are called diode pump lasers. The wavelength of diode pump Nd: YAG laser is short, about 0.85–1.65 μm. The Nd: YAG Lasers with a output power above 550 W may be used for welding and cutting.

(2) Gas lasers

Most of the gas lasers are CO_2 lasers which use the mixture of CO_2, N_2 and He as the working medium. The CO_2 laser has a wavelength of 10.6 μm, ten times that of a solid laser (Nd: YAG). There are two types of CO_2 lasers commonly used for welding and cutting: Quick axial flow type and cross-flow type.

1) Quick axial flow type CO_2 Laser. Figure 6.47 shows the structure of a quick axial flow CO_2 laser. It consists of a discharge tube, a resonant cavity, a high speed fan and a heat exchanger. The gas flows in the discharge tube at a rate close to the speed of sound and also carries away the waste heat in the laser cavity. There can be several discharge zones within the discharge tube (the 4 pairs of electrodes form 4 discharge zones, as shown in the Fig. 6.47), in which the HVDC power supply generates a uniform glow discharge. These lasers have output modes TEM00 and TEM01 and are suitable for welding and cutting. These lasers can output a power of more than 20 kW.
2) Cross-Flow CO_2 Laser. Figure 6.48 shows a schematic diagram of the structure of a cross-flow CO_2 laser. The high-speed compressor allows the mixed gas to flow perpendicular to the laser beam in the discharge area at a speed of 50 m/s. The gas directly exchanges heat with the heat exchanger, with good cooling effect. This laser can achieve a power output of more than 25 kW. The output power of the laser can be modulated by adjusting the discharge current.

At present, the lasers used for welding and cutting are mainly YAG laser and CO_2 laser, each of which has its own characteristics.

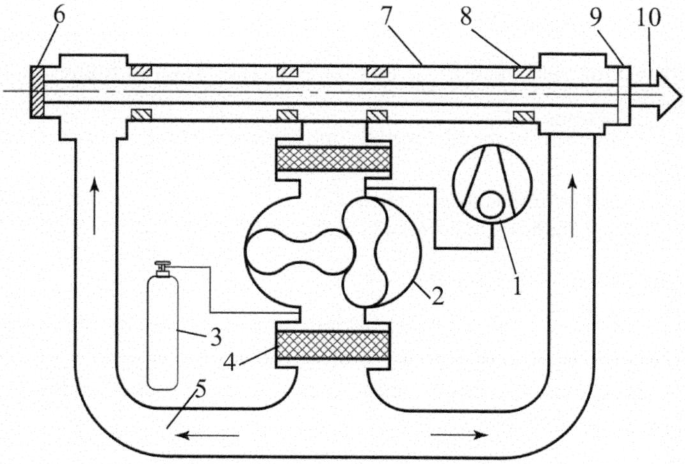

Fig. 6.47 Fast Axial Flow CO_2 Laser. 1-Vacuum system; 2-Roots blower; 3-Laser operating gas supply; 4-Heat exchanger; 5-Gas pipe; 6-Full reflection mirror; 7-Discharge tube; 8-Electrode; 9-Output window; 10-Laser beam

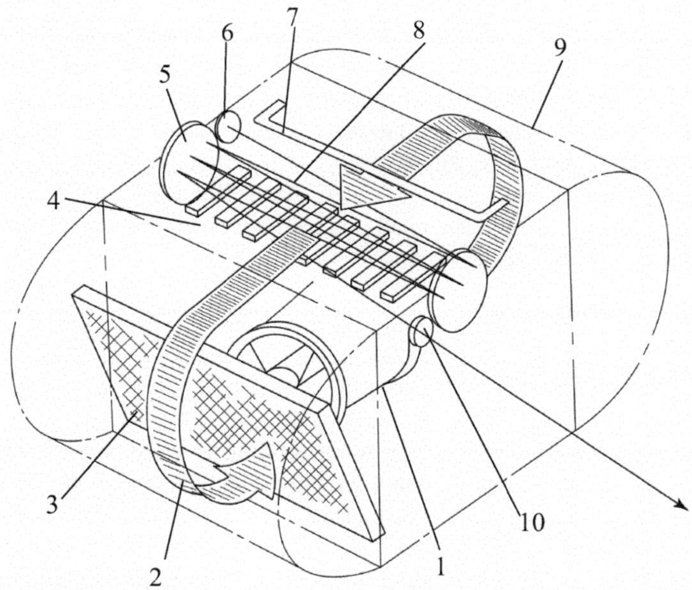

Fig. 6.48 Cross-flow CO_2 laser. 1-Compressor; 2-Gas flow Direction; 3-Heat Exchanger; 4-Anode Plate; 5-Refraction Mirror; 6-Full Reflection Mirror; 7-Cathode Tube; 8-Discharge area; 9-Sealed steel housing; 10-Semi-reflection mirror (window)

The Nd: YAG laser has the following advantages:

(a) The absorption rate of most metals for Nd: YAG laser is higher than that for CO_2 laser;
(b) Nd: YAG laser can transmit through optical fiber, which is favorable for robot welding;
(c) Nd: YAG laser beam is easy to align, delivery and split. The laser and the beam transimission systems are compact, occupying less space .

The advantages of CO_2 laser are:

(a) Large output power, high electrical-optical conversion efficiency, good focus capability, low operation cost and safety protection cost, etc.;
(b) High welding speed and penetration can be achieved when welding materials with low laser reflectivity.

6.4.2.2 Transmission and Focus Systems and Welding Torches

Beam transmission and focus systems, also known as external optical systems, are used to transmit and focus light beams onto the workpiece, with welding torches installed at the ends to provide protection or auxiliary gas flow. Figure 6.49 illustrates two types of laser transmission and focus systems. Reflectors are used to change the direction of the beam, and spherical reflectors or lens are used to focus. In solid lasers, optical glass is commonly used to make reflectors and lenses. For CO_2 lasers, because of the large laser wavelength, copper or other metals with high reflectivity are commonly used to make mirrors, and GaAs or ZnSe are used to make lens. Transmissive focusing is used for small and medium power lasers, while reflective focusing is used for high power lasers.

6.4.3 *Weld Formation Modes*

Depending on the power density, laser beam welding process can be made in two modes, i.e. melt-in (also referred to as conduction) and keyhole mode. The molten pool behavior and welding process of melt-in welding are basically similar to that of electric arc welding.

At a high power density (no less than 10^{10} W/m^2), the key-holing effect will occur during welding, as shown in Fig. 6.50. Under the radiation of laser beam, the laser beam not only melts through the workpiece, but also forms a small hole through the thickness of the workpiece underneath the laser beam as a result of the strong reaction force of metal evaporation. The walls of the hole are surrounded by molten liquid metal and the interior is filled with metal vapor. This vapor-filled hole acts as a "blackbody", sealing off the incident laser energy and significantly increasing the laser absorption rate of the workpiece. As the laser beam moves forward, the liquid

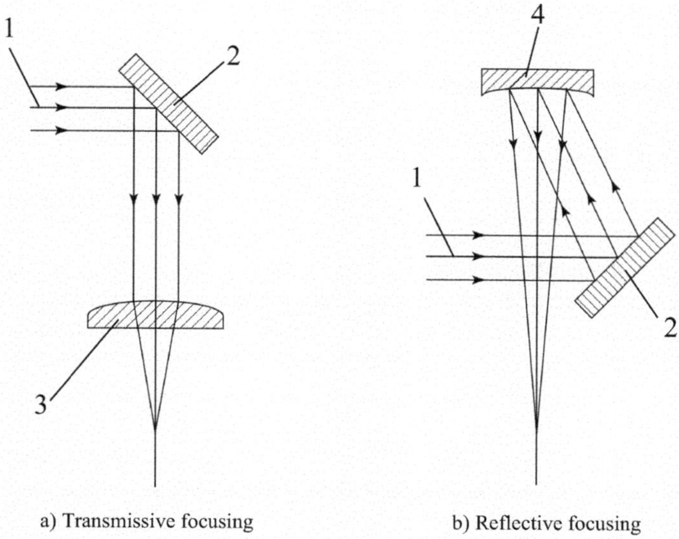

Fig. 6.49 Laser transmission and focus system. 1-Laser beam; 2-Plane mirror; 3-Lens; 4-Spherical mirror

metal flows backwards around the hole. Thereafter, the liquid metal behind the small hole is cooled down by heat transfer and solidifies gradually to form a weld.

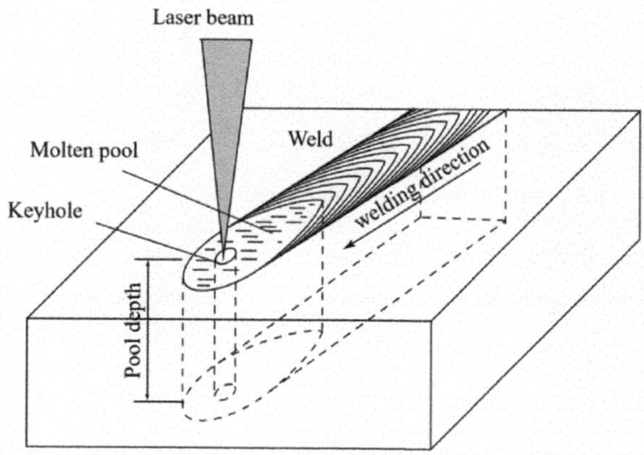

Fig. 6.50 Schematic of keyhole laser beam welding

6.4.4 Laser Beam Welding Procedure

6.4.4.1 Continuous Laser Beam Welding

(1) Type of joint

Figure 6.51 shows the joint types suitable for laser beam welding, the most common of which are butt and lap joints, with no filler metal required.

Since the diameter of the focused laser beam is generally very small, the joint must be accurately aligned and positioned laterally under the laser beam. In case of butt joint, the assembly gap shall be less than 15% of the material thickness, and the misalignment between workpieces shall not exceed 25%, as shown in Fig. 6.52a. For materials with good thermal conductivity, such as copper alloys, and aluminum alloys, etc., the error shall also be controlled within a smaller scope. In addition, when welding without wire filler, the assembly gap can result in a recessed weld surface. Although the deformation is small during laser beam welding, fixtures shall be used to clamp them.

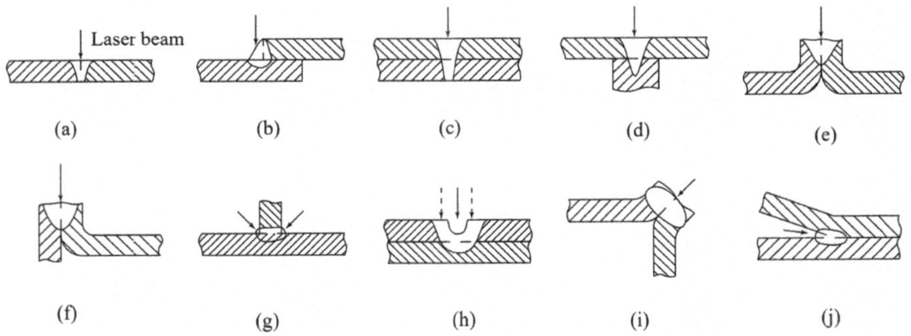

Fig. 6.51 Joint types of continuous laser beam welding. **a** Butt joint; **b** lap joint; **c** melt-through plug weld; **d** melt-through weld in T joint; **e** flange joint; **f** flange corner joint; **g** double square groove weld; **h** plug weld; **i** single flange joint; **j** flare-bevel groove weld

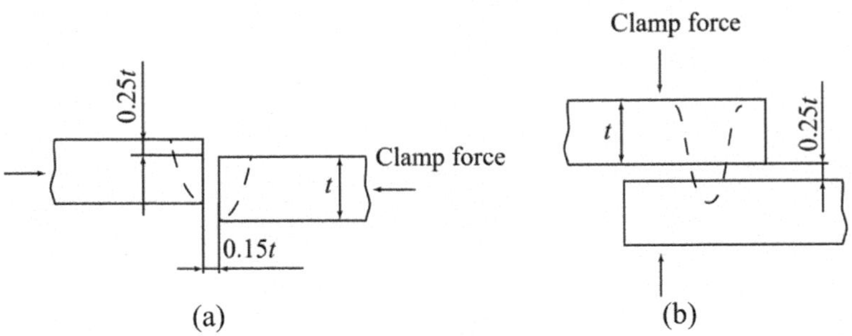

Fig. 6.52 Allowable tolerance for butt and lap joints

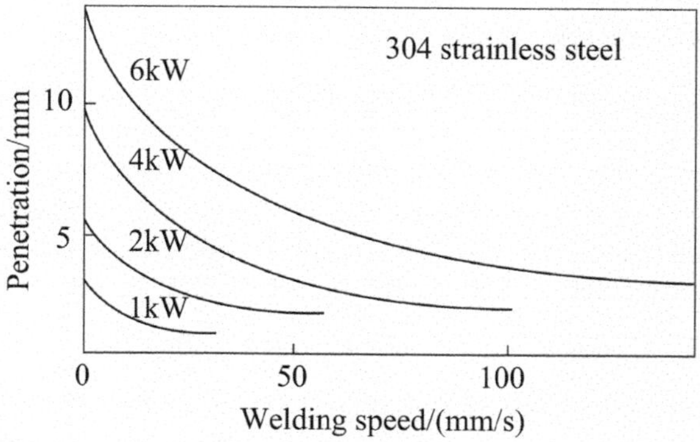

Fig. 6.53 Dependence of weld penetration on laser power and welding speed

In the case of lap joint, the allowable tolerance of the assembly gap is 25% of the thickness of the thinner part, as shown in Fig. 6.52b. If it is too large, the upper part will burn through and the lower part will not be fused.

(2) Welding parameters

1) Laser beam power

The laser beam power is the main parameter affecting the penetration of a weld. With a certain spot diameter, increasing laser power can increase the penetration depth or welding speed. Figure 6.53 shows the dependence of weld penetration on laser power and welding speed.

2) Laser wavelength

Wavelength affects absorption rate, and the shorter the wavelength, the higher the absorption rate. For example, aluminum and copper have higher absorption rate for solid lasers and lower absorption rate for gas lasers.

3) Beam diameter and defocusing distance

With a given incident laser beam power, the smaller the beam diameter, the higher the energy density, resulting in a narrower and deeper weld pool.

The light spot center on the focal plane has the maximum power density, thus excessive evaporation is likely to occur when the focal plane is on the surface of the workpiece. Therefore, laser beam welding usually requires a certain defocusing distance. When the focal plane is above the surface of the workpiece, it is considered to be positive defocusing, while the opposite is considered to be negative defocusing. In practice, negative defocusing is used when the penetration is required to be large. When welding thin materials, positive defocusing is preferably used. When the defocusing distance is large, the penetration ability is small, and the weld forming method is melt-in type. When the defocus amount decreases to a certain critical value, the penetration depth increases significantly, and the weld forming method becomes keyhole mode.

4) Welding speed

The welding speed influences the weld penetration and width. In keyhole mode, the weld penetration is almost inversely proportional to the welding speed. At a given beam power, appropriate welding speed is required for a certain penetration, and excessive welding speed can result in defects such as incomplete penetration or undercut. Too slow welding speed will result in a sharp increase in weld width and even collapse, burning-through or other defects.

5) Composition and flow rate of shielding gas

Shielding gas is used to protect the the weld pool and hot weld bead from oxidation and to suppress the production of large amount of plasma during high-power welding.

He can significantly improve the penetration of the laser, because He has a high ionization potential and is not easy to generate plasma, while the ionization potential of Ar is low and plasma is easily generated. Adding 1% (volume fraction) H_2 of double-atomic molecule to He will enhance the penetrating capability of the laser beam, thus increasing the weld penetration. With CO_2 as the protective and side-blowing gas, the penetration capacity of a laser beam is between those with helium and argon.

The weld penetration increases with the increase of flow, but remains substantially constant when it exceeds a certain value. As the flow rate increases, the shielding gas gradually improves its ability to remove the plasma above the molten pool, thereby reducing the absorption and scattering of the beam by the plasma. However, excessive gas flow may cause concave weld surface and gas waste.

6.4.4.2 Pulsed Laser Beam Welding

(1) Joint type

Pulsed laser beam welding has tiny heating spots (on the order of microns), making it suitable for welding thin sheets (0.1 mm thick), foils (a few microns to tens of microns), and wires (as small as 0.02 mm in diameter). Figure 6.54 shows several types of welded joints for pulsed laser beam welding of metal wires. Figure 6.55 shows the joint type used for pulsed laser beam welding between wire and block parts.

(2) Process parameters

The pulsed laser beam weld consists of a series of welding spots. Each laser pulse creates a welding spot in the workpiece. This method is mainly used for welding miniature, precision components and some microelectronic components.

When the power density reaches 10^{10} W/m^2, keyholing effect will occur, resulting in welding spots with depth-width ratio greater than 1 and slight gasification of metal. However, excessive power density may cause excessive gasification of metal. The resultant metal vapor form cavity in the molten metal in the weld pool which will

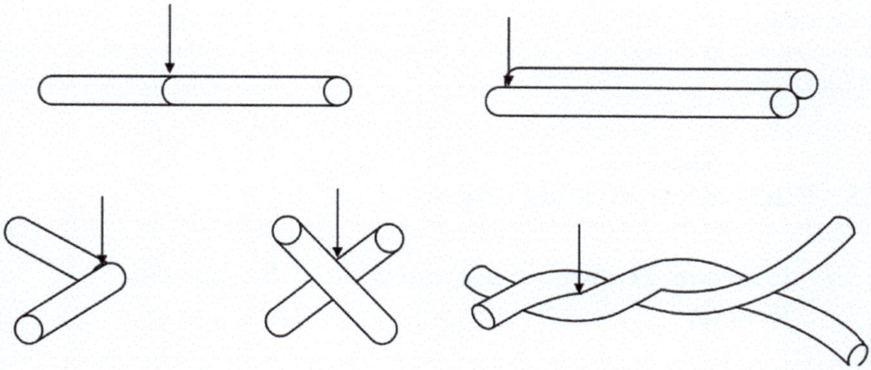

a) Butt joint b) Parallel connection c) Lap joint d) Corner joint e) Cross joint f) Twisted joint

Fig. 6.54 Joint types for laser beam welding of metal wire (Arrow denoting laser beam)

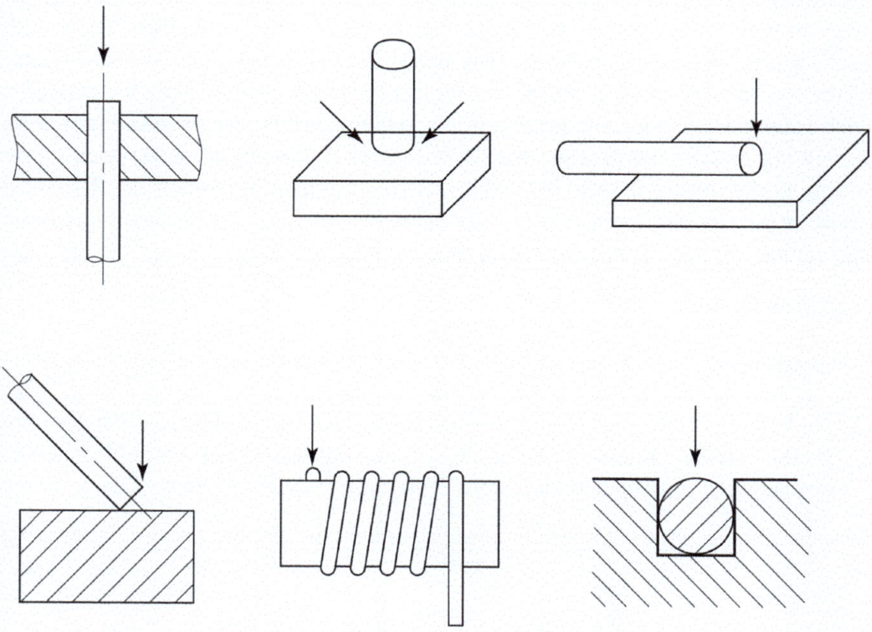

Fig. 6.55 Joint types for laser beam welding of metal wire and block parts (Arrow denoting laser beam)

evolve into porosity in the weld if it is not filled with molten metal. For a workpiece of a certain thickness, there is an optimum power density which increases with the increase of the workpiece thickness.

In pulsed laser beam welding, the pulse energy mainly affects the melting amount of metal, while the pulse width affects the penetration. Each material has an optimum pulse width under which the penetration capability is maximum, e.g., (1–5) × 10^{-4} s for copper and its alloys, (0.5–2) × 10^{-2} s for aluminum and its alloys, and (5–8) × 10^{-3} s for steel.

6.5 Friction-Stir Welding Process

6.5.1 Principle, Features and Application of Friction-Stir Welding

(1) Principles

Friction-stir welding is a solid phase welding process in which material coalescence is achieved by the heating the plasticized material transfer caused by a rotating and travelling probe and the workpieces, as shown in Fig. 6.56. First, the welding probe rotates at high speed, and the pin drills into the joint of the workpiece. Then, both the friction between the pin and the base metal and that between the probe shoulder and the workpiece surface generates heat under the pressure applied by the shoulder, which causes a thermoplastic layer to form in the vicinity of the welding probe. As the welding probe advances, the plasticized metal is transferred to the back of the welding probe, filling the gap behind it and forming a continuous weld. The weld formation is a complicated process involving metal extrusion, friction heating, plastic deformation, migration, diffusion, and recrystallization, etc.

(2) Characteristics

1) Advantages

(a) High quality weld joint. Friction-stir welding is a solid welding which does not produce defects related to the melting and solidification of materials, such as porosity, segregation and inclusions. The grains in various areas of the joint are

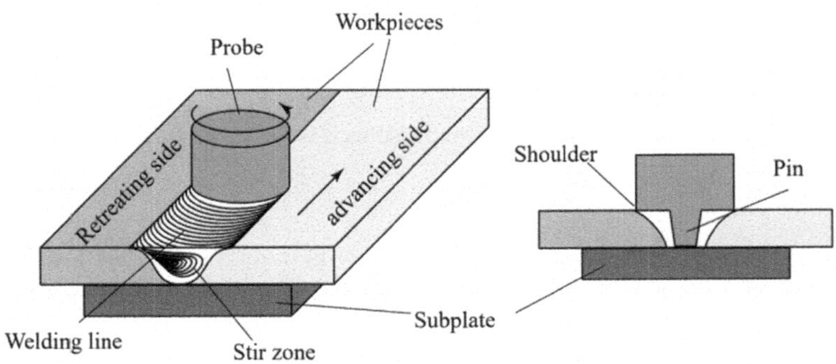

Fig. 6.56 Schematic of friction-stir welding process

small, the structure is dense, and the inclusions are dispersed. The joints have good performance, stable quality and good repeatability.

(b) High productivity and low production cost. Friction-stir welding requires no filler material, welding flux or shielding gas, little or no reserved allowance for workpiece size, and no special cleaning or bevelling before welding. There is no need to deburr the joints after welding, and costs can be reduced by about 30% compared with arc welding.

(c) High dimensional accuracy of weldement. Due to the low welding temperature and low welding deformation, friction-stir welding can achieve high precision welding.

(d) High degree of automation. The whole welding process is controlled by an automatic welding machine or robot, which avoids defects caused by human factors, and the quality of the weld is not dependent on the skill level of the operator.

(e) Environment friendly. No smoke, dust, arc radiation and other harmful substances will be generated during welding, so no smoke exhaust and ventilation devices are required.

2) Weaknesses and disadvantages of friction-stir welding

(a) At present, friction-stir welding is only applicable to butt and lap welding of light metal materials (such as aluminum and magnesium alloy, etc.), it's still difficult to weld high strength materials, such as steel, titanium alloy and powder metallurgical materials.

(b) As welding equipment is relatively complicated and requires large one-off investment, the production cost can only be reduced with mass production. The welding probe has a short service life due to the severe wearing.

(3) Applications

Friction-stir welding is only suitable for materials with good plasticity and is mainly used for welding of aluminum and aluminum alloys. The joint types suitable for friction welding include butt, lap, fillet and T-shaped welding, as shown in Fig. 6.57. At present, it is mainly used in aerospace and aviation, high-speed rail, aluminum pressure vessel and yacht manufacturing, etc.

6.5.2 Friction-Stir Welder

Figure 6.58 shows the schematic diagram of the drive system of the welding head of the friction-stir welder, which mainly consists a drive motor, a gear box, a spindle box, a clamp holder for the welding probe. During friction-stir welding, the two to-be-welded workpieces do not rotate, but the welding probe needs to be rotated, so the transmission mechanism for drives the welding probe to move is relatively simple. However, the mechanism that drives the welding probe to move in the three

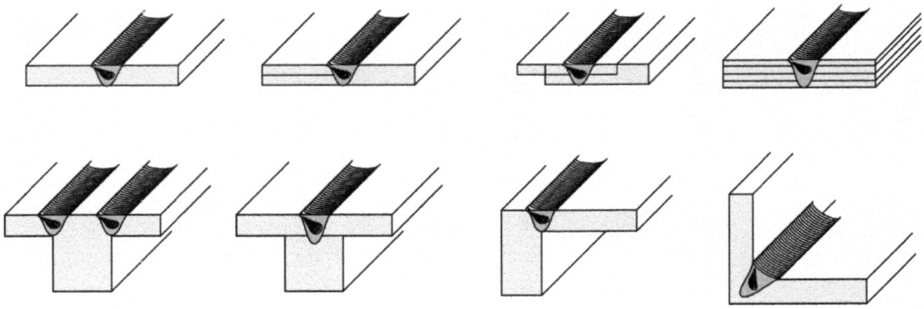

Fig. 6.57 Joint types suitable for friction-stir welding

Fig. 6.58 Schematic diagram of welding probe drive

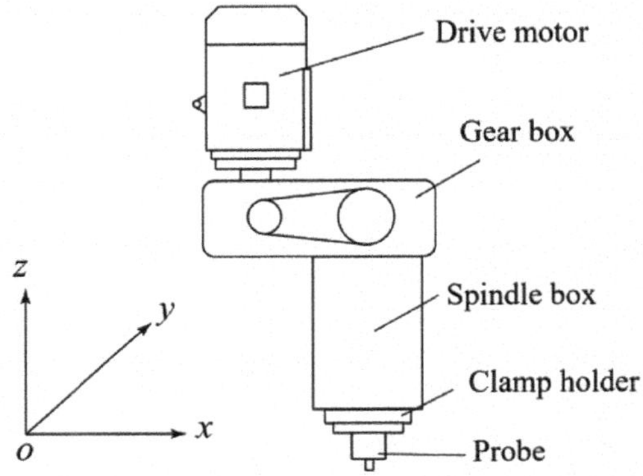

coordinate directions of *x*, *y* and *z* relative to the workpiece is more complicated. The movement along *z* axis provides welding pressure and insert depth control for the welding probe, which can be achieved by changing the distance between the welding probe and the workpiece. The movement along the *x* and *y* axes enables the welder to weld straight seams as well as planar curved seams, which is usually achieved by the relative movement between the frame supporting the driving mechanism of the welding probe and the workbench supporting and clamping the workpiece.

6.5.3 Welding Parameters

The welding parameters for friction-stir welding include inclination angle of welding probe, rotation speed, welding speed, insertion depth, insertion speed, insertion residence time, welding pressure, retraction residence time and retraction speed of welding probe.

1) Inclination angle of welding probe

Generally, the welding probe shall be inclined at a certain angle to reduce the forward resistance and allow the back edge of the welding probe shoulder to exert a certain upsetting force on the weld. For sheets with a thickness of 1–6 mm, the inclination angle of the welding probe is generally 1°–2° (the welding probe points to the welding direction); for medium-thickness plates with a thickness greater than 6 mm, it is generally taken as 3°–5°.

2) Rotation speed

As one of the main welding parameters, the rotation speed of the welding probe needs to match with the welding speed. For any base metal, there is a suitable rotation speed range at a given welding speed, within which a good quality joint can be obtained. When the rotation speed is too low, the frictional heat is insufficient to form a good thermoplastic layer, causing cavity defects to be formed in welds. When the speed is too high, the temperature of base metal near the pin will be over-high, and the high temperature base metal will be bonded to the welding probe, making it difficult to form a good joint.

According to the rotation speed of welding probe, friction-stir welding procedures can be classified into cold, weak and strong types. See Table 6.9 for the classification of welding procedures for various aluminum alloy materials.

3) Welding speed

The welding speed is the rate at which the welding probe moves along the seam between the workpieces. It is usually selected based on the workpiece thickness, in additon, factors such as production efficiency and flexibility of welding procedure shall also be taken into consideration.

Table 6.10 shows the commonly used welding speeds for friction-stir welding of aluminium alloys of different thickness.

Table 6.9 Classification of welding procedure for aluminum alloy materials

Type of procedure	Rotation speed of welding probe (r/min)	Weldable aluminium alloy material
Cold procedure	<300	2024, 2214, 2219, 2519, 2195, 7005, 7050, 7075
Weak procedure	300–600	2618, 6082
Strong procedure	>600	5083, 6061, 6063

Table 6.10 Commonly used welding speed for friction-stir welding of aluminum alloys of different thicknesses

Thickness of plate (mm)	Welding speed (mm/min)	Weldable materials
1–3	30–2500	5083, 6061, 6063
3–6	30–1200	6061, 6063
6–12	30–800	2219, 2195
12–25	20–300	2618, 2024, 7075
25–50	10–80	2024, 7075

4) Insertion depth

The insertion depth of the welding probe generally refers to the depth to which the pin is inserted into the material to be welded. However, considering that the length of the pin is generally a fixed value, the insertion depth of the welding probe can also be expressed by the depth to which the back edge of the probe shoulder is lower than the surface of the plate. Generally, the insertion depth is 0.1–0.3 mm for thin gauge sheet; and shall not exceed 0.5 mm for medium and thick plates.

5) Insertion speed

It is the speed at which the pin inserting in the workpieces. It is generally selected based on the type of pin and plate thickness. Excessive insertion can cause great damage to the spindle of the machine before the workpieces have fully attained its thermoplastic state. If the insertion speed is too slow, it will cause overheating and affect the weld quality.

When the pin is tapered, the insertion speed is usually set to be about 15–30 mm/min; when the pin is cylindrical, the insertion speed shall be reduced to about 5–25 mm/min. When welding thick plates is >12 mm, the insertion speed is about 10–20 mm/min; and the insertion speed is about 15–30 mm/min for the welding of thin plates (thickness 0.8–12 mm).

6) Insertion residence time

It is time interval after the pin is inserted into the workpiece to the predetermined depth and before the welding probe moves horizontally. It shall be selected based on the workpiece material and plate thickness. If the residence time is too short, the pin begins to move horizontally before the temperature of the weld reaches equilibrium, tunnel-shaped hole defects might occur. If the residence time is too long, the base materials will be overheated, which will easily lead to chemical segregation, burr on the surface of the weld, and S-shaped black line defects.

The insertion residence time is generally selected to be 5–20 s. For thin sheets, materials with good plastic flow or heat-sensitive materials, the insertion residence time should be shorter.

7) Welding pressure

It is the axial upsetting pressure exerted by the welding probe to the workpiece during welding, which is usually selected based on the strength and rigidity of workpiece, the shape of welding probe, and the pressing depth of welding probe, etc. During welding, the welding pressure is generally kept constant.

8) Retraction residence time

It is the time interval in which the pin stays in the workpiece after the welding probe stops moving forward. If the retraction residence time is too short, the welding probe will be pull out of the workpiece before thermoplastic flow reaches equilibrium state. This will cause cavities around the tail hole. If the residence time is too long, the weld will be overheated, resulting in chemical segregation and deteriorated weld quality.

9) Retraction speed

It is the speed at which the pin is extracted from the weldment. The value is determined mainly based on the type of pin and the thickness of base metal. If the retraction speed is too rapid, the thermoplastic metal in the base metal will inertially move upwards with the retraction of the pin, resulting in metal loss at the root of the weld and the resultant cavity.

For conical needles, the retraction rate is usually 15–30 mm/min; for cylindrical pins, the retraction speed shall be moderately reduced to about 5–25 mm/min.

10) Shape of welding probe

The shape of the welding probe has a great effect on the weld formation. For aluminum alloys with a thickness less than 12 mm, a cylindrical thread welding probe is generally used, as shown in Fig. 6.59a. For aluminum alloys with a thickness greater than 12 mm, a tapered or claw-shaped threaded welding probe is usually used, as shown in Fig. 6.59b, c. In addition, the latter two types of welding probes allow for higher welding speeds.

(a) Cylindrical threaded probe (b) Conical threaded probe (c) Claw-shaped threaded probe

Fig. 6.59 Welding probe used in Al alloy welding

Chapter 7
Operation Technologies of Welding Robots

7.1 Operation Instructions for TP

7.1.1 Teach Pendant and Its Functions

TP (abbreviation of Teach Pendant) is also referred to as teaching box. It is an interface device between the application tool software and the user. It can control the robot to accomplish certain movement and perform most robotic operations. In addition, the TP also has certain monitoring and information display functions. The TP is connected to the robotic controls by cables, and it is important to understand the functionality of the TP and how to use the individual keys before using it. Different robotic systems have different TP models wtih different component layouts, but basically the same functions. The structure and functions of the model are illustrated using FANUC robotic system as an example as below. Figure 7.1 shows the front side and Fig. 7.2 shows the rear side.

The TP consists of the following components:

- Liquid crystal display with 640 × 480 pixels;
- 2 LED indicators;
- 68 operation keys (4 of which are dedicated to various application tools);
- TP switch;
- safety switch;
- Emergency stop button.

The TP is mainly used for the following operations:

- Jogging-forward of robots;
- Program creation;
- Testing implementation of procedures;
- Operational implementation;
- Status confirmation.

M. Chen et al., *Technologies of Robotic Welding*, Advanced and Intelligent Manufacturing in China, https://doi.org/10.1007/978-981-97-3149-7_7

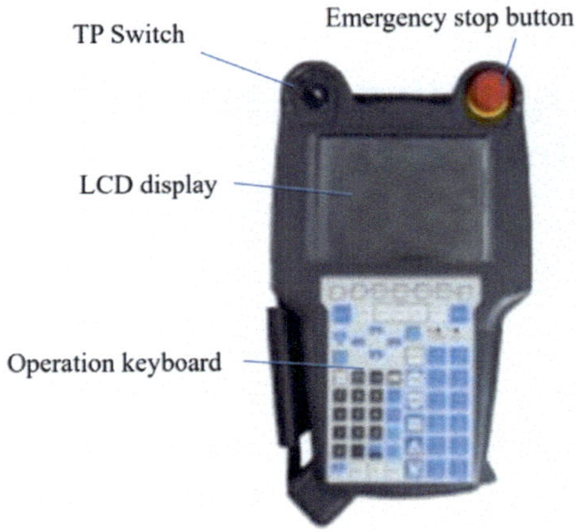

Fig. 7.1 Front side of robotic TP

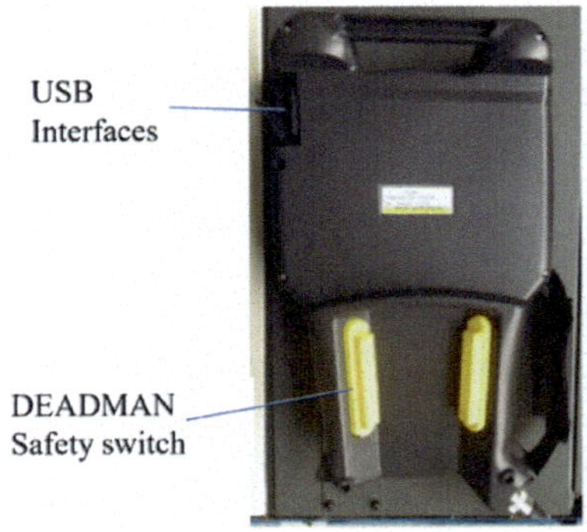

Fig. 7.2 Back side of robotic TP

7.1.1.1 TP Switch Function

(1) Emergency stop button

The red button in the upper right corner of the front of the TP is the emergency stop button, as shown in Fig. 7.1, which is used for emergency stop of robot in case of emergency. When this button is pressed, the servo switches are immediately disconnected, and all movements of the robot and external shaft stop immediately. At the same time, the button will be locked automatically. After danger or alarm is canceled, turn the emergency stop button clockwise, and it will automatically rebound and unlock.

(2) Safety Switch

The safety switch is located at the bottom of the back side of the TP, as shown in Fig. 7.2, to ensure the safety of the operator during operation. When TP is active, the servo switch will be closed by slightly pressing one or two safety switches, and the robot can be manually operated; when both switches are released or pressed down forcefully at the same time, the servo switch will be disconnected, the robot will stop its movement immediately and an alarm will be triggered.

(3) TP Switch

The TP switch is located in the upper left corner of the front side of the TP, as shown in Fig. 7.1, and is used to control the state of the TP. Turn the switch to "ON", the TP is in working status; turn the TP switch to "OFF", the TP is deactivated (TP is locked and out of service).

7.1.1.2 TP Display

(1) Display window

The TP is usually equipped with a 7-inch liquid crystal display. The display window presents status bar, function menu, edit box, information box and title bar as shown in Fig. 7.3.

(2) Status bar

The status bar of the TP display window is shown in Fig. 7.4. There are 8 LEDs used to display the current working status of robot, alarm signal, running program and line number, working mode, execution status, current coordinate, running motion coordinate motion, speed multiplication value, etc. The meanings of the LED displays are shown in Table 7.1.

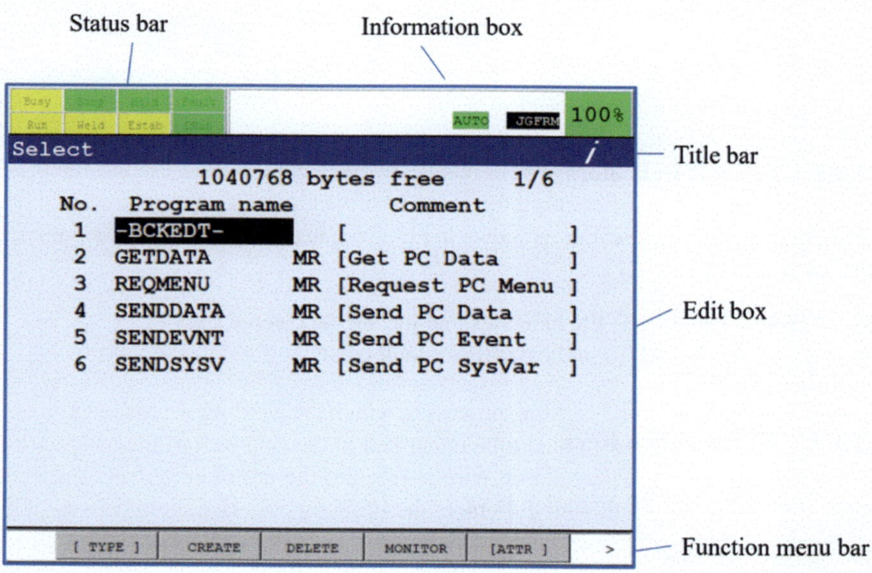

Fig. 7.3 TP display window

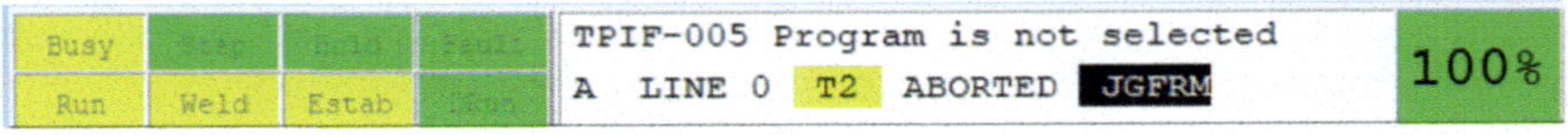

Fig. 7.4 Status window

Table 7.1 Meaning of the eight LED displays

Display LED	Meaning
Busy	Green indicates that the robot is performing a job
Step	Yellow indicates that the program is in single-step execution mode
Hold	Red indicates that the HOLD button has been pressed or the HOLD signal has been entered and the robot is in HOLD status
Fault	Red indicates an anomaly
Run	Green indicates that the procedure is in progress
Weld	Green indicates that the welding function is enabled
Estab	Green indicates welding in progress
DRun	Green indicates idling status

Fig. 7.5 LED indicators

7.1.1.3 TP LED Indicators

The TP has two LED indicators as shown in Fig. 7.5 (On the third line of the keyboard area, as shown in Fig. 7.1).

(a) POWER: Light POWER indicates that the TP is powered on.
(b) FAULT: Light FAULT indicates that a fault exists.

7.1.1.4 TP Operation Keys

Figure 7.6 shows the TP operation keys.

(1) TP key function

See Table 7.2 for the brief functions of the keys on the TP.

Fig. 7.6 Operation keys

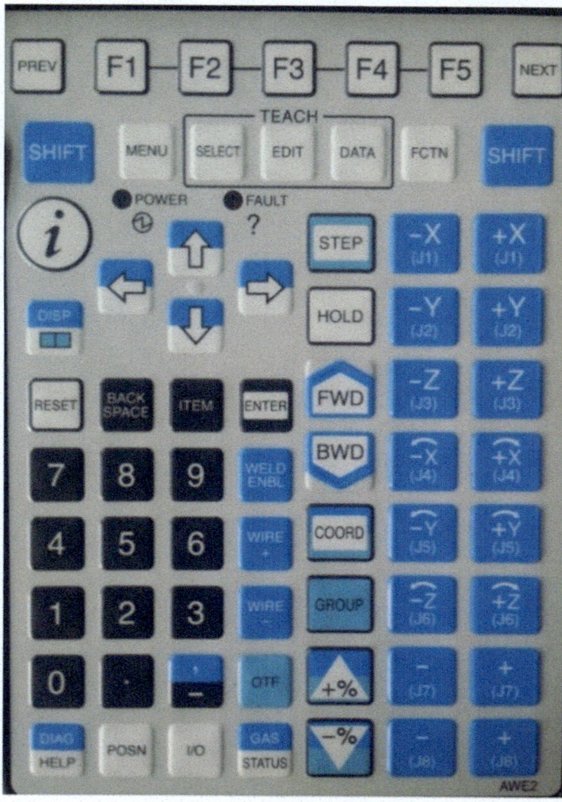

(2) Detail functions of commonly-used keys

1) F1, F2, F3, F4, F5

F1-F5 are used to select the functions in the function menu at the bottom of the screen and correspond sequentially to each function menu item, as shown in Fig. 7.7.

2) NEXT

Page keys, used to switch the function menu to the next page, as shown in Fig. 7.8, where the function menu is displayed.

3) MENU, SELECT

Press the MENU key to display the main menu interface, as shown in Fig. 7.9. Press the SELECT key to display the program overview interface, as shown in Fig. 7.10.

4) DISP/□

Press the DISP/□ key and SHIFT key at the same time to enter the screen-spliting interface, as shown in Fig. 7.11. You can select the number of display interfaces

Table 7.2 TP key functions

Button	Function
F1, F2, F3, F4, F5	Function keys, used to select the function key menu on the bottom line of the screen
PREV	Return key, used to return the screen to the previously displayed interface
NEXT	Page key, used to switch the function key menu to the next page
SHIFT	When the SHIFT key is pressed together with other keys, JOG feed, position data demonstration and program startup can be performed
MENU	Menu keys, used to display the menu interface
SELECT	List key, used to display the program list
EDIT	Edit key, used to display the program editing interface
DATA	Data key, used to display data interface
FCTN	Auxiliary key, used to display the auxiliary functions menu
DISP/	Interface changeover key, pressed simultaneously with SHIFT key to divide screen (single screen, double screen, triple screen, status/single screen)
↑, ↓, Σ, →	Cursor movement key, used to move the cursor (The cursor is the part of the display interface of the TP that can be moved and displayed in inverse way.)
RESET	Alarm canceling key, also known as reset key
BACK SPACE	Delete key, used to deletes a character or number preceding the cursor position
ITEM	Item selection key, used to move the cursor after entering the line number
ENTER	Confirm key, used to confirm the entry of value and menu selection
WELD ENBL	Welding enabling key. By pressing this key alone, the test execution and welding interface will be displayed. By pressing this key and the SHIFT key at the same time, the welding function can be switched on/off at the same time
WIRE+	Manual wire feeding
WIRE-	Manual tapping
OTF	Display micro adjustment interface of welding
DIAG/ HELP	Diagnostic/help key, press this key separately to switch to the alarm interface; press this key and the SHIFT key simultaneously to display the system version
POSN	Position display key, used to display the coordinates of the current robot location
I/O	I/O key, used to display I/O interface
GAS/ STATUS	Gas inspection/status key, press this key alone to display the weld status interface; press this key and the SHIFT key simultaneously to check the gas
STEP	Program execution mode switching key, press this key to switch the program execution mode between single step and continuous execution
HOLD	Pause key, used to pause the program execution
FWD, BWD	The forward and backward keys, used to start the program (the SHIFT key needs to be pressed at the same time)
COORD	Switch-over coordinate system

(continued)

Table 7.2 (continued)

Button	Function
+%, -%	Multiplier key, used to change the speed ratio
+X, +Y, +Z, -X, -Y, -Z	JOG key, used to move the robot manually (the SHIFT key needs to be pressed at the same time)

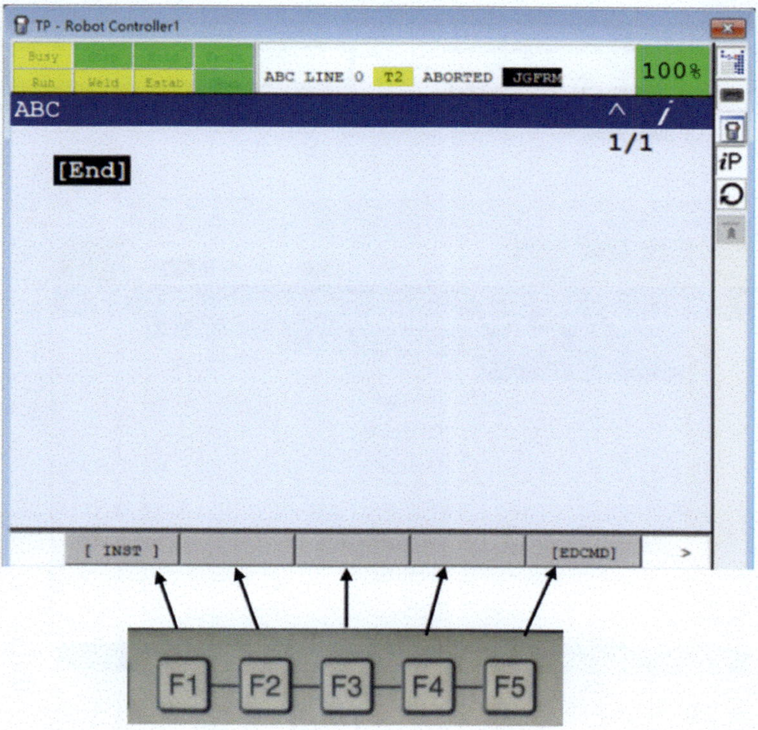

Fig. 7.7 F1-F5 function keys

(1–3), as shown in Fig. 7.11. When "Double" is selected, the window display is as shown in Fig. 7.12, and when "Triple" is selected, the window display is as shown in Fig. 7.13.

5) WELD ENBL

Welding startup key. When the switch of the "WELDING" LED indicator in the upper left corner of the screen is yellow, the welding function of robot is switched off. After pressing the WELD ENBL and SHIFT keys at the same time, the robot

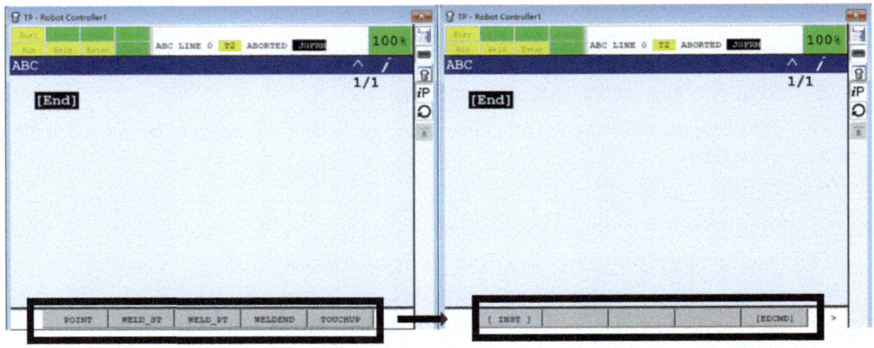

Fig. 7.8 NEXT page keys

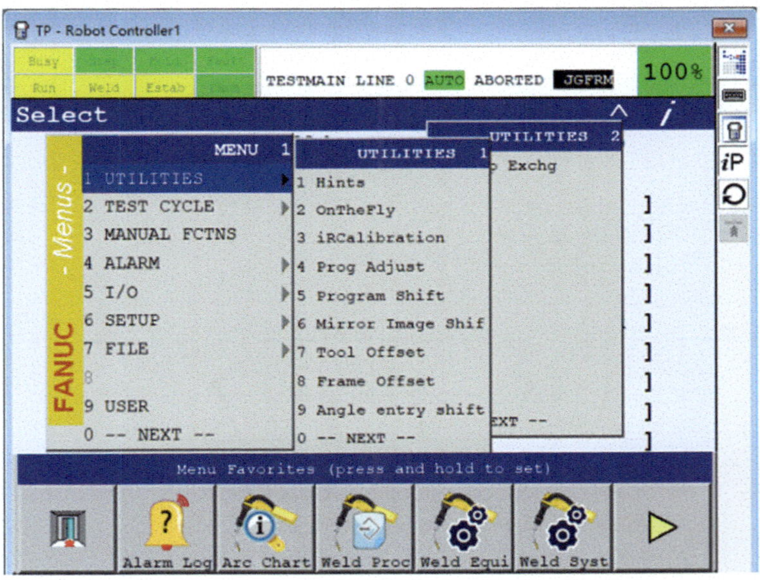

Fig. 7.9 MENU key

switches to the status of welding function enabled, and the "Welding" LED indicator in the upper left corner of the screen turns green, as shown in Fig. 7.14.

6) STEP

Program execution mode switching keys. When the "Single Step" indicator in the upper left corner of the screen is green, the robot is in the program continuous execution mode. Press the STEP key to switch to the program single-step execution mode, and the "single-step" LED indicator turns yellow, as shown in Fig. 7.15.

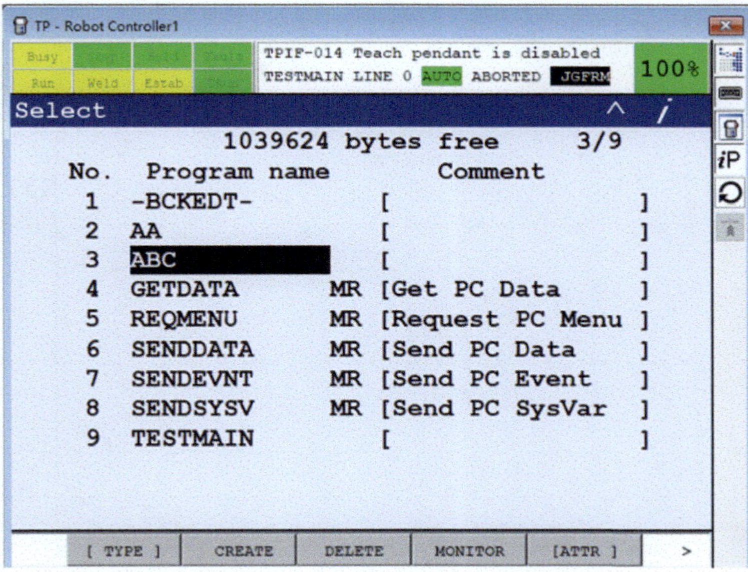

Fig. 7.10 SELECT key

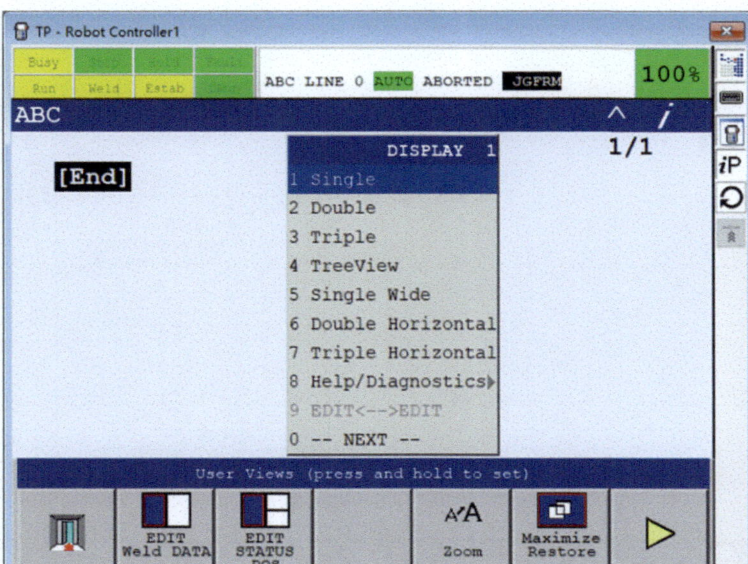

Fig. 7.11 Screen-splitting operation interface

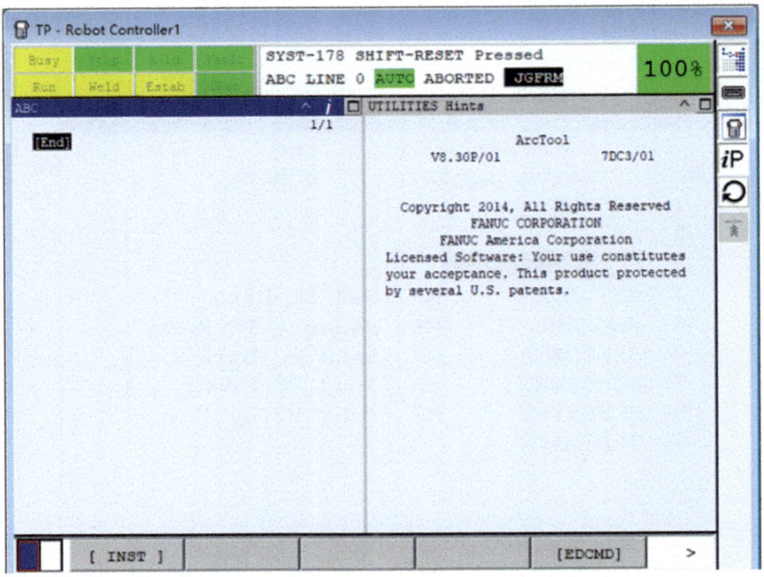

Fig. 7.12 Double-interface display

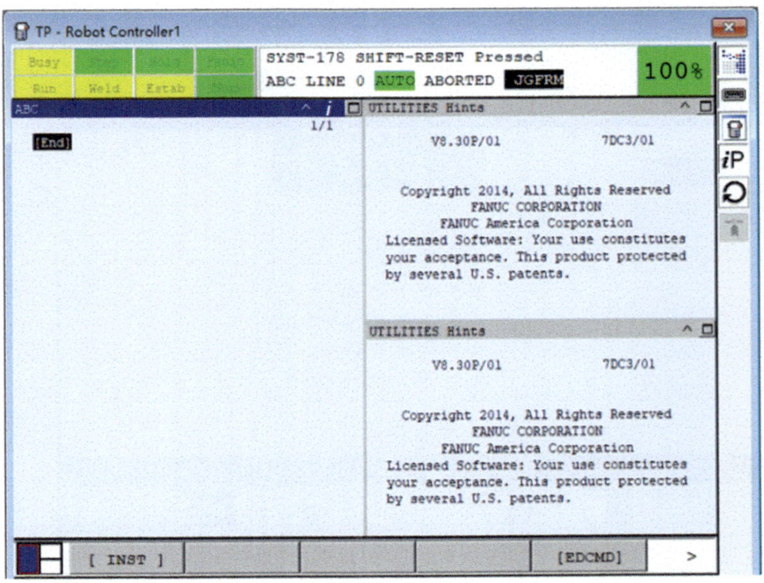

Fig. 7.13 Triple-interface display

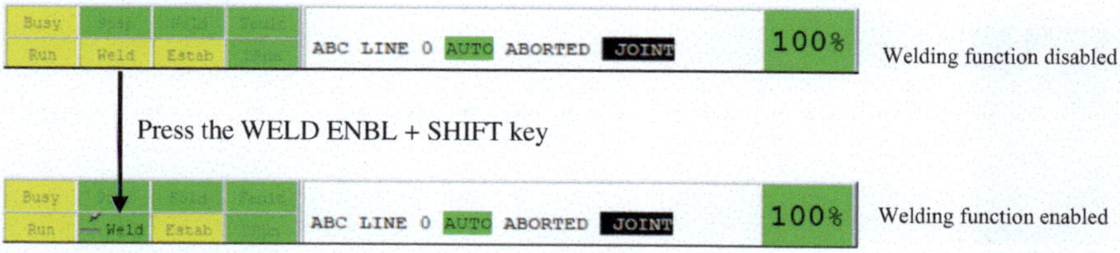

Fig. 7.14 Switching of welding functions

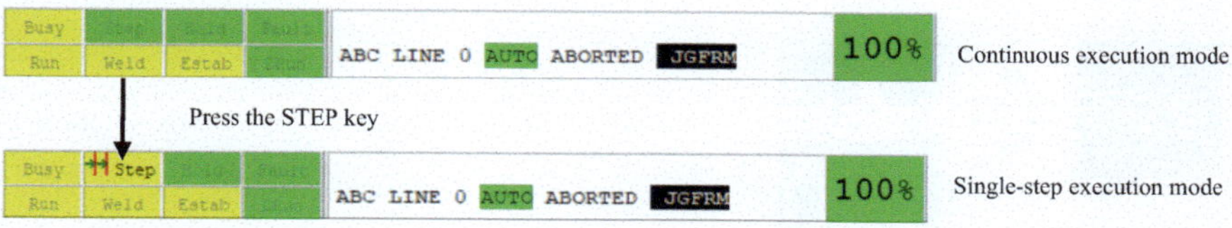

Fig. 7.15 Changing the program execution mode

7) COORD

Coordinate system switch key. When clicking the COORD key, switching is made in sequence as follows: "JOINT" → "WORLD" → "TOOL" → "USER" → "JOINT", as shown in Fig. 7.16.

8) +%, -%

Multiplier key, used to change the multiplier of robot operation speed. Click the +%, the following switch will occur: "VFINE" → "FINE" → "1%, 2%, 3%, 4% 5% → 10% → 15% → 100%", as shown in Fig. 7.17; press +% and SHIFT, the following switch will occur: VFINE → FINE → "5% → 25% → 50% → 100%", as shown in Fig. 7.18.

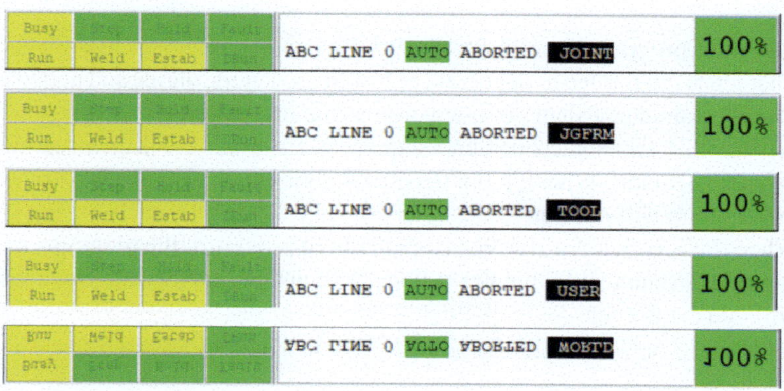

Fig. 7.16 Switch-over of coordinate systems

"VFINE" → "FINE" → "1%, 2%, 3%, 4% 5% → 10% → 15% → 100%"

Fig. 7.17 Change of speed multiplier through clicking the multiplier key

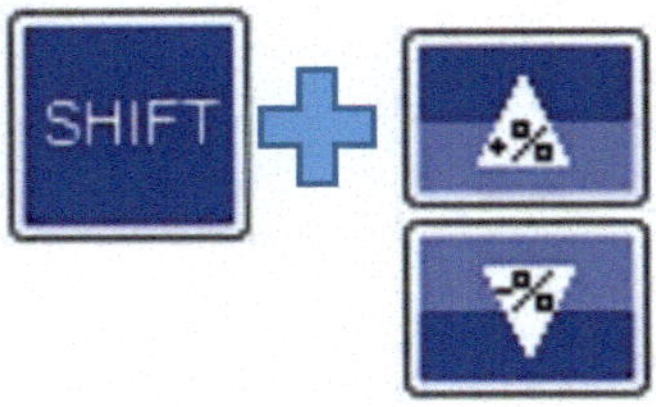

VFINE → FINE → "5% → 25% → 50% → 100%"

Fig. 7.18 Speed multiplier key changes when pressing the multiplier key and SHIFT key simultaneously

7.1.2 Programing

Programing mainly includes program creation, deletion and replication.

(1) Program creation

To create a program, first determine the program name, which is used to distinguish the programs stored in the control system memory.

1) Naming rules of program name:

(a) Two or more programs with the same name cannot be created within a control system;
(b) Length of program name: 1–8 characters;
(c) Characters are limited to uppercase or lowercase letters, digits, and underscores;
(d) The program name shall not use a digit as the first character;
(e) Where RSR automatic operating sequences are used, the sequences must be named RSRnnnn. Where, nnnn denotes 4 digits, such as RSR0001; otherwise the program will not run;
(f) Where PNS is used for automatic operation of programs, the programs must be named PNSnnnn. Where nnnn denotes 4 digits, e.g. PNS0001; otherwise the program cannot run.

1) Procedures for program creation:

(a) Press SELECT to enter the main interface of program list, as shown in Fig. 7.19.
(b) Press F2 to open the New Program window, as shown in Fig. 7.20. The "Alpha Input" (naming mode) menu pops up.
(c) Press "↓" to select the naming mode. After "Uppercase case" is selected, select the letter by pressing F1-F5 (input can be selected with the numeric key), and enter the program name "ABC", as shown in Fig. 7.21.
(d) After entering the program name, press ENTER to end the naming, as shown in Fig. 7.22.
(e) Press "F3" to open the program editing window for program edit (Fig. 7.23).

(2) Program deletion

Programs that are not needed can be deleted, but those that are not terminated cannot be deleted. Programs need to be terminated before deletion.

Procedures for program deletion are as follows:

(a) On the Program Selection page, move the cursor to the program "ABC" to be deleted, as shown in Fig. 7.24.
(b) Press "F3" to delete. The system prompts the user to delete or not, as shown in Fig. 7.25.

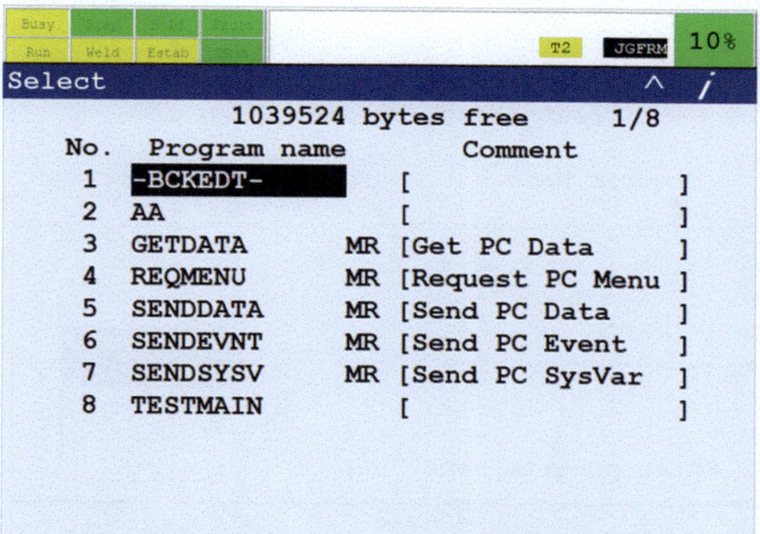

Fig. 7.19 Program list main interface

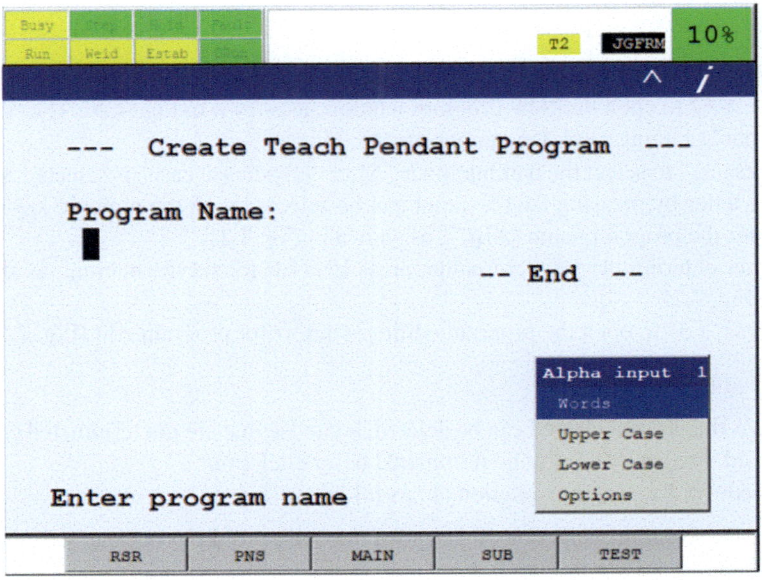

Fig. 7.20 New program creation window

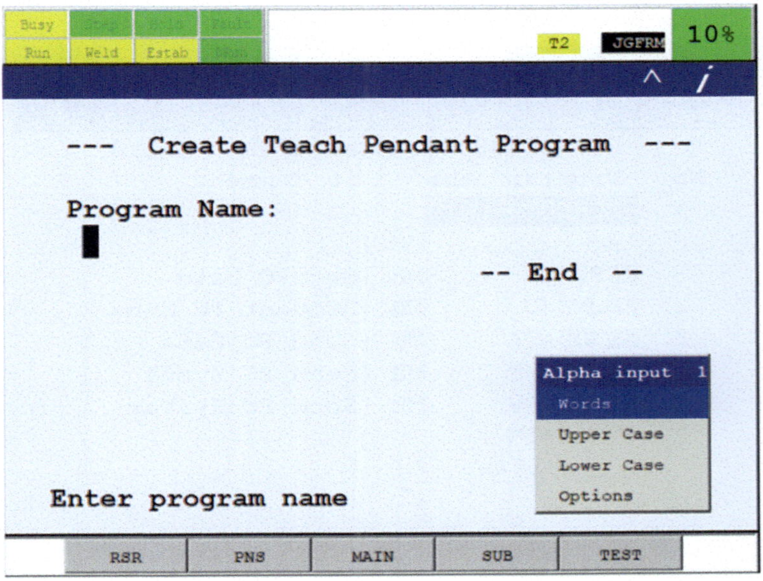

Fig. 7.21 Program name entry window

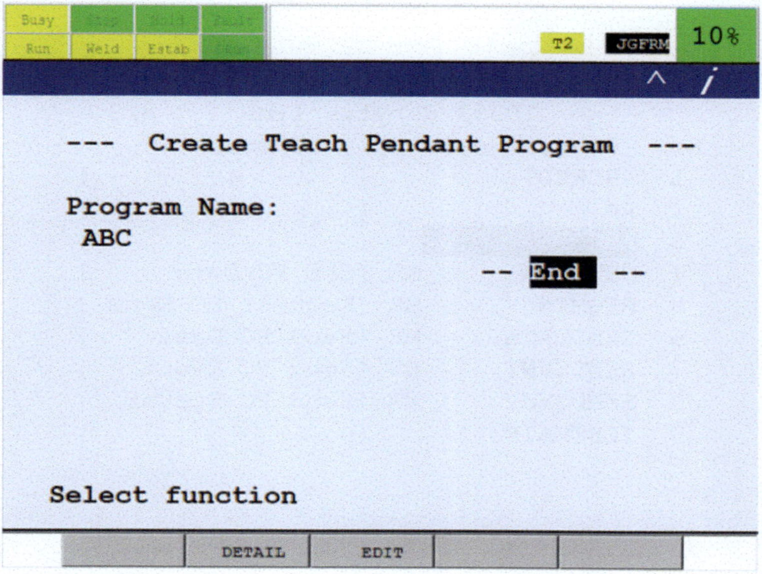

Fig. 7.22 Program naming end window

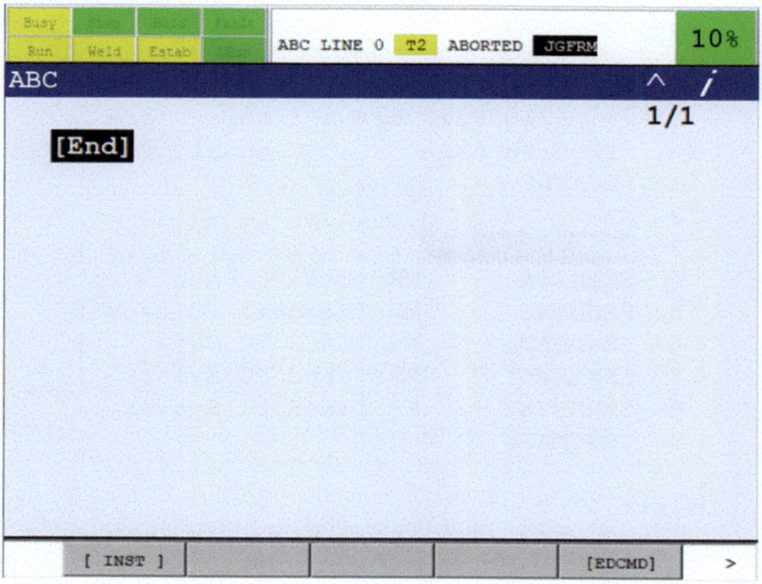

Fig. 7.23 Program editing window

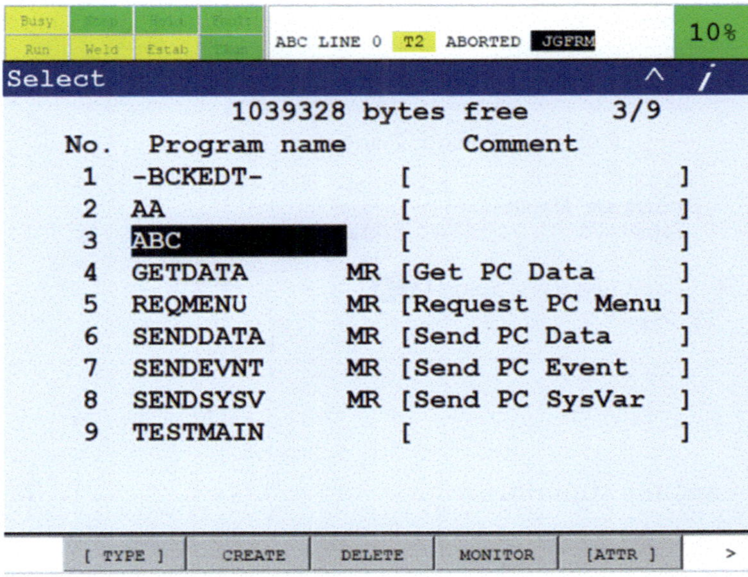

Fig. 7.24 Select the program to be deleted

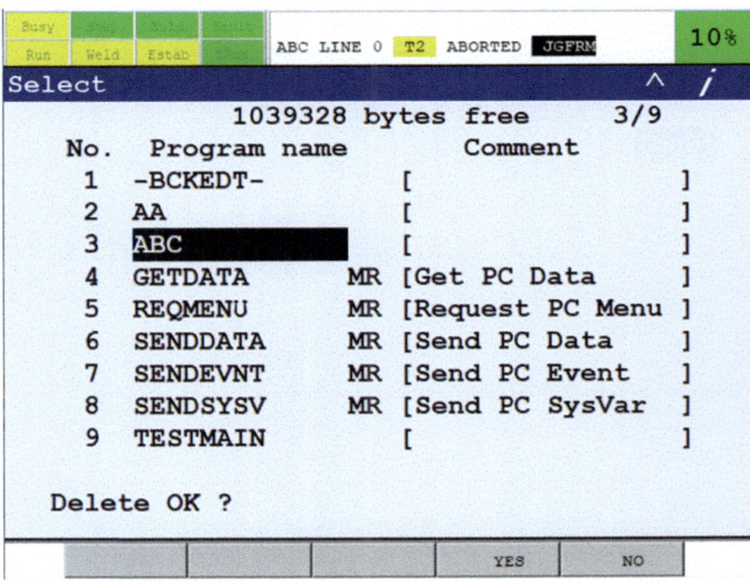

Fig. 7.25 Screen after performing deletion operation

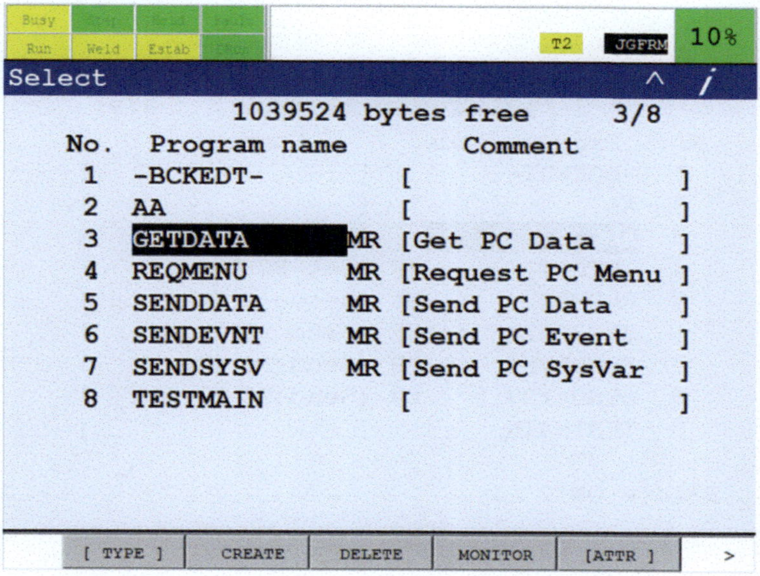

Fig. 7.26 Screen after confirmation of deletion

(c) Press “F4” to confirm the deletion, and the program “ABC” will be deleted (Fig. 7.26).

(3) Program copy

You can copy the same contents to programs with different names.

Procedures for program copying:

(a) On the Program Selection page, move the cursor to the name of the original program to be copied, as shown in Fig. 7.27.
(b) Press NEXT to display the next menu bar, as shown in Fig. 7.28.
(c) Press “F1” to copy the original program ABC to the paste board and create a new program name, as shown in Fig. 7.29.
(d) Press ENTER to display the copy confirmation window, as shown in Fig. 7.30.
(e) Press “F4” to select Yes to confirm the copy. The name of the target program ABC1 is added to the program list display window, and the contents of program “ABC” are completely copied to program “ABC1”, as shown in Fig. 7.31.

7.1.3 Programming Instructions

The programming instructions most commonly used during teach programming of the robot include motion instructions, arc instructions, and weaving welding instructions, etc.

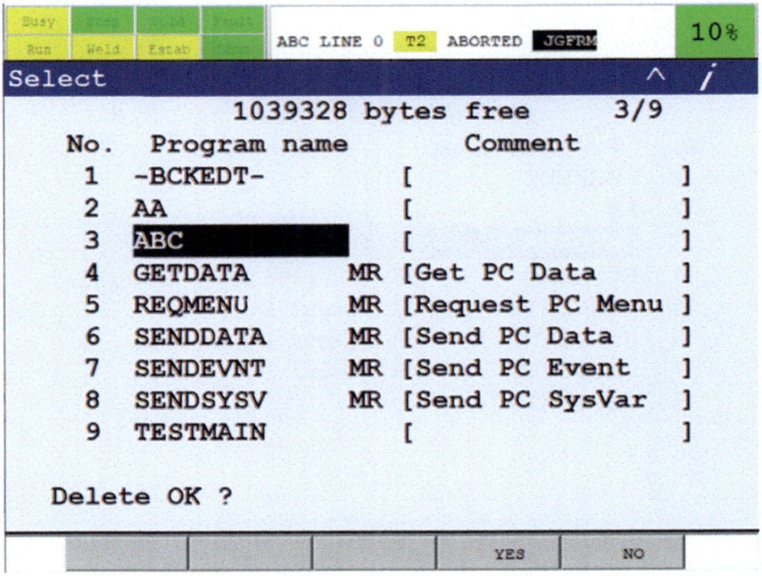

Fig. 7.27 Select the program to be copied

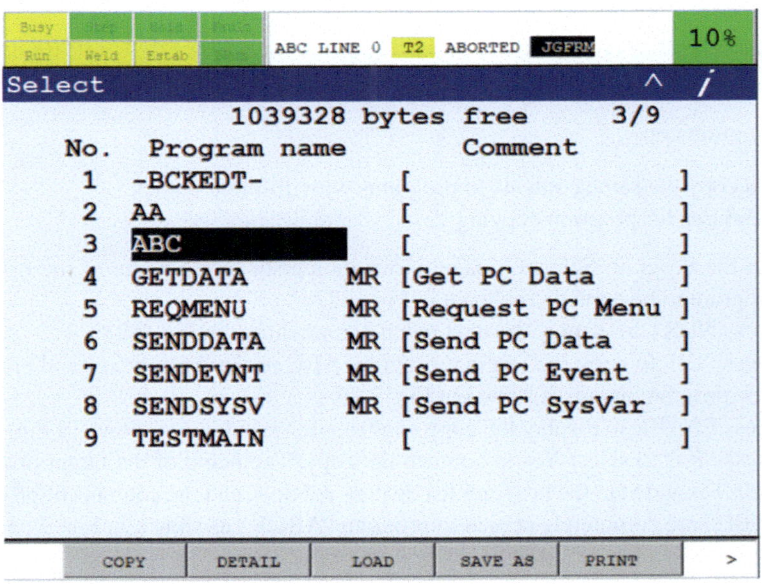

Fig. 7.28 Shows the next page of the function menu

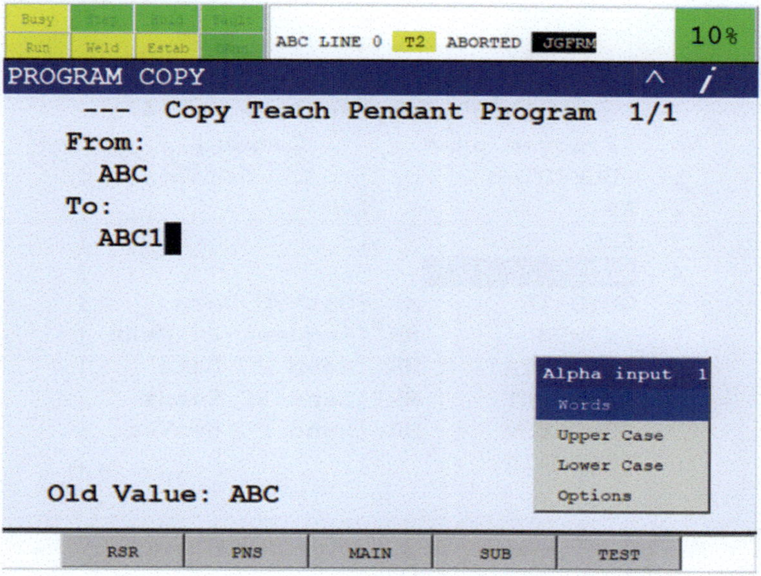

Fig. 7.29 Copy the original program and create the program name of the target program

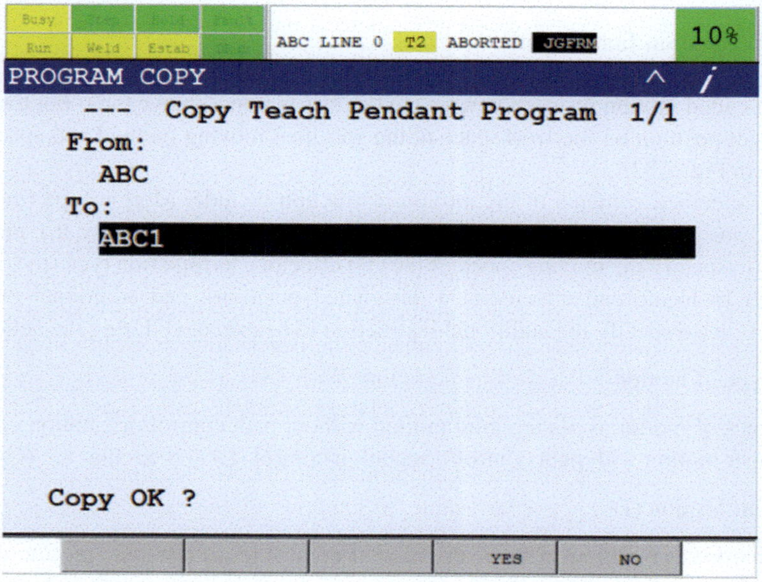

Fig. 7.30 Copy confirmation window

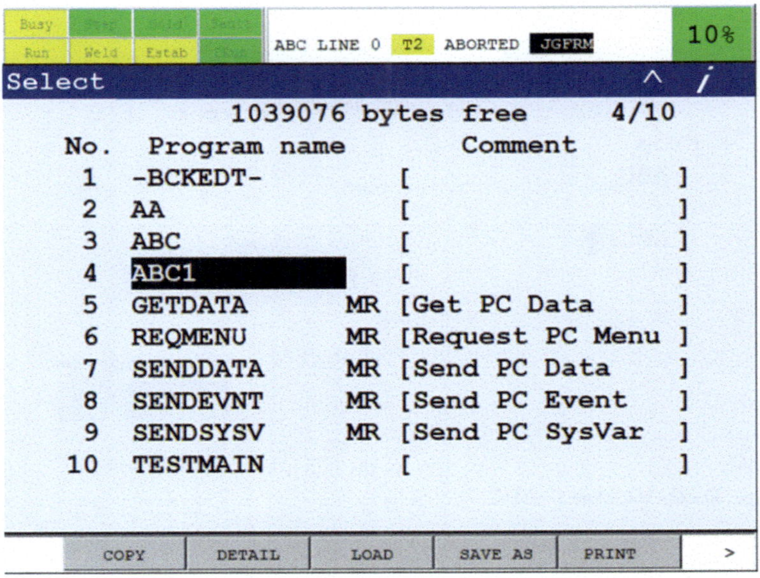

Fig. 7.31 Target program ABC1 added to the program list display window after copy completion

7.1.3.1 Motion Instructions

The so-called motion instruction refers to the instruction to move the robot towards the target position on the workspace at the specified moving method and speed, as shown in Fig. 7.32.

The specified contents of the motion instruction include: Motion type (moving mode), position data (to teach the position for the end effector to move to), moving speed (to specify the moving speed of the end effector), termination type (to specify whether to locate end effector at a designated position), and additional motion instruction (to specify the additional instruction to be executed during the action).

(1) Type of motion

The types of motion types are joint motion without path control/orientation control (J), linear motion with path control/orientation control (L) and circular arc (C).

1) Joint Motion (J)

Joint motion (J) without path or orientation control is an arbitrary motion of the robot, during which the end effector moves arbitrarily between two designated points without controlling the path and orientation, as shown in Fig. 7.33. The joint moving speed is described as a percentage of the maximum moving speed, with the unit being %.

The type of motion and speed are described in the teaching instructions for the target point.

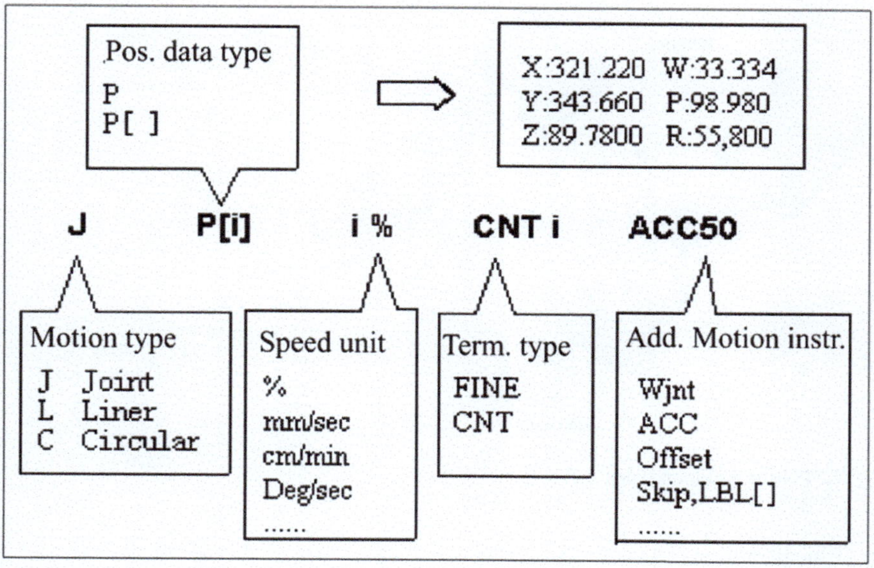

Fig. 7.32 Overview of robot motion instructions

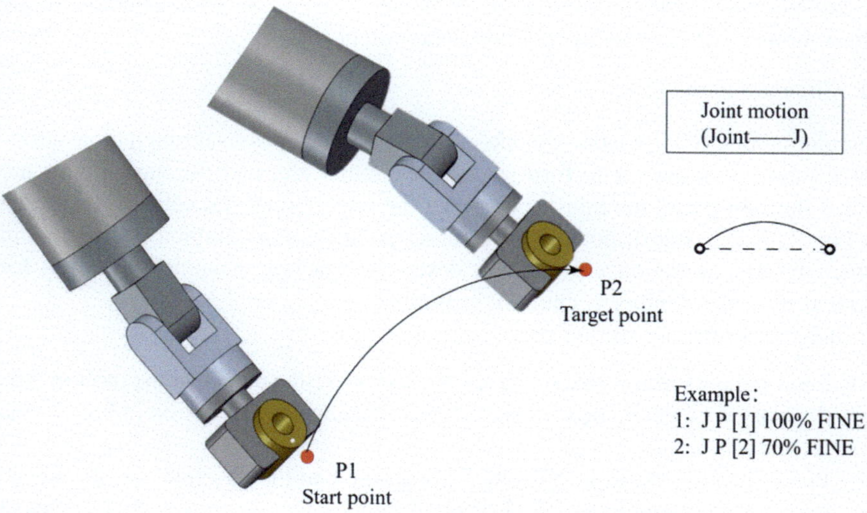

Fig. 7.33 Joint motion (J) without path control/orientation control

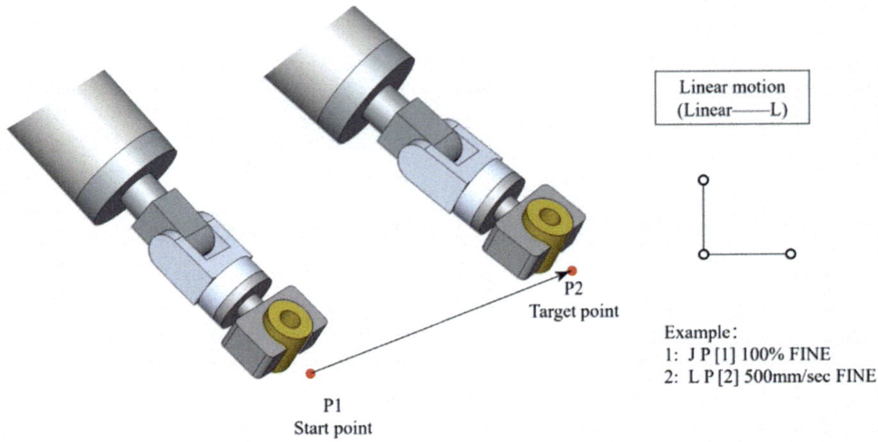

Fig. 7.34 Linear motion types

2) Linear motion (L)

Linear motion (L) is a motion mode in which a robot moves along a linear path from the start point to the target point, as shown in Fig. 7.34. The motion speed is described in the teaching instructions for the target point. The optional units for the linear motion include mm/sec, cm/min, inch/min or sec.

3) Circular motion (C)

Circular motion (C) is a motion mode in which the robot passes through three designated points in sequence to form a path of circular arc, and the three points are called the start point, the transit point, and the target point, respectively, as shown in Fig. 7.35. The transit point and the target point share a common instruction, in which the type of motion and speed are described. The optional units of velocity for circular motion are mm/sec, cm/min, inch/min.

Steps for switching motion type:

(a) Press the cursor movement key ⇦, ⇨, and move the cursor to the motion type designated by the motion instruction. The current type of motion in Fig. 7.36 is L (linear motion).
(b) Press “F4” to select the “CHOICE” function, and the “Motion specify menu” pops up, as shown in Fig. 7.37. Press the cursor movement key ⇧, ⇩ to select the motion mode “Joint (J)”.
(c) Confirm with ENTER to modify the action mode, as shown in Fig. 7.38.

(2) Position data

During teach programing an motion instruction, the position data is automatically written to the program. Position Data includes both position and orientation data:

(a) Position (X, Y, Z) represents the position of the tool cusp (the origin of the tool coordinate system) in the Cartesian coordinate system as a 3D coordinate value.

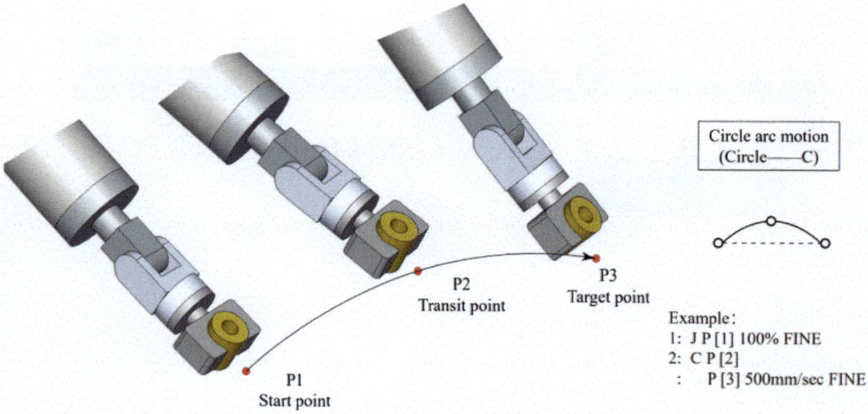

Fig. 7.35 Types of circular arc motion

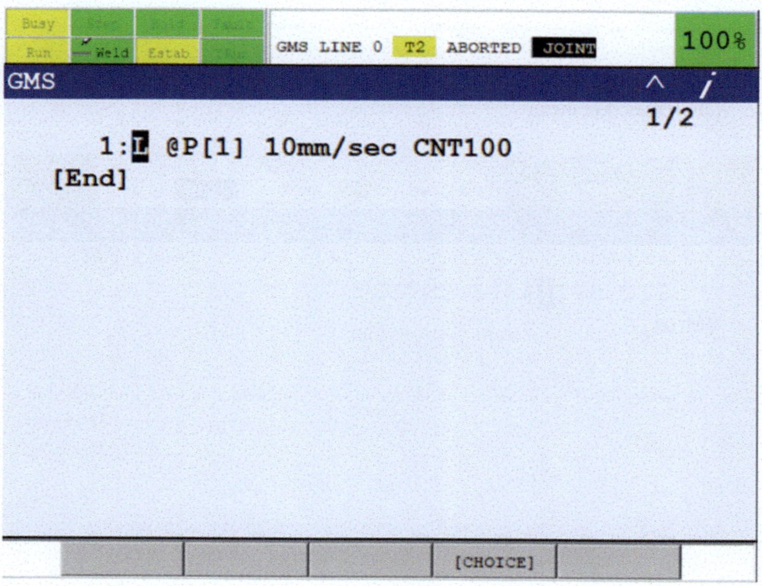

Fig. 7.36 Cursor moving to action type

(b) Orientation (*W, P, R*) is expressed in terms of three rotation angles relative to the X, Y, and Z axes in the Cartesian coordinate system.

In an motion instruction, the position data is presented as a position variable (P [i]) or position register (PR [i]). The position variable is used in the default configuration. The position variable and position register are compared as shown in Fig. 7.39.

To view/modify position data:

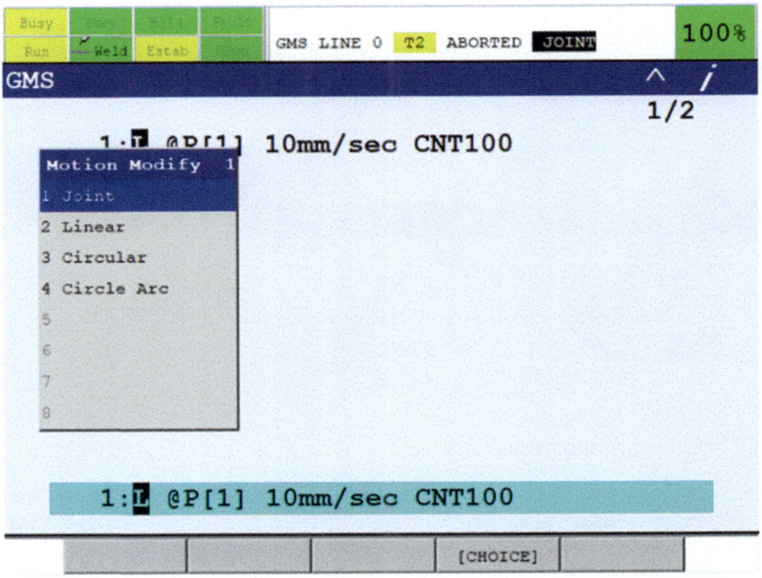

Fig. 7.37 Action selection menu

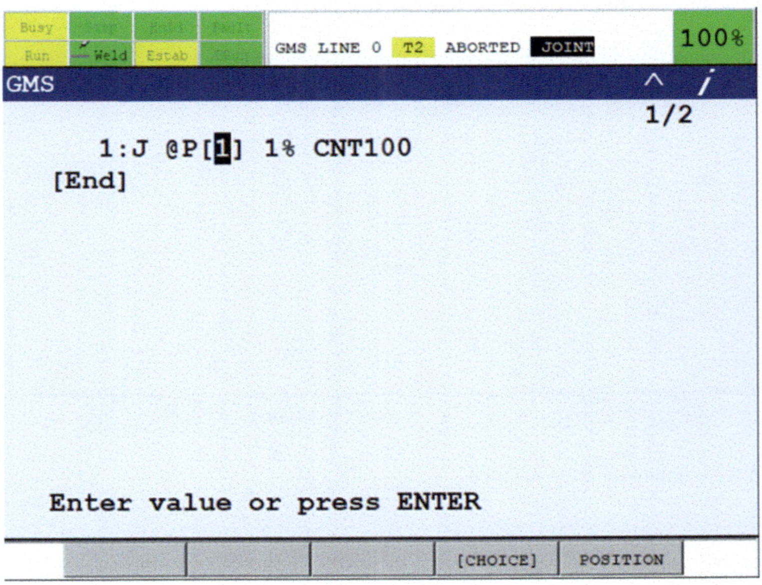

Fig. 7.38 Modified motion instructions

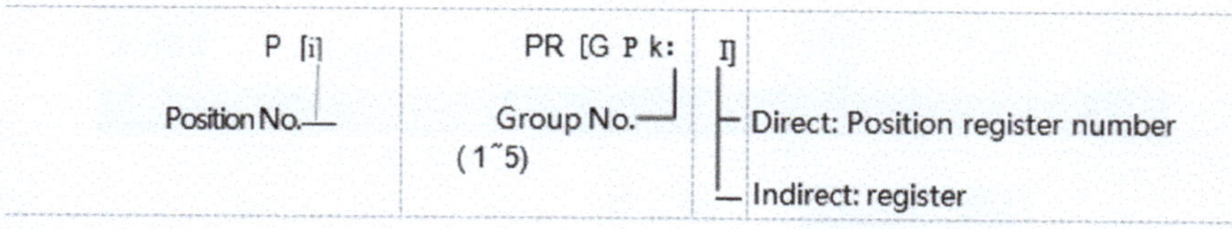

Fig. 7.39 Position data types

(a) Press the cursor movement button ⇦, ⇨, to move the cursor to the "position data" bit of the motion instruction, as shown in Fig. 7.40.
(b) Press F5 to select the position function and display the position data in *X, Y, Z* and *W, P, R* world coordinate systems, as shown in Fig. 7.41.
(c) Press "F5" to select the "REPRE" function, and the Format menu pops up. Press the cursor to move ⇧, ⇩ to select the joint and press "ENTER", as shown in Fig. 7.42.
(d) The window displays position data in the joint coordinate systems J1, J2, J3 and J4, J5, J6, as shown in Fig. 7.43.
(e) To change coordinates, move the cursor to the corresponding value, enter the new value and press ENTER.
(f) After viewing or modifying, press the PREV key to exit and return to the action command display window, as shown in Fig. 7.44.

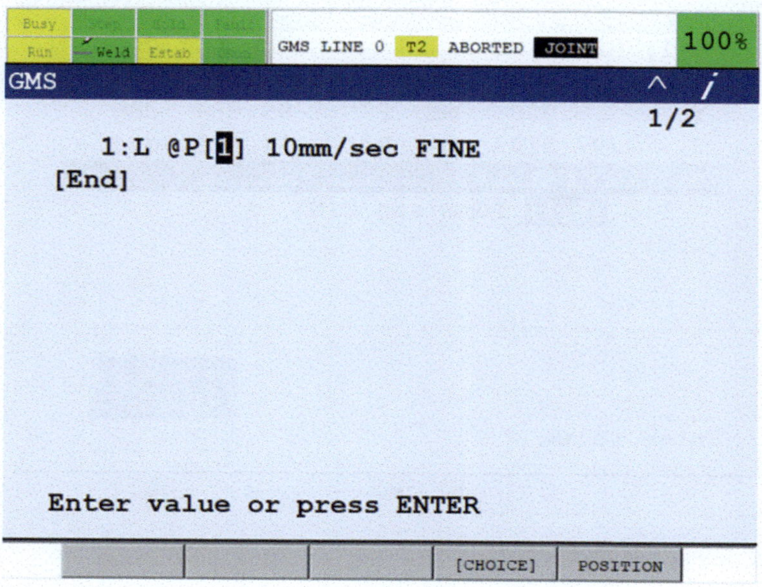

Fig. 7.40 Moving the cursor to the position data of an motion instruction

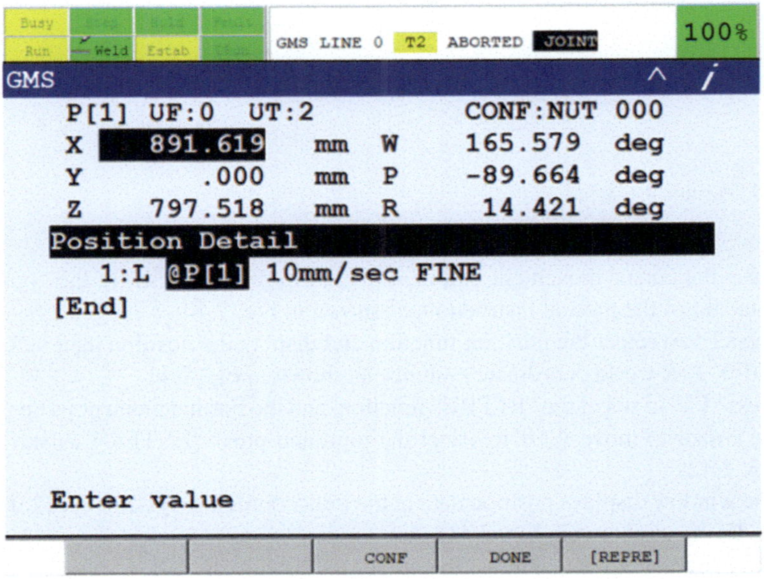

Fig. 7.41 Position data under the *X, Y, Z,* and *W, P, R* world coordinate systems

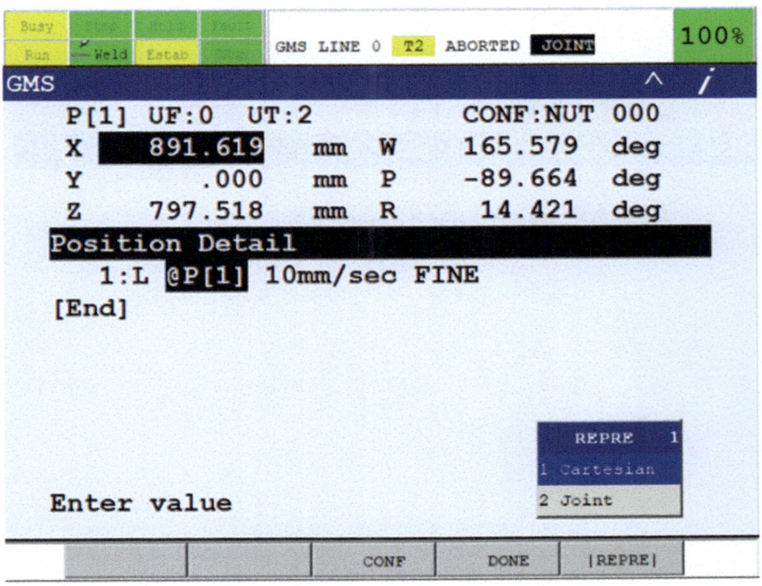

Fig. 7.42 REPRE menu

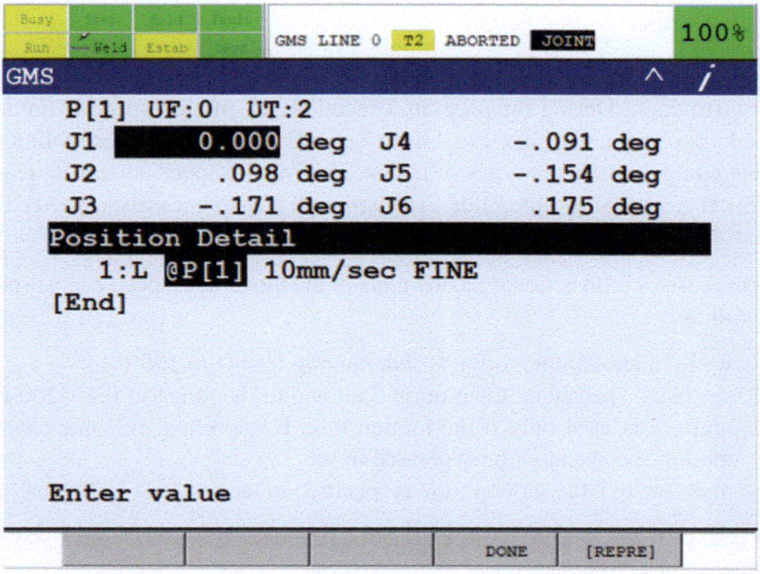

Fig. 7.43 Position data under joint coordinate systems J1, J2, J3, and J4, J5, J6

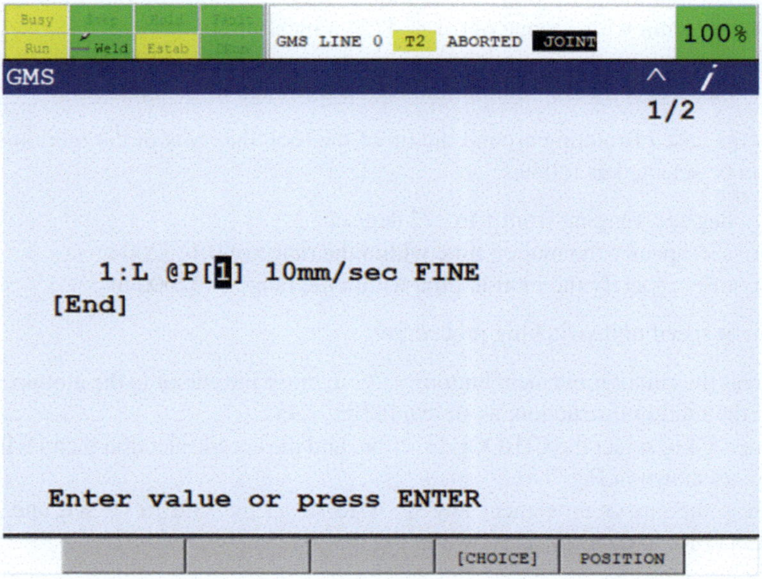

Fig. 7.44 Motion instruction display window

(3) Motion speed

The motion speed of the robot can be set or modified at the "motion speed" bit of the motion instruction. During the program execution, the motion speed is limited by the speed ratio, which can be selected from 1 to 100%. The units of the motion speed vary depending on different types of action. The motion speed set during teaching must not exceed the allowable value. An alarm will be given if the set speed exceeds the allowable range.

(a) In the case of joint action type, the units of the movement speed can be selected as follows:

   a) Ratio to maximum motion speed, ranging from 1 to 100%;
   b) sec (sec), specifying the motion time within the range of 0.1–3200 s. The unit sec is used only if the motion time is important. In some cases, the motion speed cannot be expressed in sec.
   c) msec (msec), the motion time is specified in the range 1–32000 ms.

(b) In case of linear action or circular action, the unit of the movement speed can be selected as follows:

   a) mm/sec, ranging from 1 to 2000 mm/sec;
   b) cm/min, ranging from 1 to 12000 cm/sec;
   c) inch/min with a range of 0.1 to 4724.4 inch/min;
   d) sec, specify the motion time within the range of 0.1–3200 s;
   e) msec, specify the motion time within the range of 1–32000 ms.

(c) In the case of rotating around the tip of the tool, the units of the motion speed can be selected as follows:

   a) deg/sec, ranging from 1 to 272 deg/sec;
   b) sec, specify the motion time within the range of 0.1–3200 s;
   c) msec, specify the motion time within the range 1–32000 ms.

Motion speed unit switching procedure:

(a) Press the cursor movement button ⇦, ⇨, to move the cursor to the motion speed of the motion instruction, as shown in Fig. 7.45.
(b) Press F4 to select the CHIOCE function, and the speed selection menu will pop up, as shown in Fig. 7.46.
(c) Press the cursor movement key ⇧, ⇩ to select the unit to be set, and press "ENTER" to confirm, as shown in Fig. 7.47.

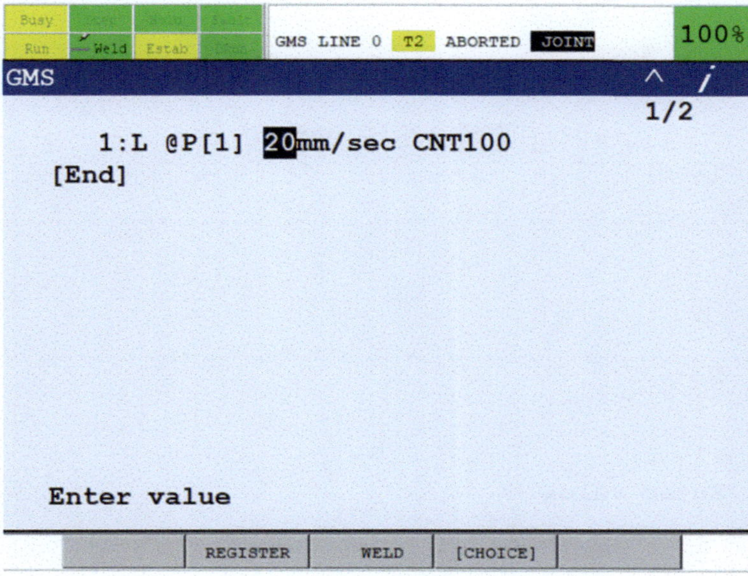

Fig. 7.45 Moving the cursor to the motion speed position of the motion instruction

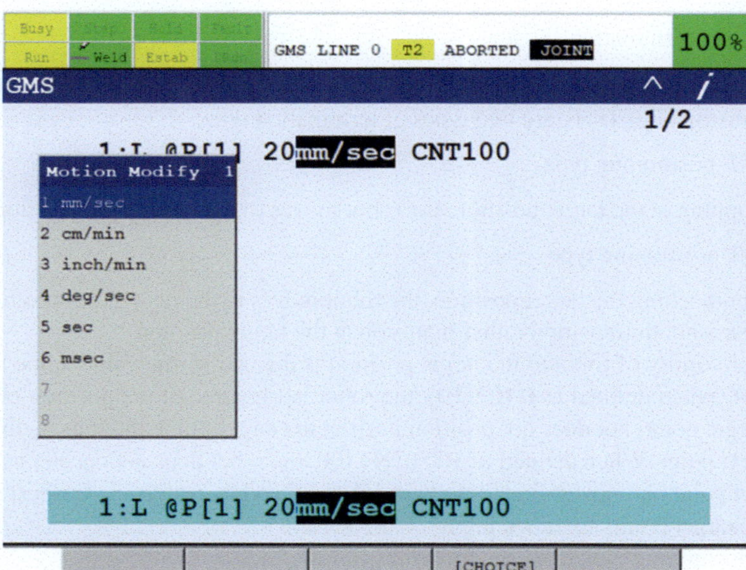

Fig. 7.46 Speed unit selection menu

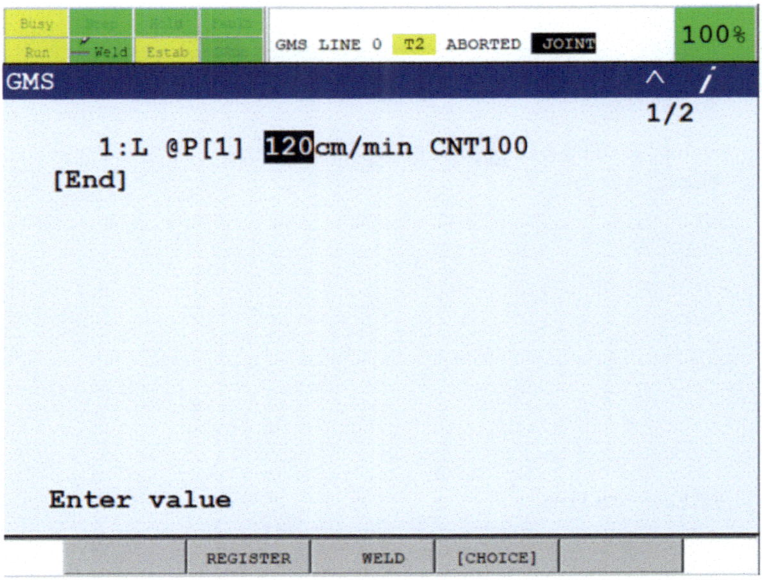

Fig. 7.47 Modified display window

(4) Type of termination

Termination type defines the method, for the robot to terminate the motion, in the motion instruction. There are two types of termination type:

(a) FINE positioning type

After stopping at the target position, the robot moves to the next target position.

(b) CNT positioning type

When appraoching the target position, the robot moves to the next target position in non-linear smooth transition rather than stop at the target position.

The proximity of robot to the target position is defined by the value in the range of 0–100. When defined as 0 (CNT0), the robot is triggered at the position closest to the target point, but does not position itself at the target point and moves directly to the next point. When defined as 100 (CNT100, the robot does not decelerate near the target point but moves directly to the next point, and its path is the furthest away from the target point, as shown in Fig. 7.48.

The positioning type can be switched as follows:

(a) Press the cursor movement key ⇦, ⇨, to move the cursor to the "positioning type" bit of the motion instruction, as shown in Fig. 7.49.
(b) Press "F4" and select the CHOICE function, and the Termination Type Selection menu pops up. Use the cursor movement key to move the cursor to the desired termination type, as shown in Fig. 7.50.

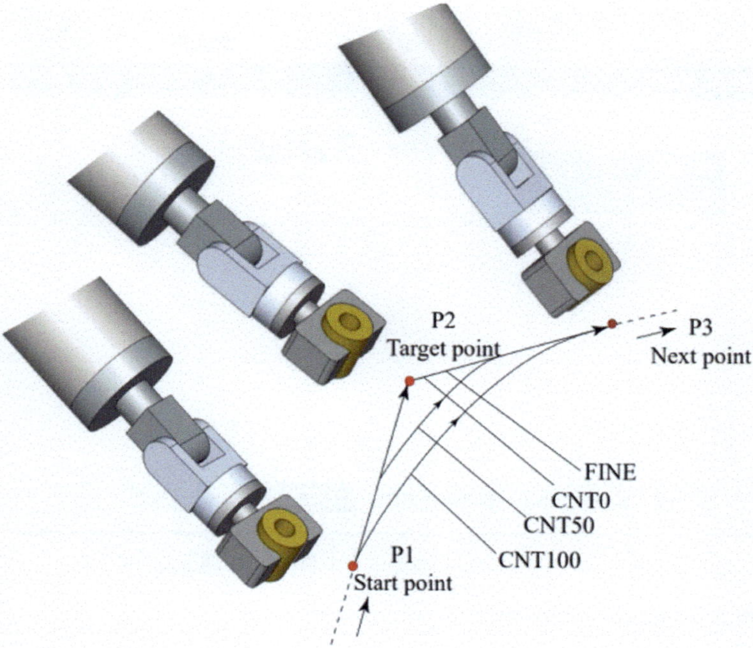

Fig. 7.48 Termination types

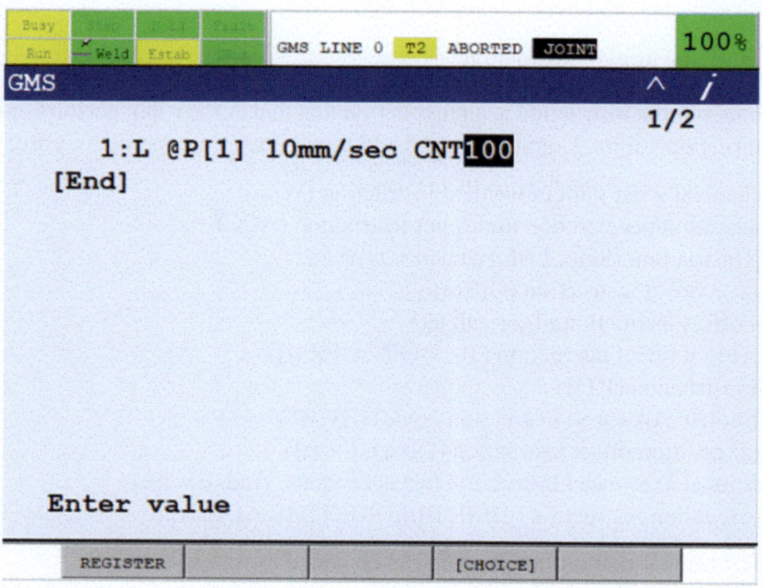

Fig. 7.49 Moving the cursor to the positioning type bit of the motion instruction

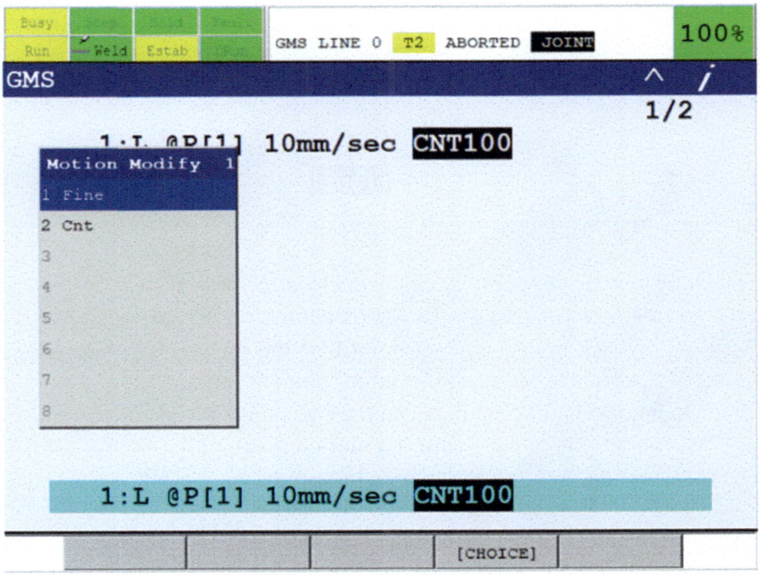

Fig. 7.50 Selecting termination type from motion modify menu

(c) Confirm by pressing the ENTER key, and the positioning type of the motion instruction is modified, as shown in Fig. 7.51.

(5) Additional motion instruction

Additional motion instruction is an instruction that makes the robot perform specific work during operation. Commonly used additional operating instructions are:

- Mechanical wrist joint movement instruction (Wjnt)
- Acceleration/deceleration multiplier instruction (ACC)
- Skip instruction (Skip, LBL [i])
- Position offset instruction (Offset)
- Tool offset instruction (Tool_Offset)
- Direct tool offset instruction (Tool_Offset, PR [i])
- Path instruction (PTH)
- Additional axis speed instruction (sync) (EV i%)
- Direct position offset instruction (Offset, PR [i])
- Additional axis speed instruction (asynchronous) (Ind. EV i%)
- Pre-execution instruction (TIME BEFORE/TIME AFTER).

The additional motion instructions can be added as follows:

(a) Press the cursor movement key ⇦, ⇨, and move the cursor to the blank after the motion instruction, as shown in Fig. 7.52.

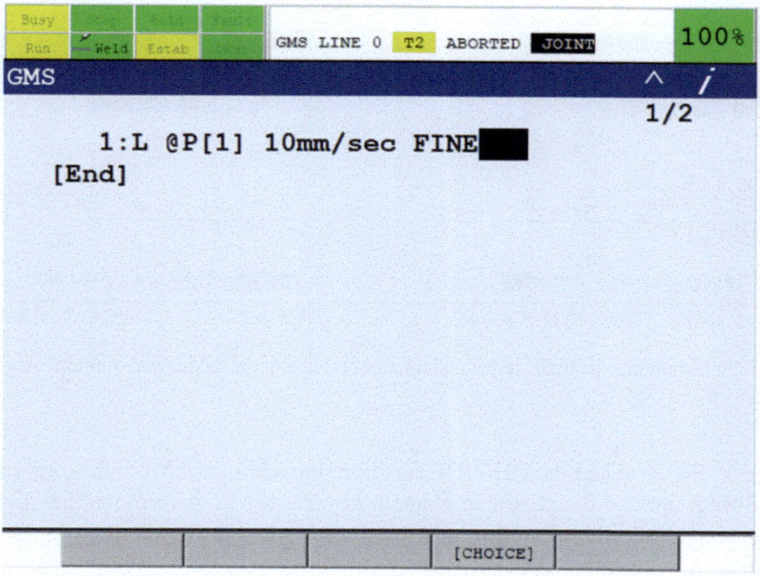

Fig. 7.51 Motion instruction with modified positioning type

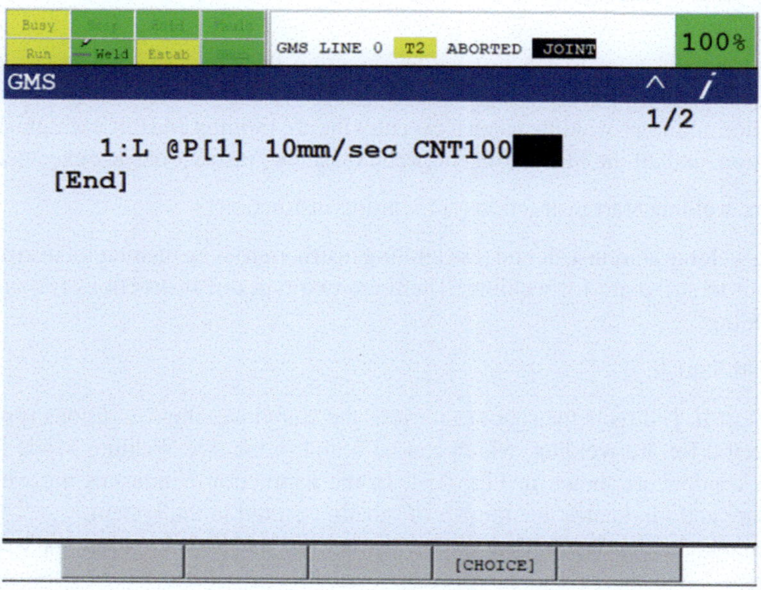

Fig. 7.52 Moving the cursor to the blank after the motion instruction

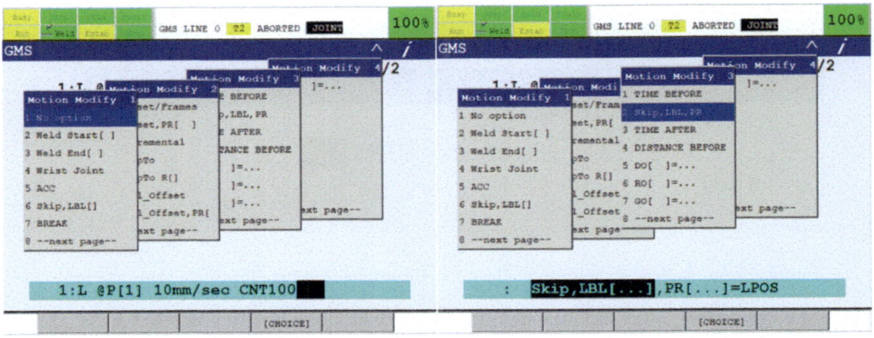

Fig. 7.53 Selecting the desired additional instruction from the additional motion instruction selection menu

(b) Press "F4" to select the CHOICE function, the additional Motion Modify menu will pop up; use the cursor movement key ⇦, ⇨, ⇧, ⇩ to move the cursor to the desired additional motion instruction, as shown in Fig. 7.53.

(c) Confirm by pressing ENTER.

7.1.3.2 Arc Instructions

Arc instructions refer to the instruction directing the robot to start and stop arc welding. The robot starts welding when it executes the arc welding start (arc igniting) instruction and stops welding when it executes the arc welding end (arc extinguishing) instruction, and all the instructions between these two instructions are executed.

(1) Arc welding start instruction (arc igniting instruction)

The arc welding start instruction (arc igniting instruction) is an instruction instructing the robot to strike arc for welding. There are two types of instructions for starting arc welding:

1) Weld Start *[i, j]*

Weld Start [*i, j*] directs the robot to retrieve the stored welding conditions (process parameters) for arc welding, which can be found in the Arc Welding Mode Information window, as shown in Fig. 7.54. In the instruction *i* indicates the welding procedure and *j* indicates the number of schedule stored in the system.

Example: In the Weld Start [1, 1] instruction, the first 1 represents welding procedure 1 and the second 1 represents Schedule 1 (voltage: 20 V, current: 200 A).

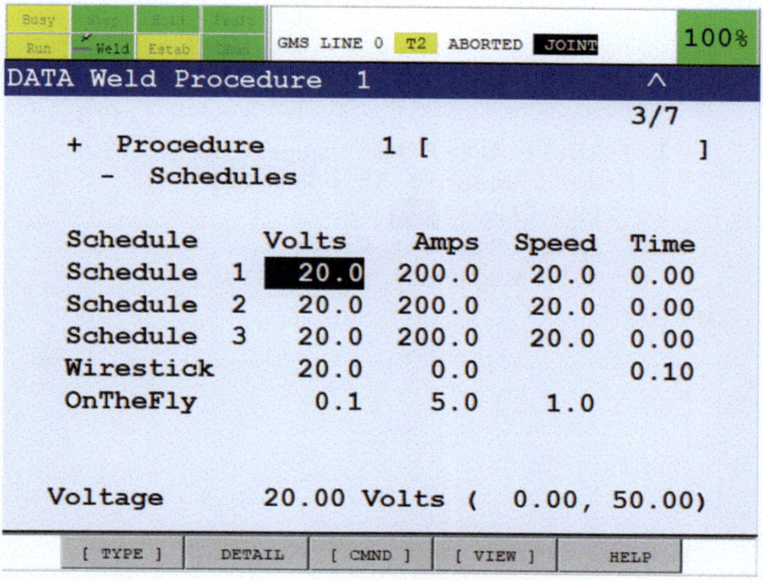

Fig. 7.54 Available welding schedule for WeldStart [*i*, *j*]

2) Weld Start [*i*, V, A, …]

The Weld Start [*i*, V, A, …] instruction instructs the robot to start welding by using the arc voltage and welding current (or wire feeding speed) set directly in this instruction, without retrieving the stored welding schedules, as shown in Fig. 7.55. The specified welding parameters depend on the specific welding process.

Example: Weld Start [1, 18.0 V, 190.0 A]: 1 represents welding procedure 1; 18 V means the arc voltage; 190.0 A refers to welding current.

(2) Arc welding end instruction (arc extinguishing instruction)

The arc welding end instruction (arc extinguishing instruction) is the instruction to indicate the robot to end the arc welding. The arc welding end instructions is given in two forms:

1) Weld End [*i*, *j*]

The Weld End [*i*, *j*] instruction instructs the robot to retrieve the arc extinguishing parameters stored in the system for extinction and to end the arc welding process. *i* indicates the welding procedure and *j* indicates the number of the extinction schedule stored in the system.

If the arc is extinguished by cutting off the welding current, the arc crater directly below the arc will not be filled in time, and an arc crater will form in the arc extinguished part of the weld. The arc extinguishing control function of the robot can be used to avoid craters.

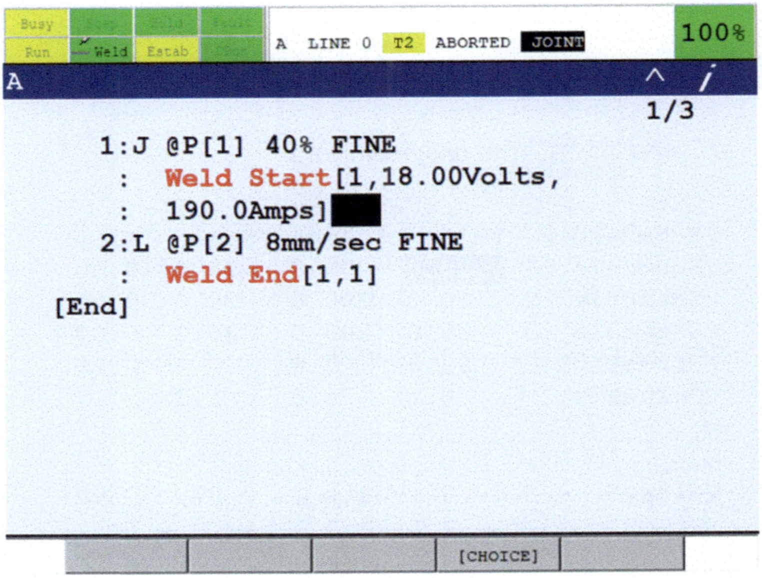

Fig. 7.55 Weld Start [V, A, …] instruction

2) Weld End [*i*, V, A, sec]

The Weld End [*i*, V, A, sec] instruction directs robots to extinguish arc in accordance with the welding conditions specified in this instruction. The arc extinction voltage, arc extinction current (or metal wire feed speed) and extinction time are directly input to the instruction, as shown in Fig. 7.56. The specified value for these parameters depend on the specific welding procedure. When no arc extinction control is needed, the extinction time is set to zero.

Example: Weld End [1, 10.0 V, 100.0 A, 0.5 s], Where 1 stands for welding procedure 1, 10 V for an arc extinction voltage of 10 V, 100 A for an arc extinction current of 100A, and 0.5 s for an extinction time of 0.5 s.

The programming procedure for the arc instruction is as follows:

(a) Select the newly established program and press Enter to enter the programming interface, as shown in Fig. 7.57.
(b) Press F2 and select "WELD-ST" function. The "Weld Start def" menu will pop up. Use the cursor movement key ⇧, ⇩ to move the cursor to the desired standard arc igniting instruction, as shown in Fig. 7.58.
(c) After pressing the ENTER key, the selected instruction appears in the program edit window, as shown in Fig. 7.59.

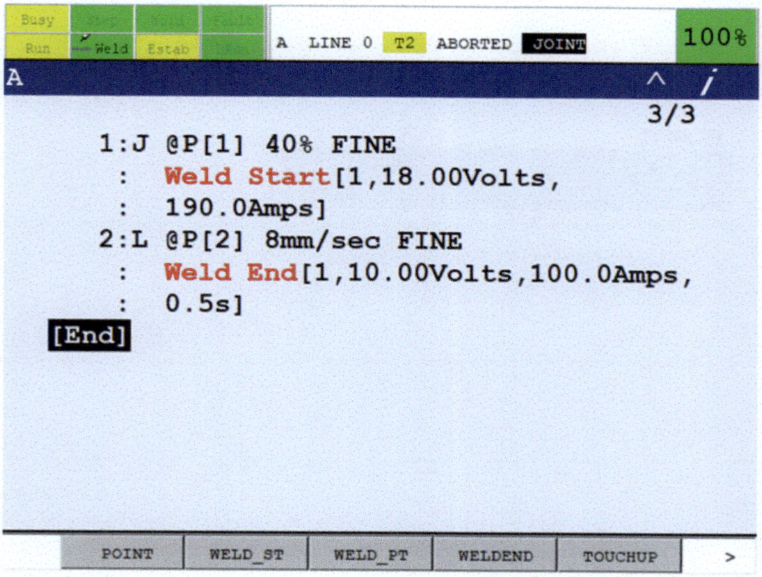

Fig. 7.56 Weld End [*i*, V, A, sec] instruction

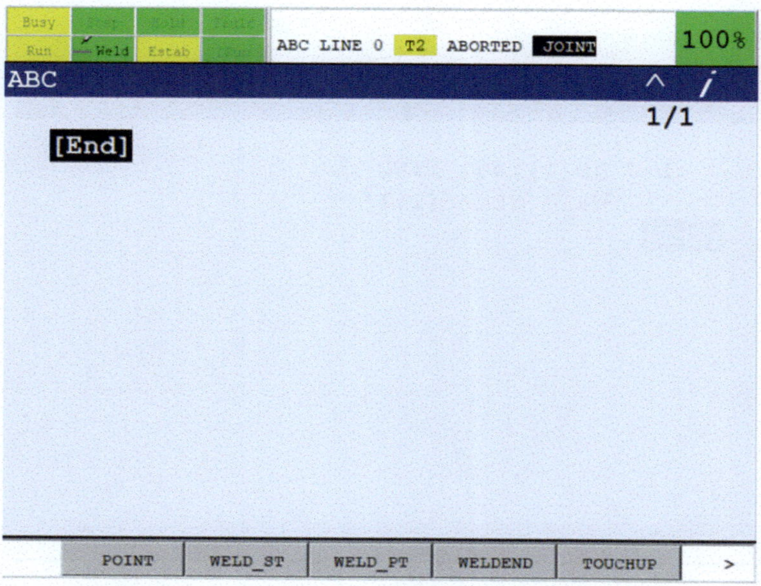

Fig. 7.57 Enter the programming interface

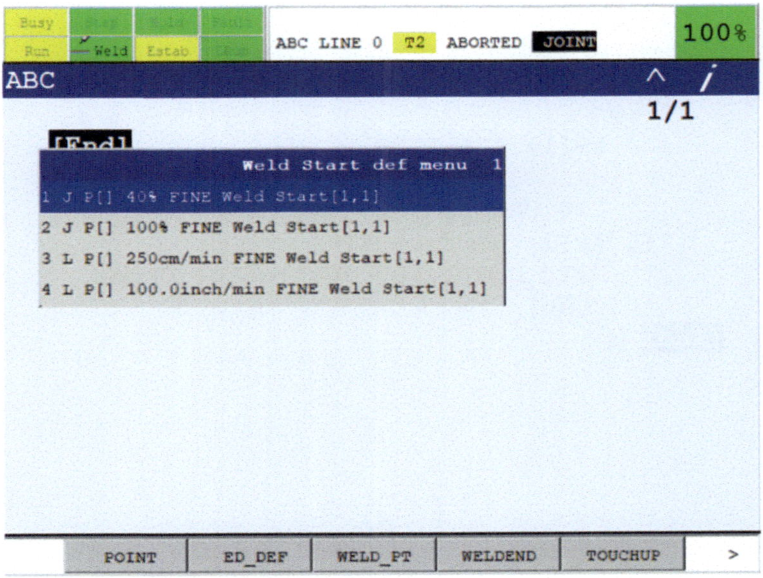

Fig. 7.58 Moving the cursor to the desired weld start instruction using the cursor movement key ⇧, ⇩

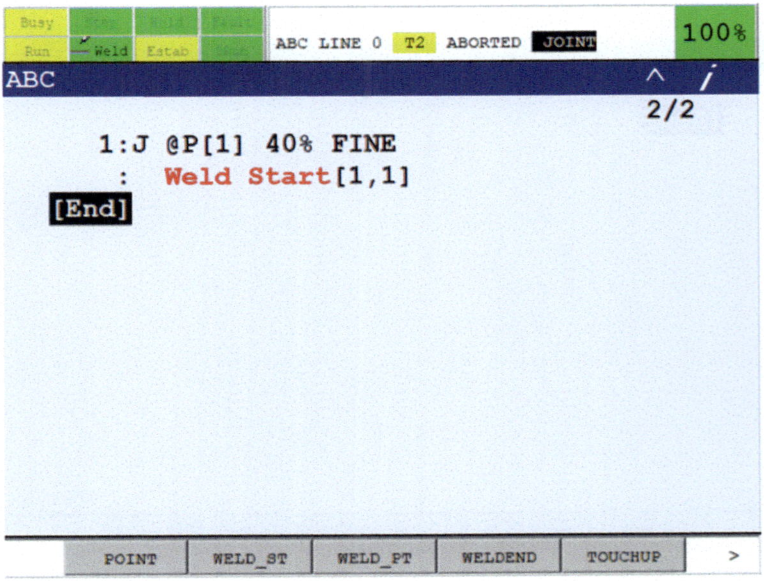

Fig. 7.59 Selected instruction in program edit window

Note:

- In the instruction for moving to the arc weld start point, the type of termination must be FINE;
- Do not use joint motion J in the motion instruction for moving to the arc weld path point and end point;
- The type of termination must be FINE in the motion instruction for moving to the arc weld end point.
- The angle between the torch and the workpiece shall be set to the correct value.

7.1.3.3 Weave Instructions

The weave instructions are a kind of instruction that command the welding torch to weave in a specified way. The weave instructions include the weave-start instruction (the instruction commanding the welding torch to start weaving) and weave-end instruction (the instruction commanding the welding torch to end weaving). The welding torch weave is performed simultaneously with all other motions in the time interval between the weave-start instruction and the weave-end instruction.

(1) Weave start instruction

The weave start instruction commands the robot to start torch weave. There are two modes of weave start instruction: Weave (mode) [*i*] and Weave (mode) [Hz, mm, sec, sec].

1) Weave (mode) [*i*]

The Weave [*i*] instruction commands the robot to start the torch weave in the specified mode by retrieving the weaving schedules stored in the system, as shown in Fig. 7.60.

Example: Weave Sine [1] instruction means that the weave mode is sinusoidal, and the weave schedule 1 is retrieved.

2) Weave (mode) [Hz, mm, sec, sec]

The Weave (mode) [Hz, mm, sec, sec] instruction commands the robot to start the torch weave with the specified mode and the weave parameters specified in the instruction. The weave parameters that can be set include frequency, amplitude, left

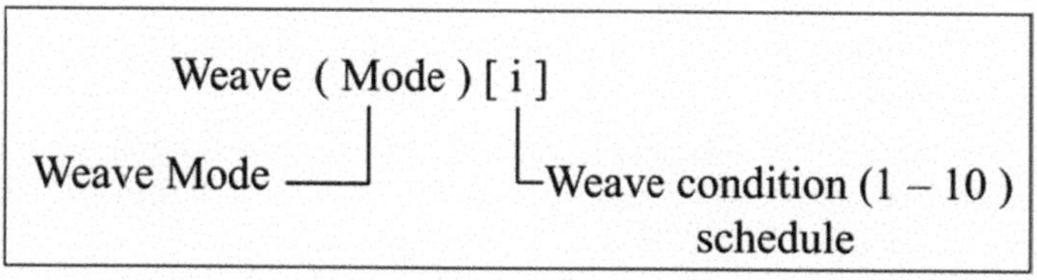

Fig. 7.60 Weave [i] instructions

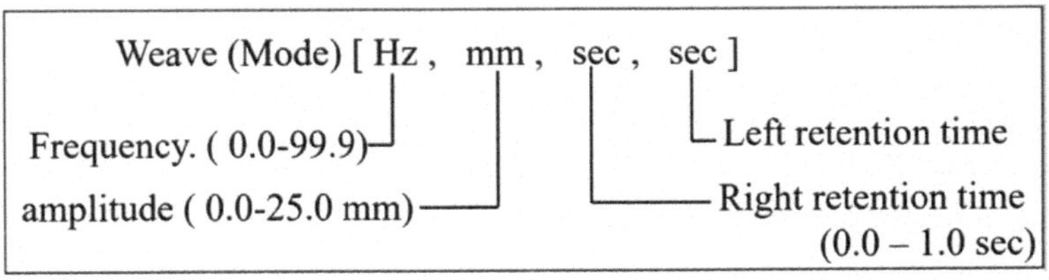

Fig. 7.61 Weave [Hz, mm, s, s] instructions

residence time, right residence time, etc., as shown in Fig. 7.61. Every parameter should be set within its allowable range.

Example: WeaveSine [5.0 Hz, 20.0 mm, 1.0, 1.0 s] instruction means that the weave mode is sinusoidal, the weave frequency is 5 Hz, and the weave amplitude is 20 mm, and the residence time at the left and right endpoints are both 1.0 s.

(2) Weave end instruction

The weave end instruction is an instruction command the robot to end the torch weave. There are two modes of weave end instruction: Weave End and Weave End [*i*].

1) Weave End

The Weave End instruction instructs the robot to terminate the torch weaves being executed.

2) Weave End [*i*]

The Weave End [*i*] instruction is used when there are no less than two motion groups and several Weave [*i*] instructions in the control program.

The weave commanded by the Weave (mode) [*i*] can be terminated by designating the same weave schedule in the Weave End [*i*] instruction with that in the Weave (mode) [*i*] instruction.

The programming procedure for the weave instruction is as follows:

(a) Select the newly established program and press Enter to enter the programming interface, as shown in Fig. 7.62.
(b) Press NEXT, the function menu bar page down to display instruction menu items, as shown in Fig. 7.63.
(c) Press "F1" keyboard to select the instruction function, and several instruction selection menus will pop up, as shown in Fig. 7.64.
(d) Move the cursor "WEAVE" to the weaving welding instruction with the cursor movement key ⇦, ⇨, ⇧, ⇩, and press "ENTER" to confirm that the weave welding instruction "WEAVE" is selected, as shown in Fig. 7.65.
(e) Weave menu pops up. Move the cursor to select the desired instruction and press ENTER to confirm (Fig. 7.66).
(f) Weave instruction appears in the edit window; move the cursor to the parameter setting position of the instruction, as shown in Fig. 7.67.
(g) Enter schedule number 1 and confirm by pressing ENTER, as shown in Fig. 7.68.

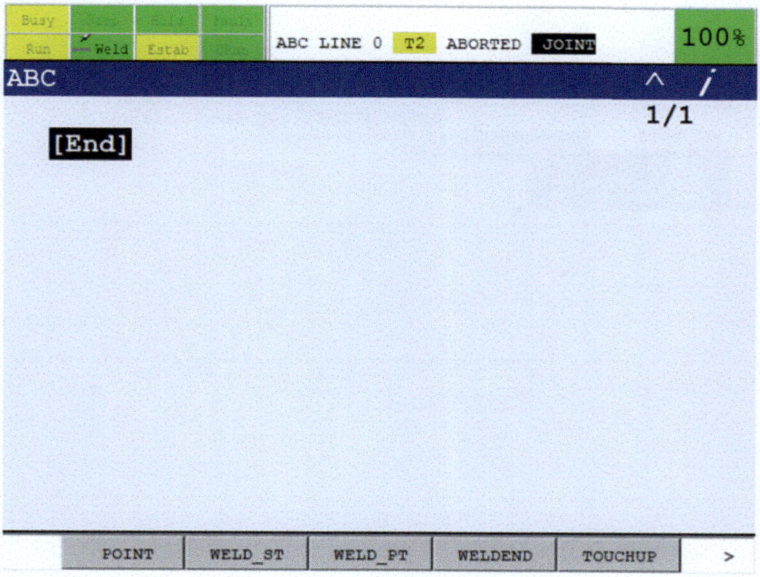

Fig. 7.62 Programming interface

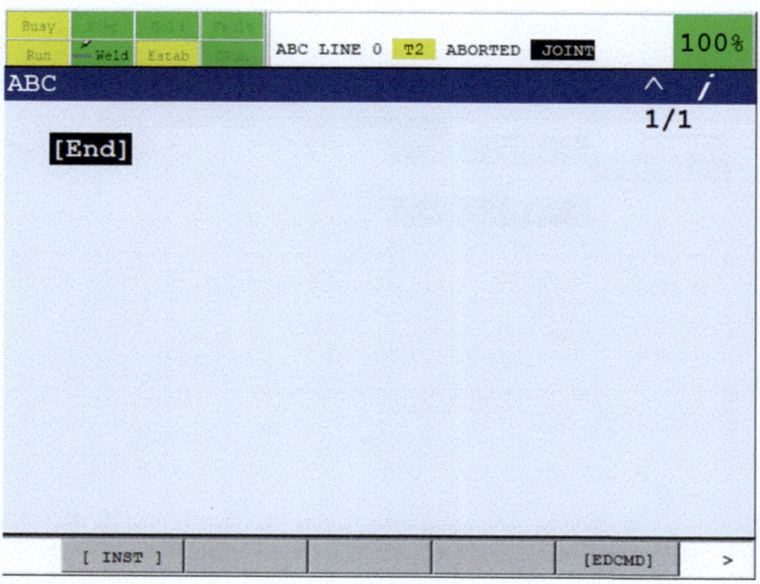

Fig. 7.63 Pressing the NEXT key to display the instruction menu items

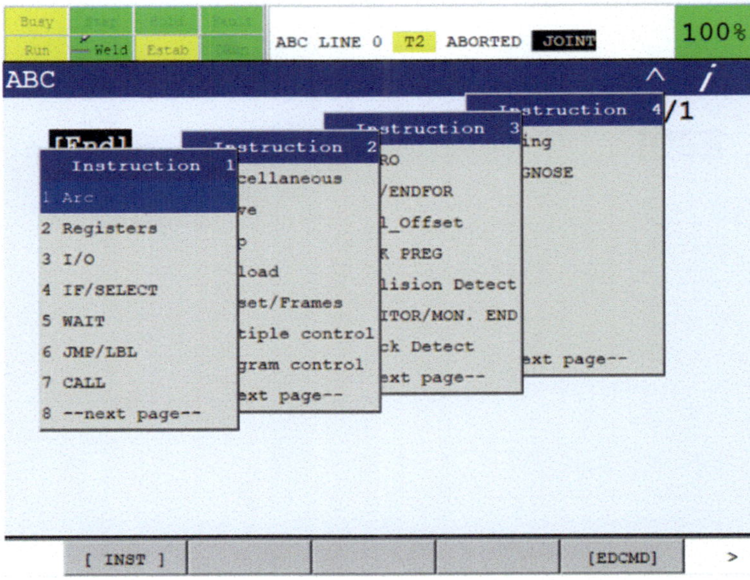

Fig. 7.64 Pressing F1 to pop up instruction selection menu

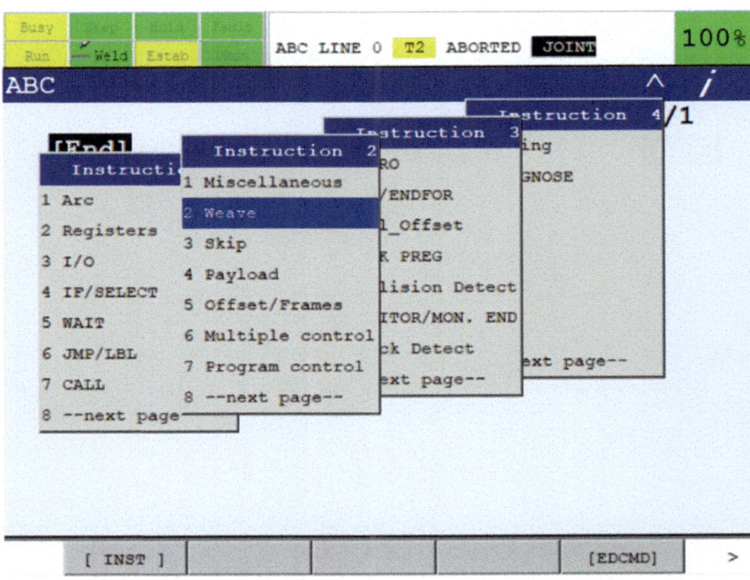

Fig. 7.65 Choosing WEAVE instruction

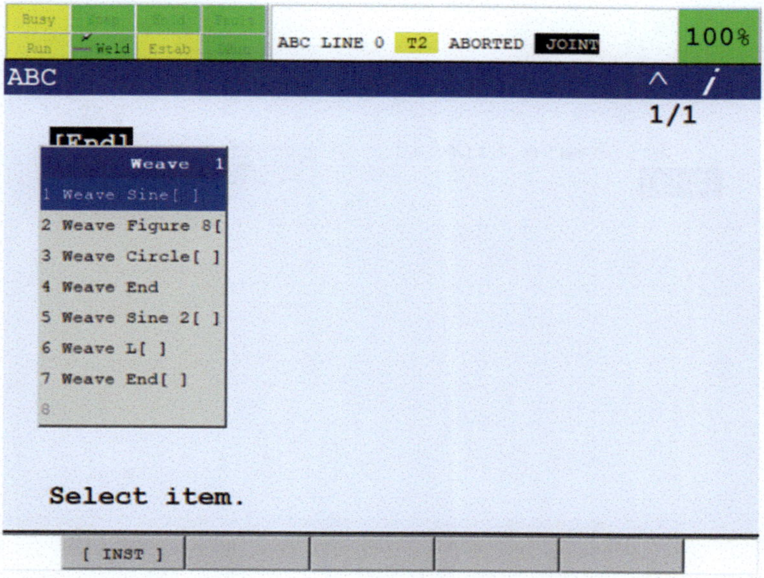

Fig. 7.66 Weave menu

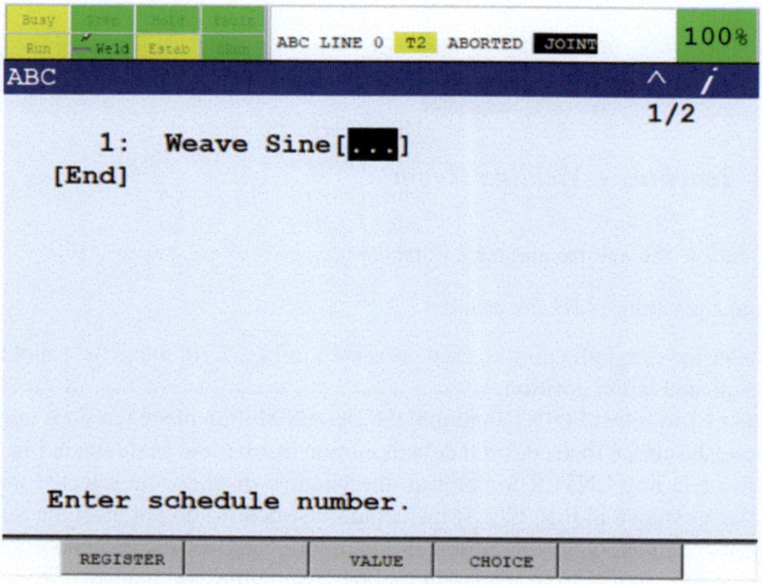

Fig. 7.67 Move the cursor to the parameter setting position of the weave instruction

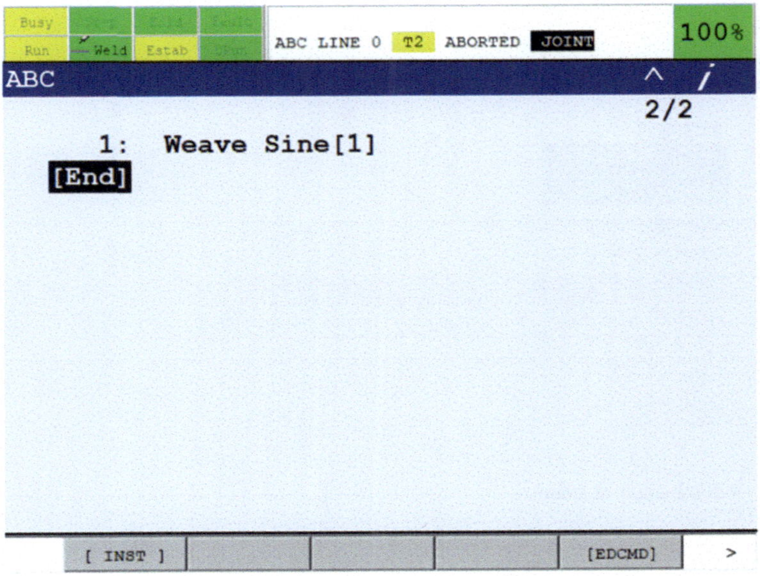

Fig. 7.68 Entering the weave schedule number

(h) Alternatively, press “F3” in step (f), select the other mode of weave start instruction, and then enter the weave parameters directly, as shown in Fig. 7.69.

7.1.4 Teaching a Welding Robot

(1) Creation of teaching-playback points

To create a teaching-playback point:

(a) Under the program edit interface, as shown in Fig. 7.70, move the robot to the designated target position.
(b) Press F1 to select POINT function, the Default Motion menu bar pops up; then, move the cursor to the desired default motion instruction, as shown in Fig. 7.71.
(c) After pressing ENTER to confirm, the window displays the selected instructions, as shown in Fig. 7.72. If the default instructions do not meet the requirements completely, a similar default instruction may be selected and modified, or modifications can be made after all the instructions are loaded.
(d) Move the robot to the next target position, repeat the above steps, and add other teaching-playback points.

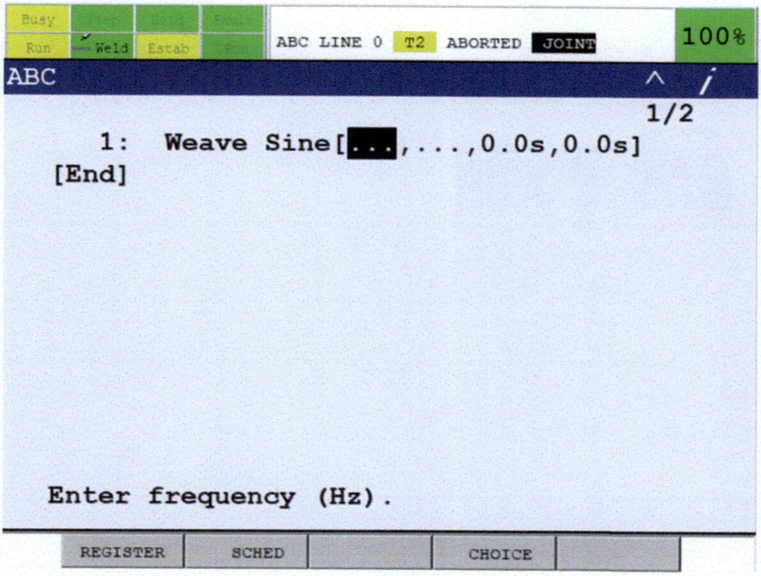

Fig. 7.69 Direct input weave parameters

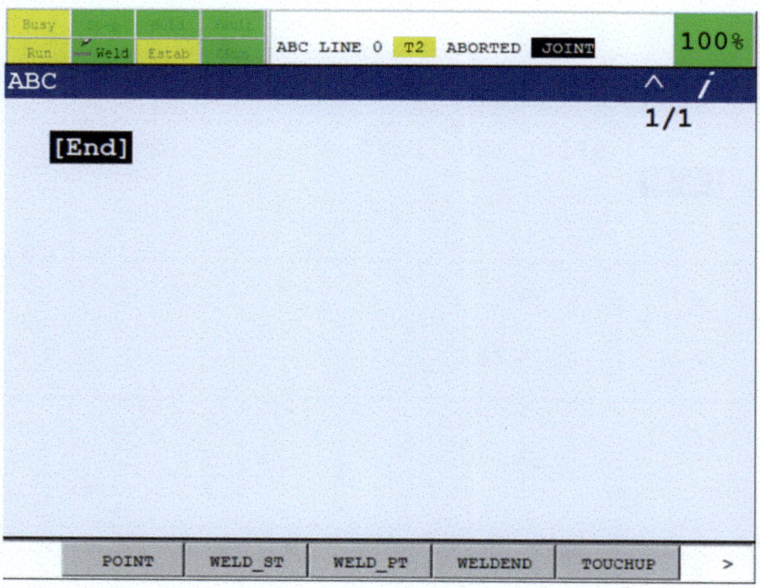

Fig. 7.70 Program edit interface

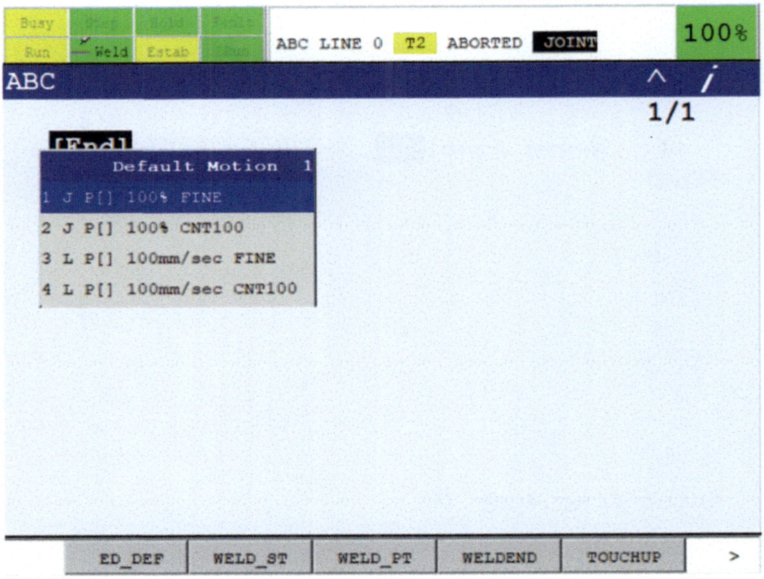

Fig. 7.71 Moving the cursor to the desired default motion instruction

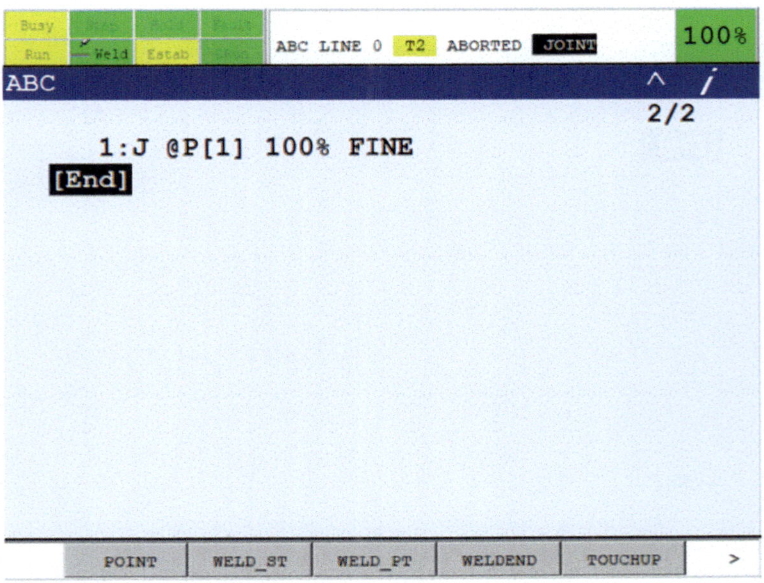

Fig. 7.72 Edit window for teaching-playback point

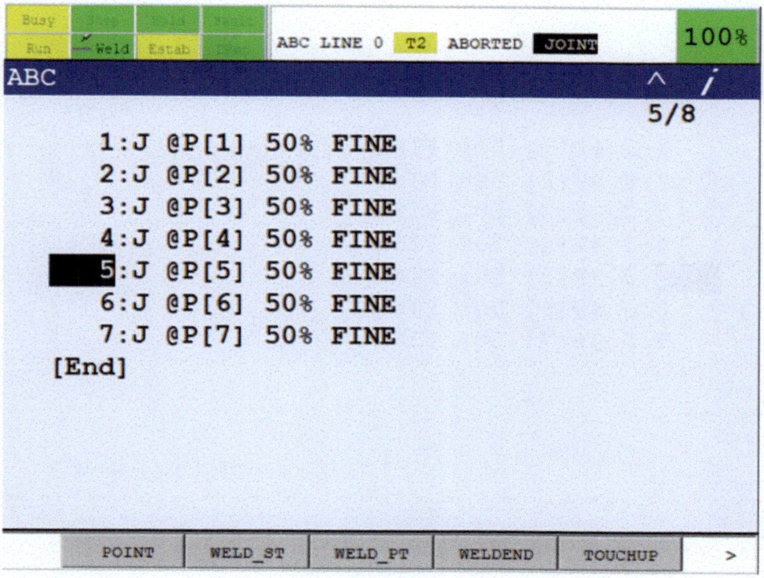

Fig. 7.73 Moving the cursor to the line number preceding the motion instruction you want to delete

(2) Delete and insert teaching-playback points

During the teaching process, it is sometimes necessary to delete the redundant teaching-playback points except the safety points, so as to save the running time and improve the efficiency.

1) Pocedure for deleting teaching-playback points:

(a) Move the cursor to the line number preceding the motion instruction you want to delete, as shown in Fig. 7.73.
(b) Press NEXT to page the function menu bar to display the edit menu item, as shown in Fig. 7.74.
(c) Press F5, select the EDCMD function, and the EDCMD bar will pop up, as shown in Fig. 7.75.
(d) Move the cursor to the "2 Delete" menu item, as shown in Fig. 7.76, and confirm with ENTER.
(e) To delete adjacent multiple lines, select multiple lines with ⇧ or ⇩, as shown in Fig. 7.77; skip this step if deleting a single line.

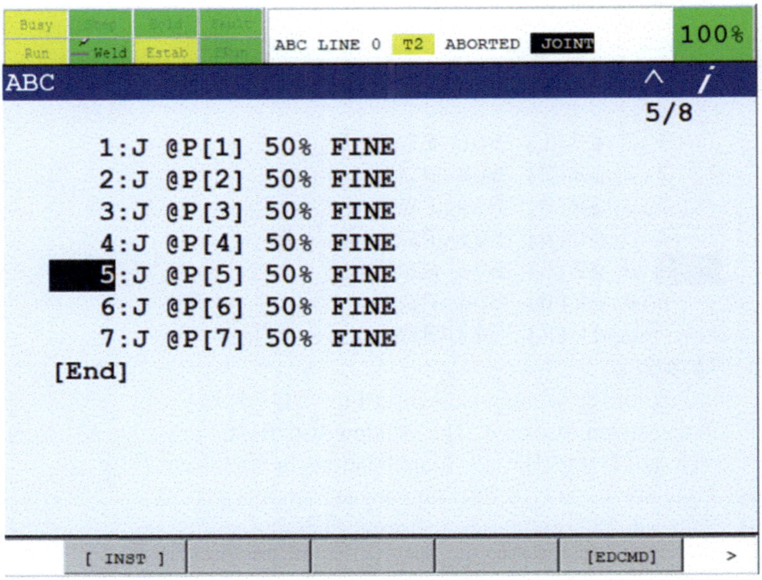

Fig. 7.74 Pressing the NEXT key to display the edit menu items

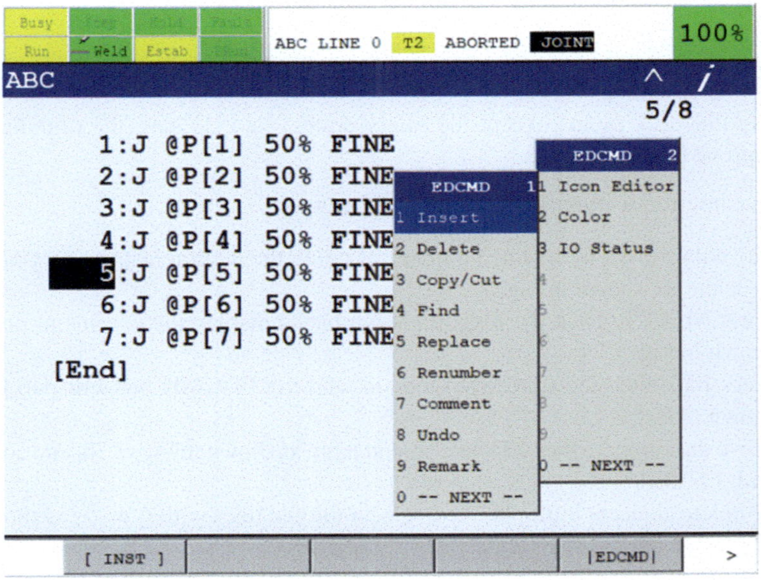

Fig. 7.75 EDCMD menu bar (1)

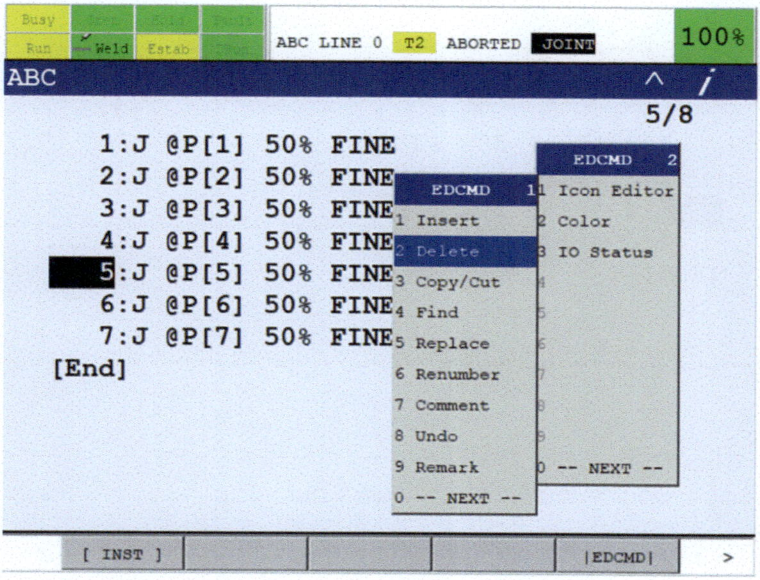

Fig. 7.76 Moving the cursor to the “2 Delete” menu item

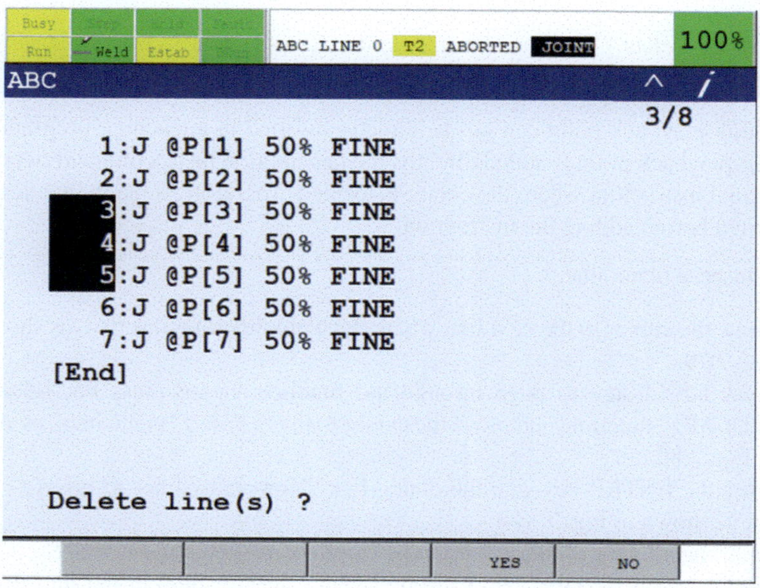

Fig. 7.77 Selecting multiple lines

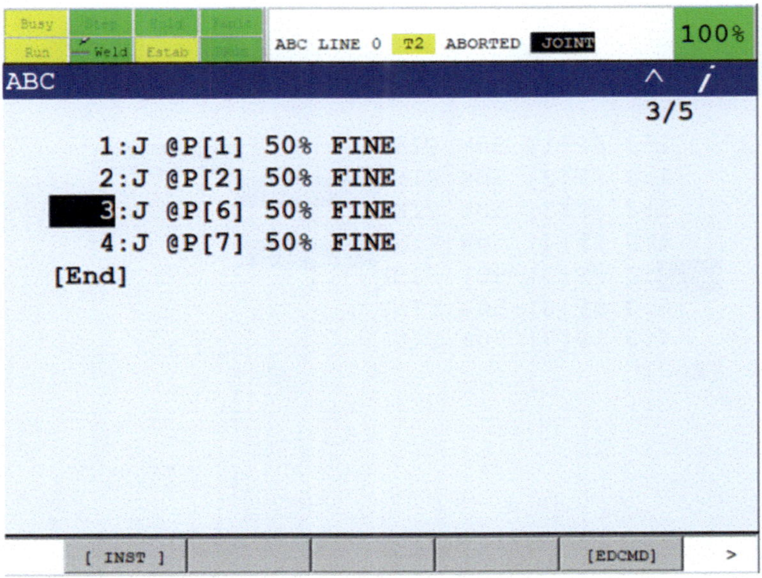

Fig. 7.78 Deletion instruction disappears from the edit window

6) Press F4, select "Yes" to confirm deletion. The deletion instruction disappears from the edit window, as shown in Fig. 7.78.

A teaching-playback point can not be directly inserted in a teaching program. If a teaching-playback point is added directly, the new motion instruction will overwrite the original instruction where the cursor is located. Therefore, a blank line needs to be inserted before adding the instruction.

2) To insert a blank line:

(a) Move the cursor to the next line where the blank lines are inserted, as shown in Fig. 7.79.
(b) Press NEXT key to page through the function menu; press F5, select the "EDCMD" function, and move the cursor to the "1 Insert" menu item, as shown in Fig. 7.80.
(c) Press the ENTER key, and the dialog box "How many lines to insert?:" pops up, as shown in Fig. 7.81.
(d) Enter the number of blank lines to be inserted, as shown in Fig. 7.82.
(e) Confirm with ENTER, the 3 inserted blank lines display in the interface, as shown in Fig. 7.83.

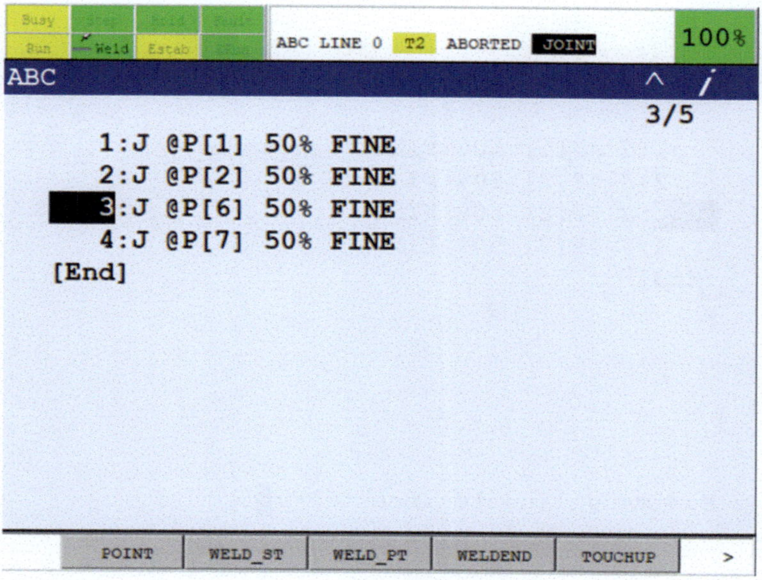

Fig. 7.79 Moving the cursor to the next line where blank lines are inserted

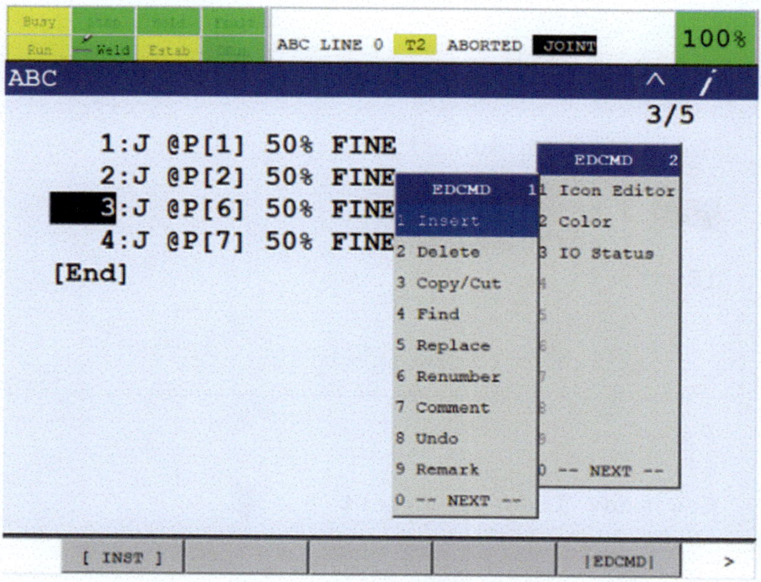

Fig. 7.80 Moving the cursor to the "1 INSERT" menu item

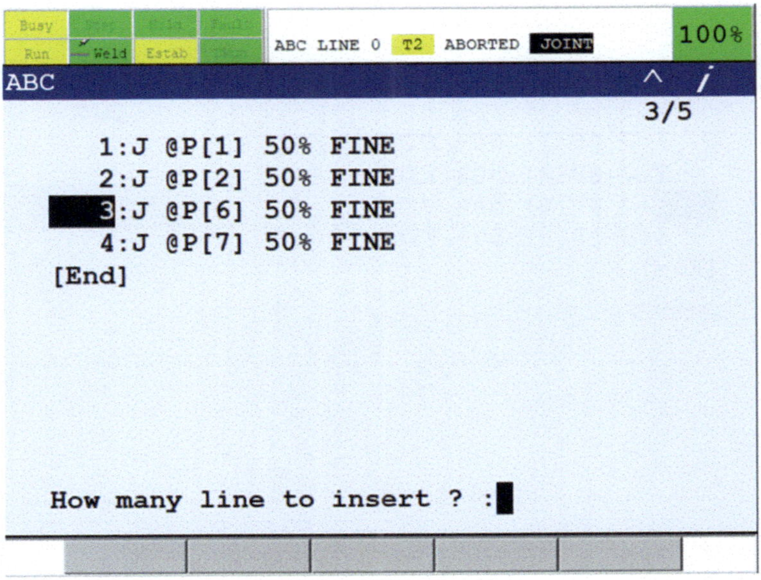

Fig. 7.81 Displaying "How many lines to inserted"

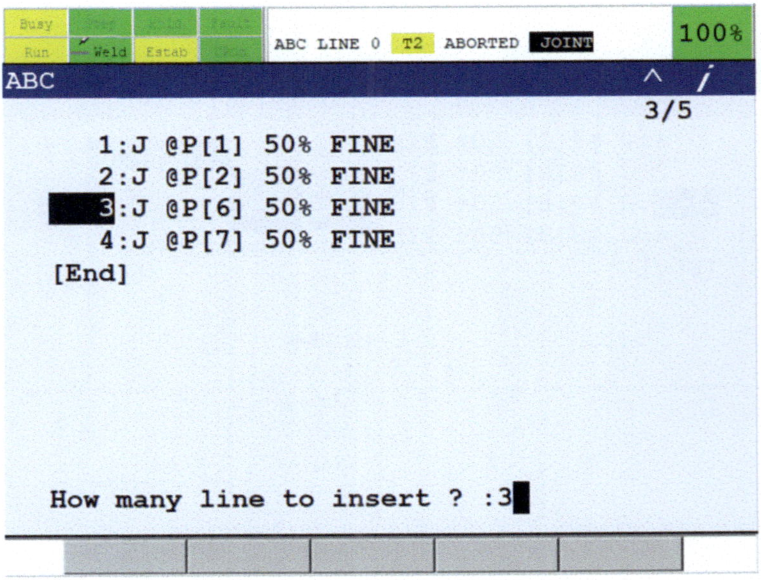

Fig. 7.82 Entering the number of lines

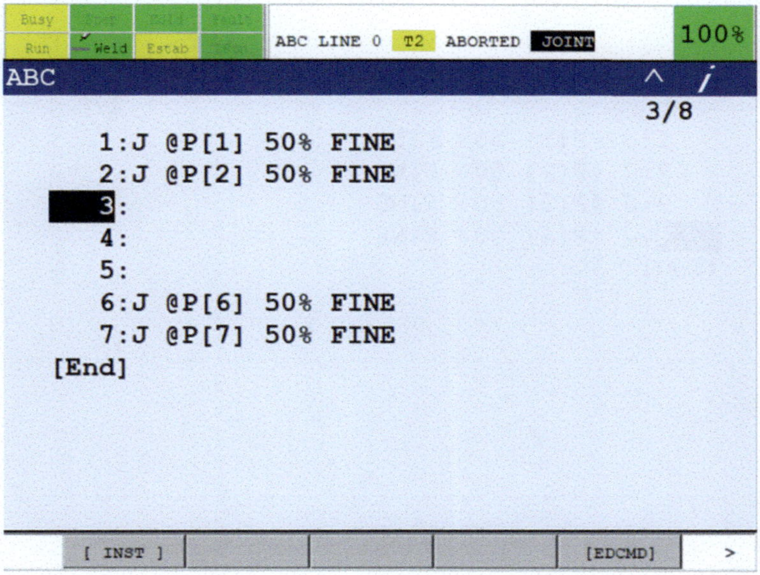

Fig. 7.83 Interface with inserted three blank lines displaying

(3) Copying, cutting and pasting of teaching-playback points

There are three ways to paste the teaching-playback point:

Logic data: no position information is pasted, only program instructions are pasted;
Location ID data: position information and position number are pasted;
Position data: position information is pasted, but position number is not.

The steps for copying/cutting and paste of teaching-playback points are as follows:

(a) Move the cursor to the line number preceding the instruction to be copied/cut, as shown in Fig. 7.84.
(b) Press NEXT to page through the function menu to display the EDCMD item in the function menu bar, as shown in Fig. 7.85.
(c) Press F5 to select the EDCMD function, and the EDCMD menu will pop up, as shown in Fig. 7.86.
(d) Select the Copy/cut menu item in the EDCMD menu, as shown in Fig. 7.87.
(e) Confirm by pressing ENTER, and the SELECT menu item appear in the function menu bar, as shown in Fig. 7.88. Press F2 to select the SELECT function.
(f) Select one or more lines by moving the cursor up or down; then COPYN or CUT button, and other menu items will be displayed in the function menu, as

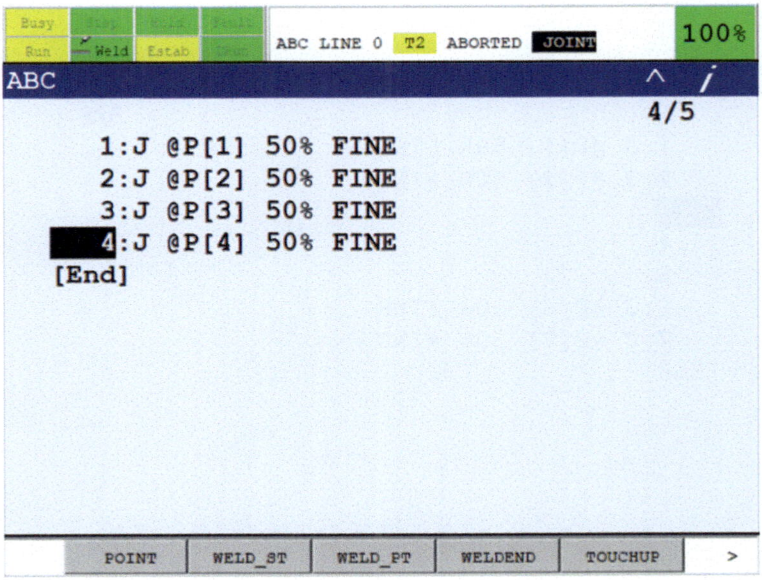

Fig. 7.84 Moving the cursor to the line number preceding the instructions to be copied/cut

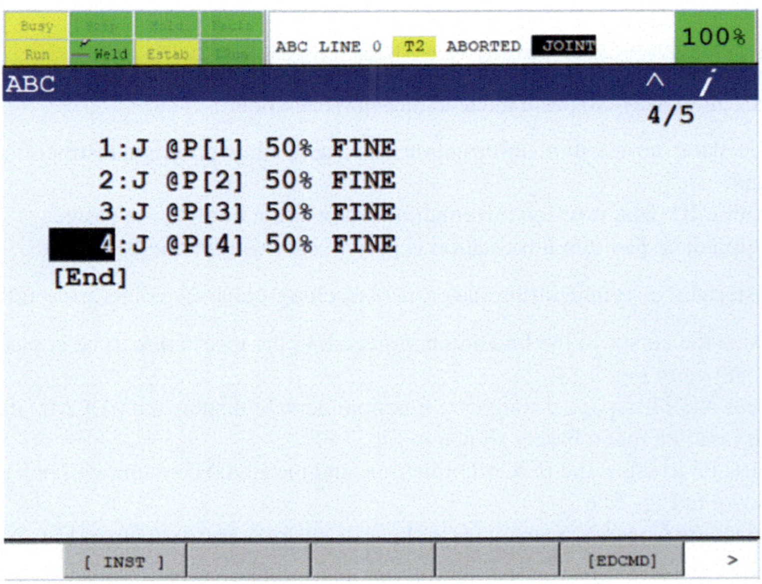

Fig. 7.85 Paging through to display EDCMD function button

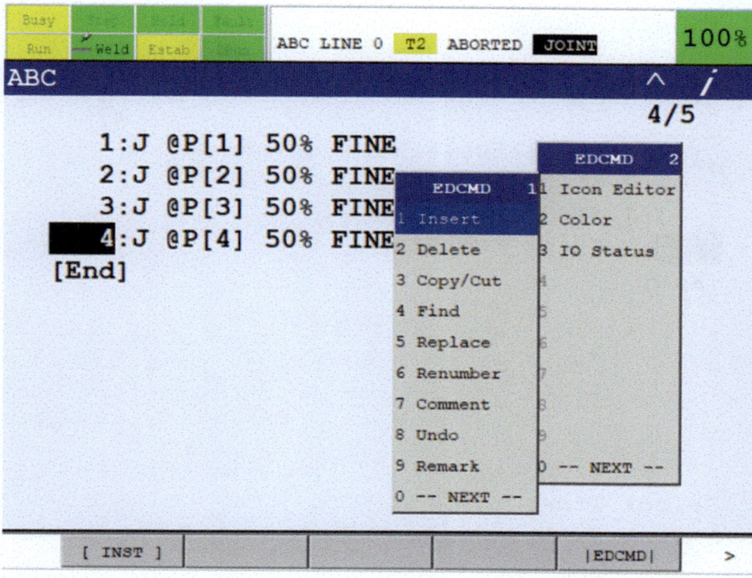

Fig. 7.86 Edit menu (2)

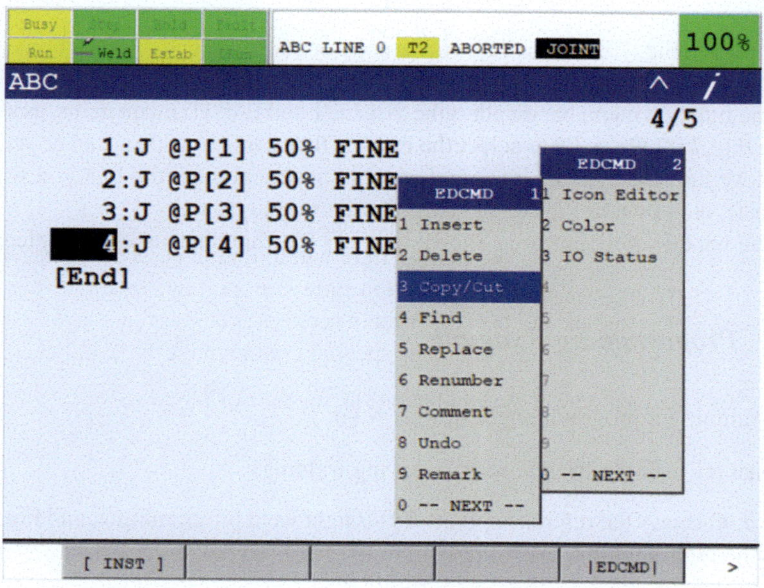

Fig. 7.87 Selecting copy/cut menu items from the EDCMD menu

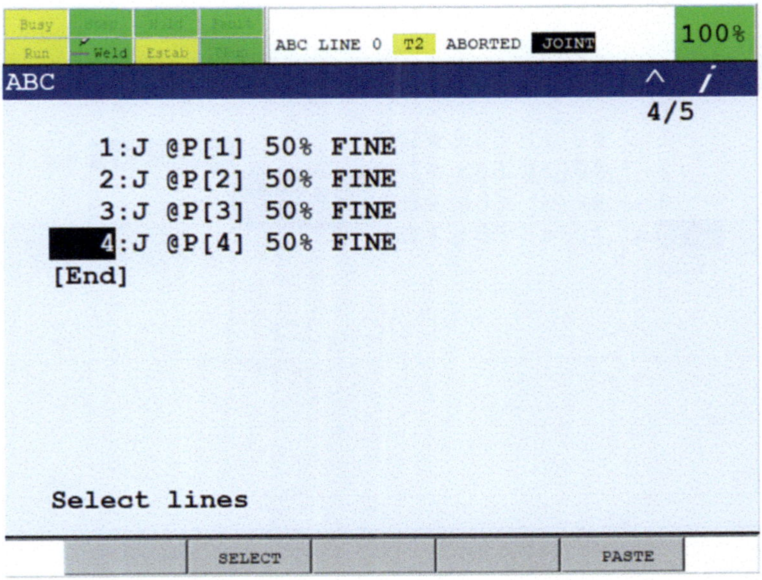

Fig. 7.88 Display SELECT menu items in the function menu

shown in Fig. 7.89. Press F2 or F3 to copy or cut the selected instructions to the paste board.

(g) The function menu bar displays the SELECT and PASTE menu items, as shown in Fig. 7.90. Press F5 to select the PASTE function.

(h) Move the cursor to the line number of the instruction next to that you want to paste, as shown in Fig. 7.91. Press F2, F3 or F4 to paste.

(i) The window will appear as shown in Fig. 7.92 when the paste is completed.

7.1.5 Programming Example

(1) Example for programming a straight weld

1) Teaching a single straight weld (weaving welding)

Figure 7.93 shows the reference points for straight weld programming, and Fig. 7.94 shows the programming.

2) Programming steps:

(a) Move the robot to the point 1, insert a motion instruction (refer to 7.1.4 Procedures for Creating Welding Robot Teaching-playback point), and set point 1 as the safety point 1 of the robot. The safety point shall be far away from the workpiece or fixture as much as possible and at a safer position so as to avoid affecting the installation and removal of workpiece.

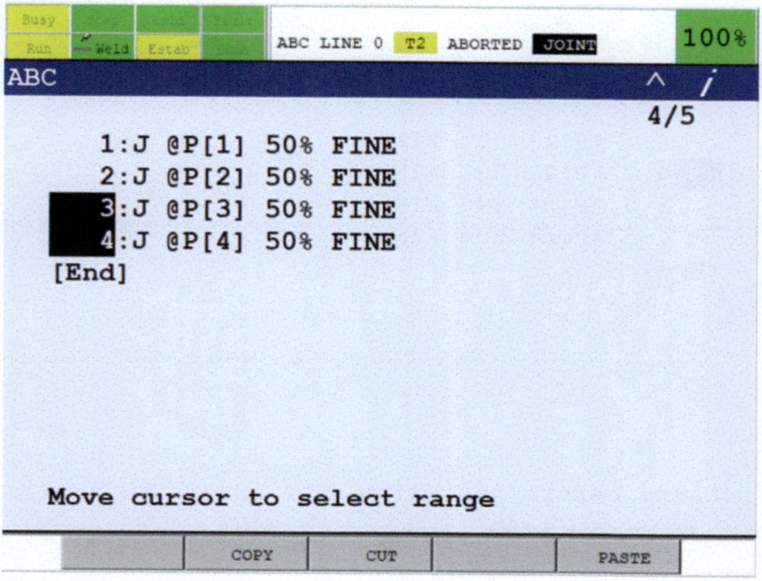

Fig. 7.89 Choose two lines

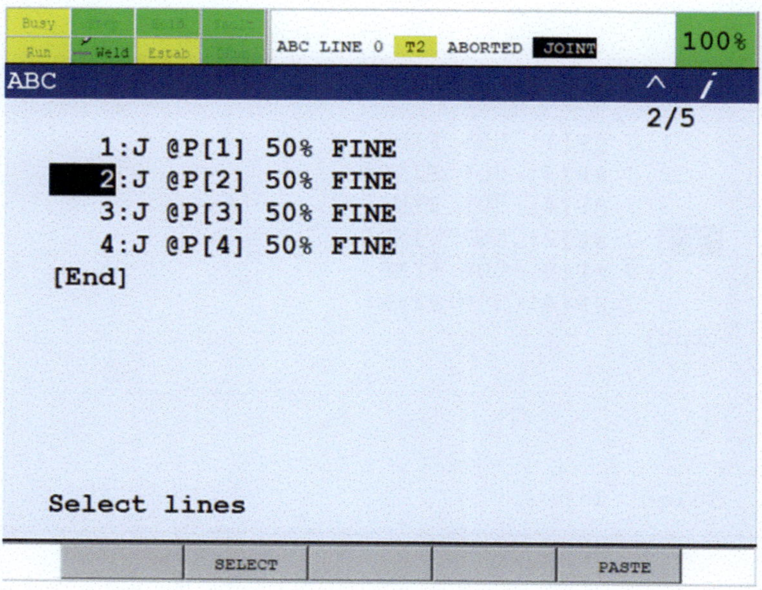

Fig. 7.90 SELECT and PASTE menu items appear in the function menu after copying to the paste board

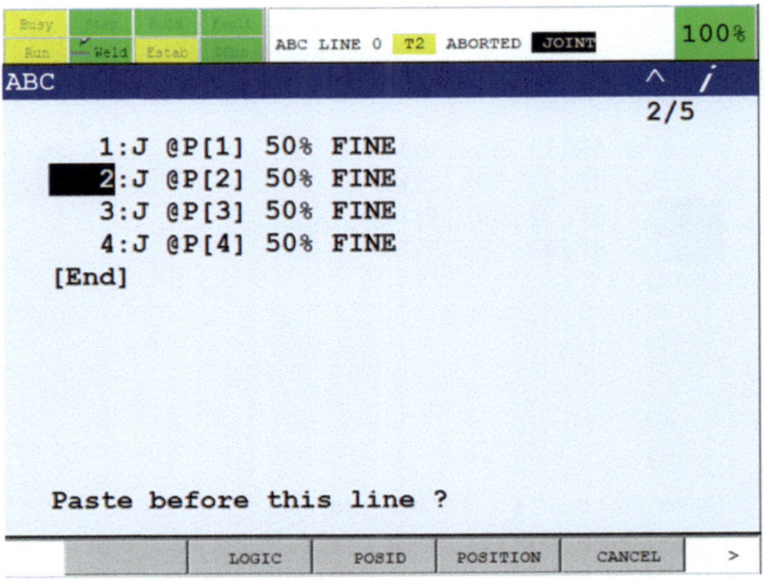

Fig. 7.91 Moving the cursor to the line number of the instruction next to that you want to paste

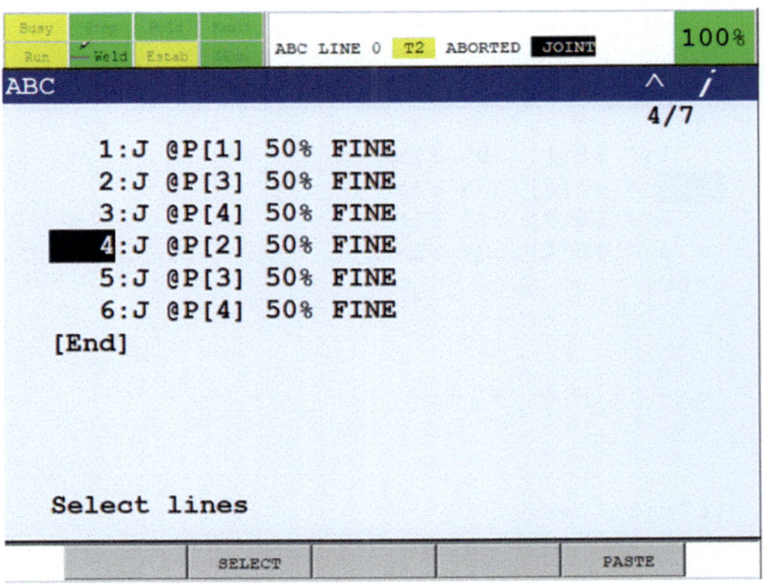

Fig. 7.92 Window display after completing paste

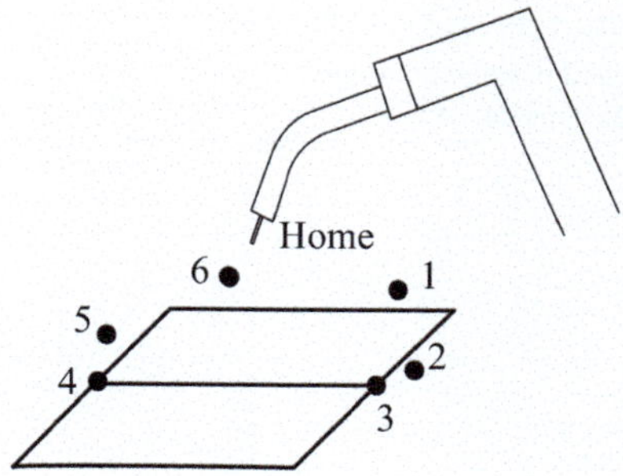

Fig. 7.93 Reference points used in a straight weld teaching

Fig. 7.94 Programming for a straight weld

```
 1:J @P[1] 50% CNT100
 2:J @P[2] 50% CNT100
 3:J @P[3] 50% FINE
 4:  Weld Start[1,1]
 5:  Weave Sine[1]
 6:L @P[4] 8mm/sec FINE
 7:  Weld End[1,1]
 8:  Weave End
 9:J @P[5] 50% CNT100
10:J @P[6] 50% CNT100
[End]
```

(b) Move the robot to the point 2 which is close to the arc-starting point, insert a motion instruction (refer to 7.1.4 Procedures for Creating Welding Robot Teaching-playback point), and set point 2 as the proximity arc-starting point. If there is an obstacle between the safety point 1 and arc-starting point, more proximate points shall be set between the two points to evade the obstacle.

(c) Move the robot to the point 3, insert a motion instruction (refer to 7.1.4 Procedures for Creating Welding Robot Teaching-playback point), and set point 3 as the arc-starting point. As shown in the 3rd instruction in Fig. 7.94, the termination type at the arc-starting point must be FINE to ensure that the robot reaches the arc-starting point accurately, and add an arc-starting (weld start) and a weave start instruction (Refer to Sect. 7.1.3).

(d) Move the robot to the point 4, insert a motion instruction (refer to Sect. 7.1.4), and set point 4 as the arc extinction point. As shown in the 6th instruction in Fig. 7.94, the motion type on the straight weld must be L to ensure that the robot runs in the same path as the weld; the termination type at arc extinction point must be FINE to ensure the robot reaches the arc extinction point accurately. Insert an arc-extinction (weld end) and a weave end instruction (Refer to Sect. 7.1.3).

(e) Move the robot to the point position 5, and insert a motion instruction (refer to Sect. 7.1.4). Set point 5 as the proximity arc extinction point to evade ovstacles.

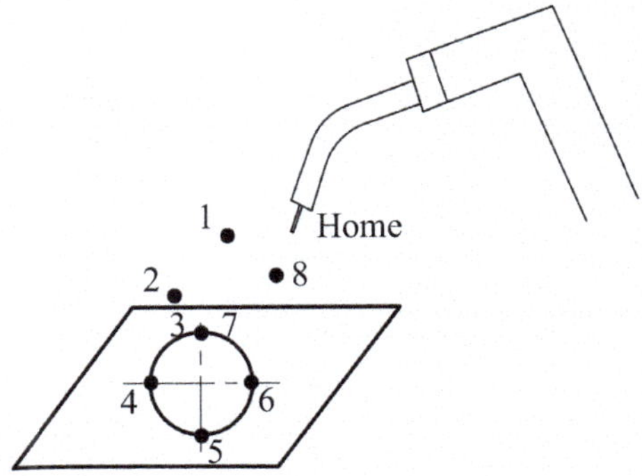

Fig. 7.95 Reference points used in the teaching of a circular-arc weld

(f) Move the robot to the point 6 and insert a motion instruction (refer to Sect. 7.1.4). Set point 6 as the safety point 2 of the robot. After the welding process is completed, move the robot to a safe position so as to avoid affecting the installation and removal of workpiece.

(2) Example for programming a circular-arc weld

1) Teaching a circular-arc weld

Figure 7.95 shows the reference points for programming a circular-arc welds, and Fig. 7.96 shows the programming.

2) Programming steps

(a) Move the robot to the point 1, insert a motion instruction (refer to Sect. 7.1.4), and set point 1 as the safety point 1 of the robot. The safety point shall be far away from the workpiece or fixture as much as possible and at a safe position so as not to affect the installation, disassembly, and loading and unloading of workpiece;

Fig. 7.96 Programming for a circular-arc weld

```
1:J @P[1] 50% CNT100
2:J @P[2] 50% CNT100
3:J @P[3] 50% FINE
4:  Weld Start[1,1]
5:C @P[4]
 :   P[5] 8mm/sec CNT100
6:C @P[6]
 :   P[7] 8mm/sec FINE
7:  Weld End[1,1]
8:J @P[8] 50% CNT100
9:J @P[1] 50% CNT100
```

(b) Move the robot to point 2 which is close to the arc-starting point, and insert a motion instruction (refer to Sect. 7.1.4). set point 2 as the proximity arc-starting point. If there is an obstacle between the safety point 1 and arc-starting point, more proximate points shall be set between the two points to evade the obstacle.
(c) Move the robot to the point 3, insert a motion instruction (refer to Sect. 7.1.4), and set point 3 as the arc-starting point. As shown in the third instruction in Fig. 7.96, the end type at the arc-starting point must be FINE to ensure that the robot reaches the arc-starting point accurately. Add arc-starting instruction (refer to Sect. 7.1.4).
(d) Move the robot to the point 4, insert a circular motion instruction, set the point 4 as the midpoint of the first arc segment. Move the robot to the point 5, insert a circular motion instruction, set the point 5 as endpoint of the first arc segment. The arc-starting point (point 3) is the starting point of the first arc segment. The termination type must be CNT to ensure smooth transition of the robot.
(e) Move the robot to the point 6, insert a circular motion instruction, set the point 6 as the midpoint of the second arc segment. Move the robot to the point 7, insert a circular motion instruction, set the point 7 as endpoint of the first arc segment. The endpoint of the first segment (point 5) is the starting point of the second arc segment. The end point of second are segment is also the arc extinction point. So an arc extinction (weld end) instruction shall be added. The termination type of the extinction point must be FINE.
(f) Move the robot to close to the point 8, insert a motion instruction (refer to Sect. 7.1.4), and set point 8 as the proximity arc extinction point to evade ovstacles.
(g) Move the robot to safety point 1 and insert a motion instruction (refer to Sect. 7.1.4). Set the safety point 1 of the robot.

7.1.6 Program Execution Mode

After the program is prepared and confirmed, the program can be executed manually or automatically.

(1) Manual execution mode

Procedure for manual execution:

(a) Hold the TP, turn the TP switch to "ON".
(b) Switch the program execution mode to "continuous". Press the STEP key, "one-step" LED on the TP changes from yellow to green, indicating that continuous mode is on.
(c) Press the multiplication key to set the speed multiplication to 100%.
(d) Set the welding status to "on". When the SHIFT key and the WELD ENBL key are pressed at the same time, the "WELD" LED on the TP changes from yellow to green, indicating that the welding status is switched on.

(e) Press the SELECT key to display the main interface of the program. Select the program to be executed and press ENTER to display the program. Move the cursor to the leftmost end of the first line of the program. Gently press the safety switch on the back side, and press the SHIFT and the FWD keys at the same time to execute the program.

(3) Automatic execution mode

Procedures for automatic execution of program:

(a) At the interface of the program to be executed, move the cursor to the leftmost end of the first line of the program, then perform the (b), (c) and (d) steps in manual execution mode.
(b) Turn the TP switch on the TP to OFF and the mode selecting switch on the operation panel of the control cabinet to AUTO, as shown in Fig. 7.97.
(c) Press the RESET key to cancel the TP alarm and press the Auto button to automatically execute the program.
(d) If the program is not executed automatically, the TP screen displays as shown in Fig. 7.98. Select Yes, press ENTER key, and then press the Auto button again to execute the program automatically.

7.2 Offline programming technology for robots

Off-line programming for robots means to establish geometric model of robots and their working environment by using computer graphics. Then, some planning algorithms are used to plan the paths off-line by controlling and operating the graphics. Then, 3D graphical animation simulation of the planned path is carried out to check the correctness of programming. Finally, the generated code is transferred to the robot controller to control the robot movement and accomplish the given task. The off-line

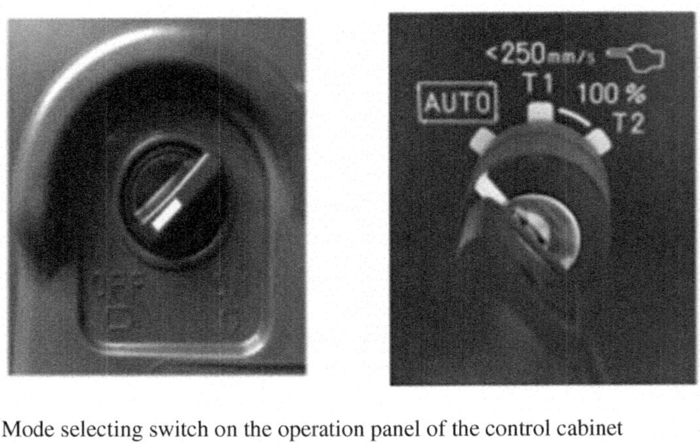

Fig. 7.97 Mode selecting switch on the operation panel of the control cabinet

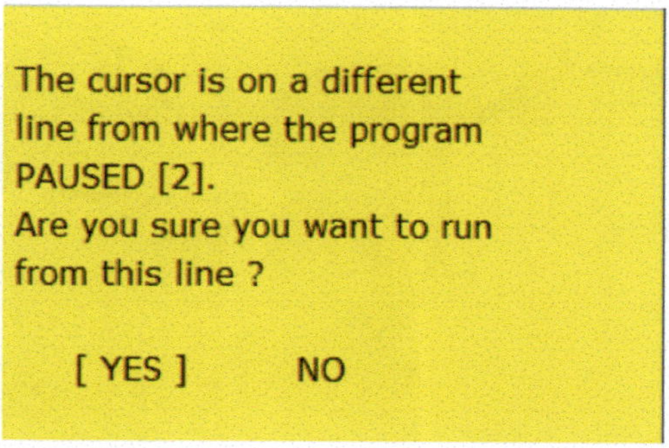

Fig. 7.98 Prompt on the TP screen when automatic program execution is not available

programming process is shown in Fig. 7.99. The off-line programming software interface of robot is as shown in Fig. 7.100.

In addition to basic programming and simulation, off-line programming allows optimization of welding sequences, verification of accessibility (Fig. 7.101).

7.2.1 Features of Off-line programming

The off-line programming system has a vast mode database. For example, its external axis system covers all commonly-used welding positioners. The off-line programming system also provides an interface for users to develop other functional modules.

The off-line programming system offers the following advantages over teach programming:

(a) The robots can still work on the production line when programming the next task, shortening the downtime and increasing productivity;
(b) The programmer stay away from dangerous working environment when programming
(c) High flexibility, allowing for programming most of the available robots and easy to implement program optimization
(d) Easy to integrate with CAD/CAM system to achieve CAD/CAM/robotic integrated welding
(e) High-level computer programming language can be used for programming complex tasks;
(f) Fast programming for small-lot and multi-variety products;
(g) The quality of teach programming depends on the skills and experience of the operator, while off-line programming provides the best path using planning algorithms;
(h) It is difficult to program complex path paths for teach programming, but easy to do so for off-line programming.

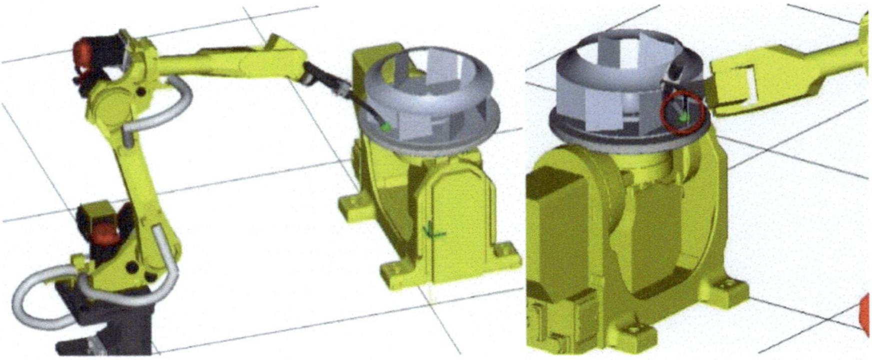

(a) Overall Equipment scheme (b) Designating Weld Path

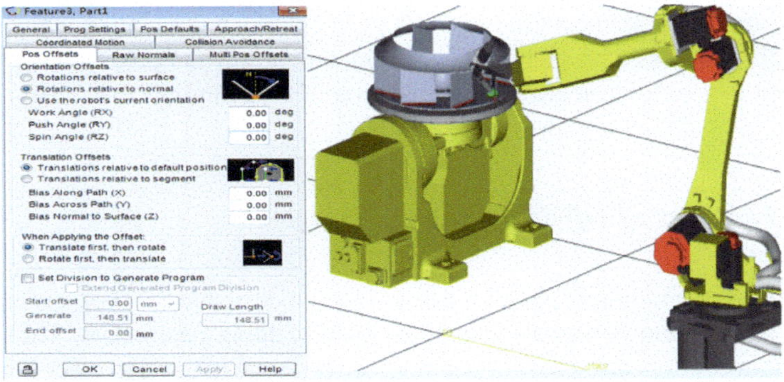

(c) Automatic Program Generation (d) Simulation and Certification of Robotic Paths

Fig. 7.99 Off-line programming process

7.2.2 Components of the Off-Line Programming System

The off-line robot programming system is required not only to create the physical model of the robot system on the computer, but also carry out programming and animation simulation of the robot system as well as post-processing of programming results. Therefore, the off-line programming system of robots mainly includes the following modules: CAD modeling, graphical simulation, sensor and post-processing, etc.

(1) CAD modelling

CAD modeling includes part modeling, equipment modeling, system design and arrangement, and graphical processing of geometric models. As there is an error between the robot models built using the existing CAD data and the theoretical structure parameters of robot and the actual model, the robot must be calibrated and its error measured, analyzed and continuously corrected to perfect the model, as shown in Fig. 7.102.

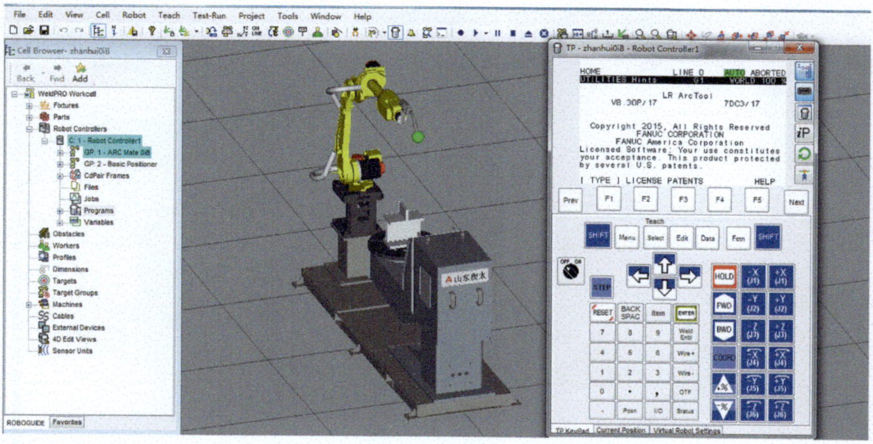

Fig. 7.100 Off-line programming software interface

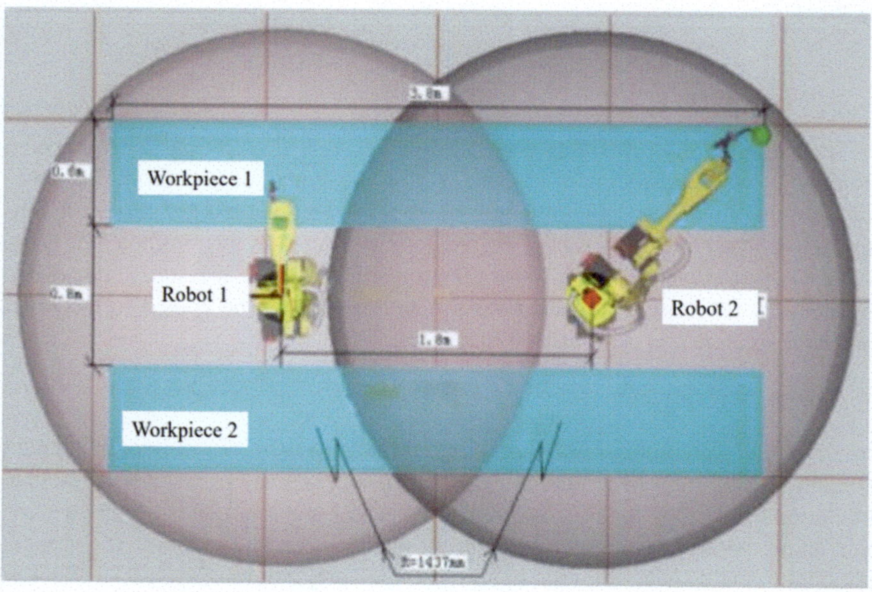

Fig. 7.101 Verification of accessibility of work scope

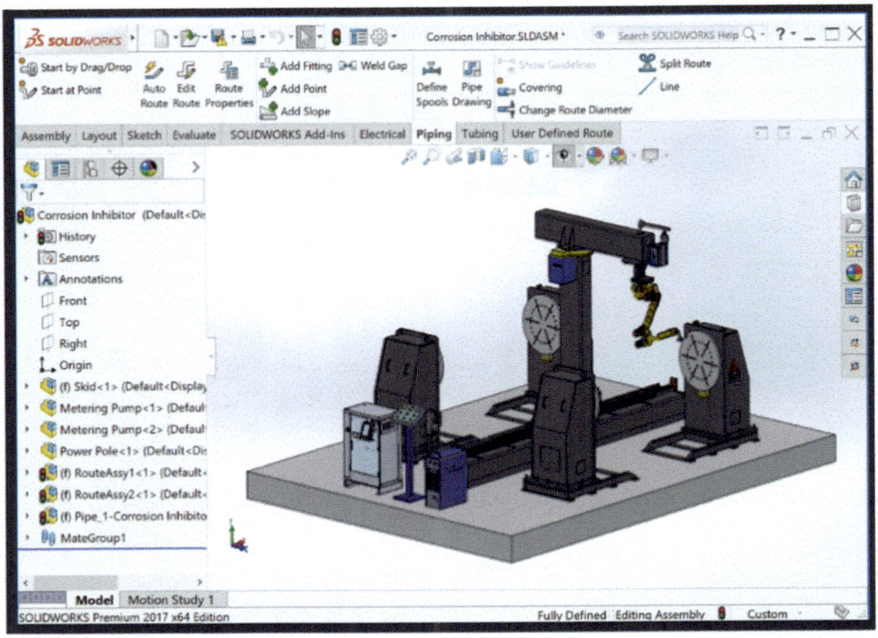

Fig. 7.102 CAD modeling

(2) Off-line programming

Off-line programming modules generally include task description for robots and equipment (including setting of path points), establishment of transformation equations, solving kinematic equations (generally inverse problems) and task programming. Graphical simulation is performed after the programming is completed. Then, the program is corrected according to the dynamic simulation results. Only when the accuracy reaches the expected value, can the program be copied to the robot controller, as shown in Fig. 7.103.

(3) Graphical simulation

An important function of off-line programming system is off-line debugging program, which allows for simulating the operation process of robot with graphic simulation technology without contacting the actual robot and its working environment and provides a virtual environment for interaction with the robot. The graphic simulation module is an important part of off-line programming system of robot. It displays the simulation results in the form of visible 3D graphics to visualize the motion of robot, thereby obtaining important information that is difficult to analyze from data curve or data. The correctness and accuracy of off-line programming are verified through this module. Figure 7.104 shows the graphic simulation.

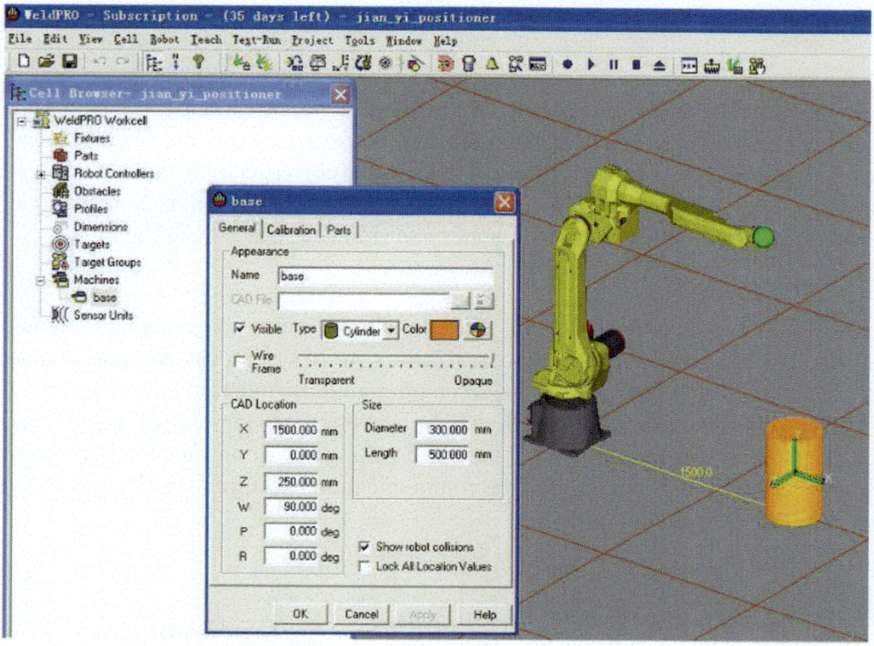

Fig. 7.103 Off-line programming

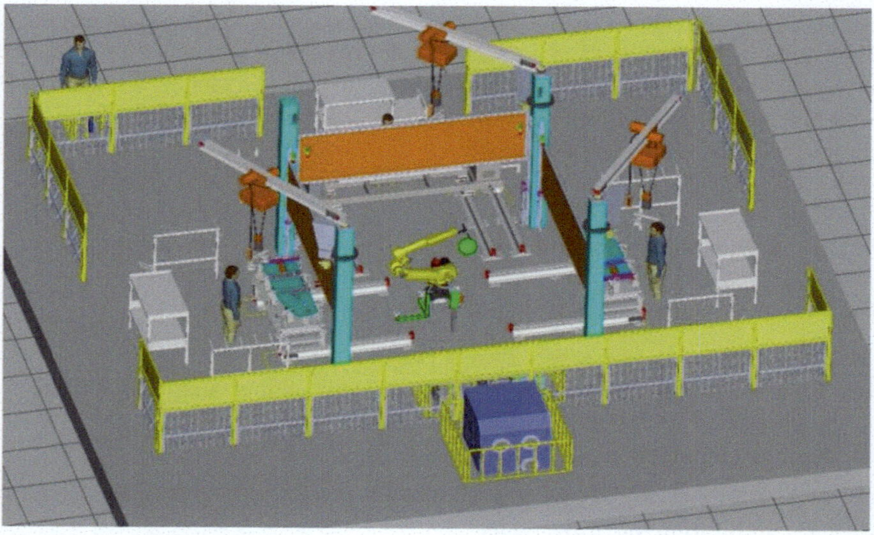

Fig. 7.104 Graphical simulation

(4) Sensors

The robots commonly adopt sensors for guidance and control to improve system stability and reliability. For a sensor-driven robotic system, the sensor-based motion is unpredictable because the signals generated by the sensor are interfered by many factors (such as light condition, physical reflectance, object geometry and unbalanced motion process). Therefore, the off-line programming system shall be able to model sensors, generate sensor control strategies and simulate sensor-based tasks.

(5) Post-treatment

The primary task of post-processing is to compile the source program that is created off-line into the target program that the robotic control system can identify. That is, when the simulation result of the operation program fully meets the requirements of the operation, the operation program shall be converted into the control program and data for the target robot, which will then be downloaded and installed to the control cabinet of the target robot through the communication interface to drive the robot to accomplish the specified tasks.

7.2.3 Off-Line Programming Simulation Software and Its Usage

The simulation software of FANUC robot Roboguide V7.7 is taken as an example. Roboguide is the core application software and is used in conjunction with other modules such as handling, arc welding, painting and spot welding. The simulation environment interface of ROBOGUIDE is a traditional Windows interface, consisting of menu bar, toolbar, and status bar, etc.

7.2.3.1 Overview of Simulation Software

The Roboguide software has the following features:

(a) The interface is concise, as shown in Fig. 7.105;
(b) There is list of rich resources, including various welding equipment resources, as shown in Fig. 7.106;
(c) Simulation of motion paths, as shown in Fig. 7.107;

7.2.3.2 Simulation Software Usage

(1) Software installation

To install Roboguide V7.7, firstly click open…\RoboguideV7.7\setup.exe, as shown in Fig. 7.108. If the linkage function of the positioner is required, the MultiRobot Arc Package shall also be installed.

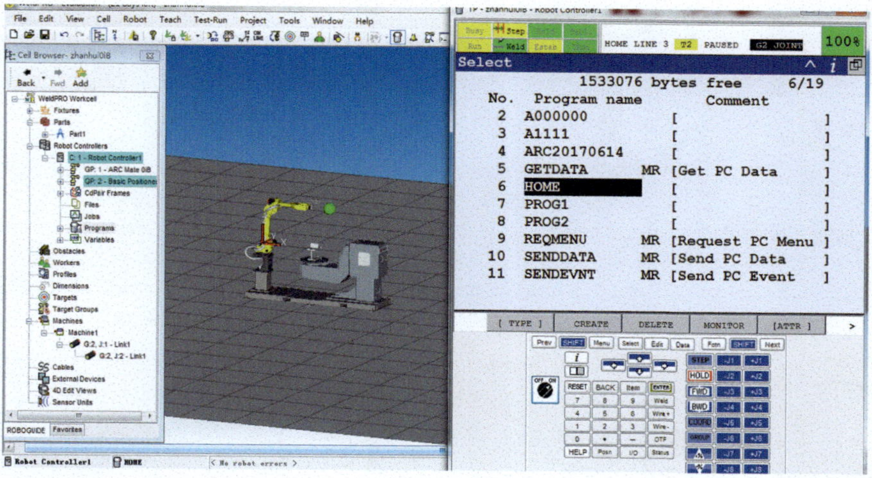

Fig. 7.105 Simulation software interface

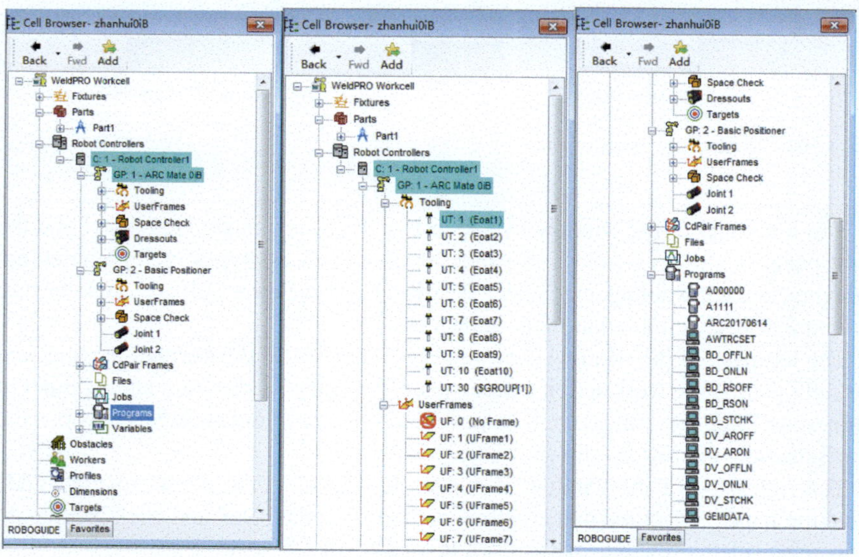

(a) Control system (b) Coordinate (c) Motion procedure

Fig. 7.106 List of resources

(2) Steps to create a new Workcell

After opening Roboguide, the window shown in Fig. 7.109 will be opened. Click the file button "File" on the toolbar and click the menu item "New Cell" to open the next window, as shown in Fig. 7.110.

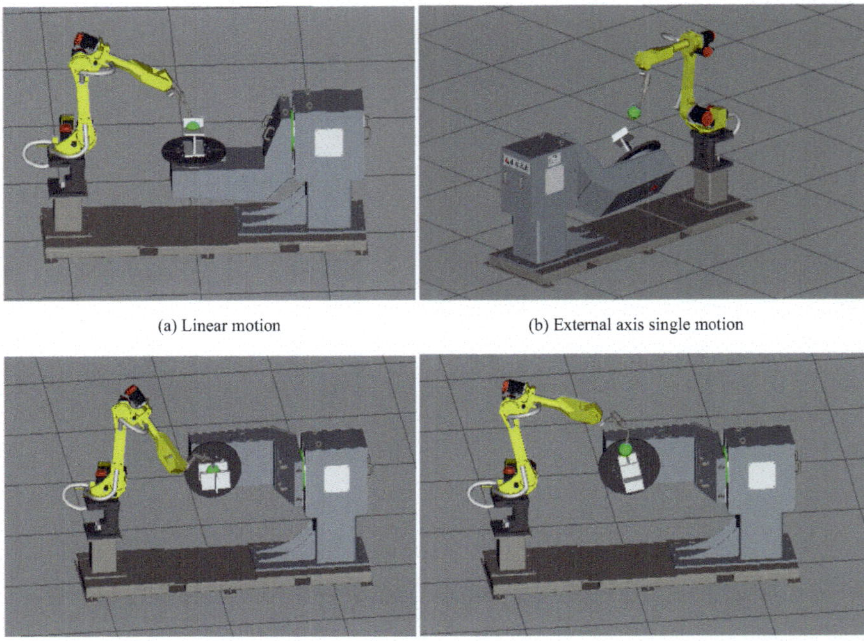

(a) Linear motion (b) External axis single motion

(c) Welding positoner motion (d) Linkage motion of the postioner

Fig. 7.107 Simulation of motion path

The left panel of the window shown in Fig. 7.110 is a navigation panel and the right panel is a process selection panel. Optional process types include handling, spraying, and arc welding, etc. Select WeldPRO-Arx welding Cell and click Next to go to the next selection window, as shown in Fig. 7.111.

The left pane of the window shown in Fig. 7.111 is still the navigation pane, and the right pane is the workstation naming pane. Enter the workstation name (either in Chinese or English) or the default name in the Name box, and click Next to open the next selection window, as shown in Fig. 7.112.

The right side of the window shown in Fig. 7.112 is the robot creation mode selection pane., Select the first creation method from the pane, and click Next to go to the next selection window, as shown in Fig. 7.113.

The right side of the window shown in Fig. 7.113 shows the pane for selecting the software version. Select the software version installed on the robot from the right pane (the higher the version, the more the functions) and click Next to enter the next selection window, as shown in Fig. 7.114.

The right side of the window shown in Fig. 7.114 shows the pane for selecting the toolkit. Options mainly include spot welding tool, arc welding tool, and handling tool, etc. Select appropriate tools according to the needs of off-line programming simulation and click Next to enter the next selection window, as shown in Fig. 7.115.

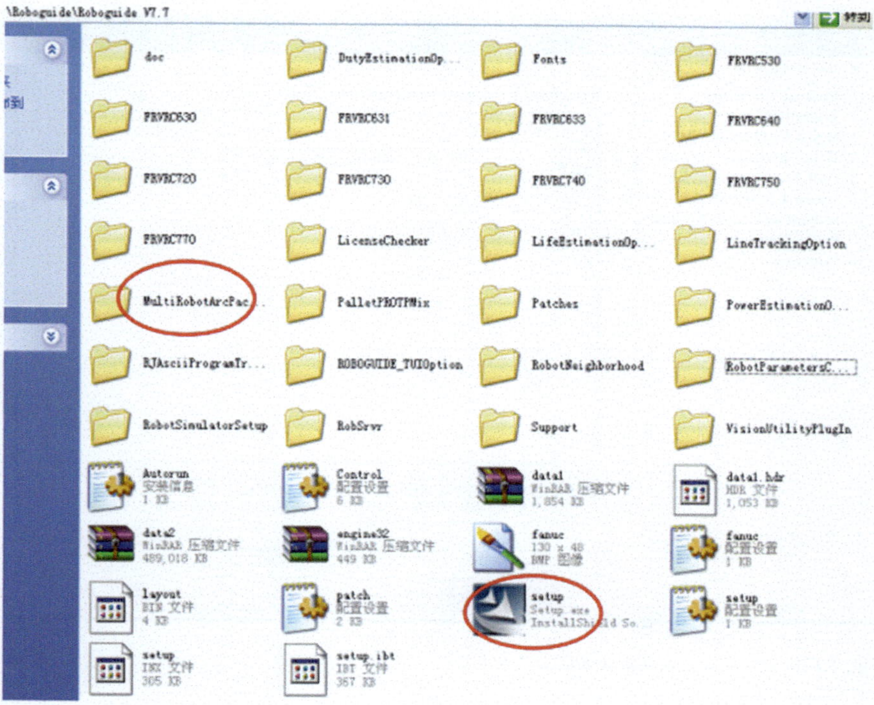

Fig. 7.108 Software installation

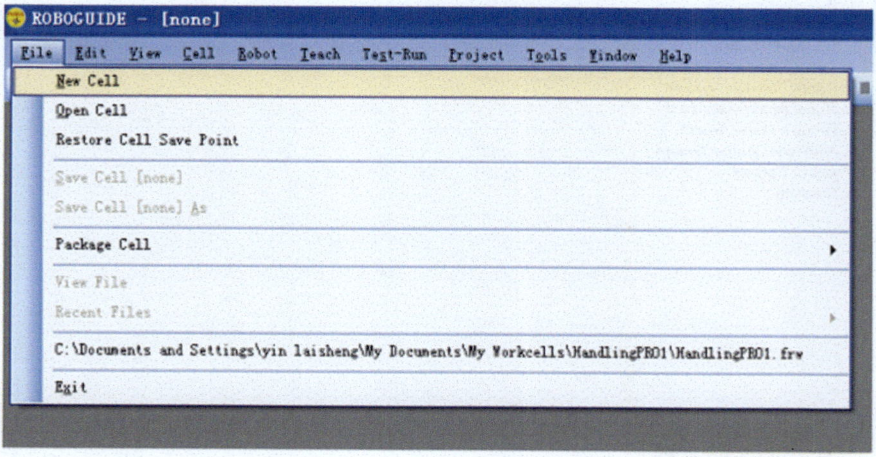

Fig. 7.109 File menu bar

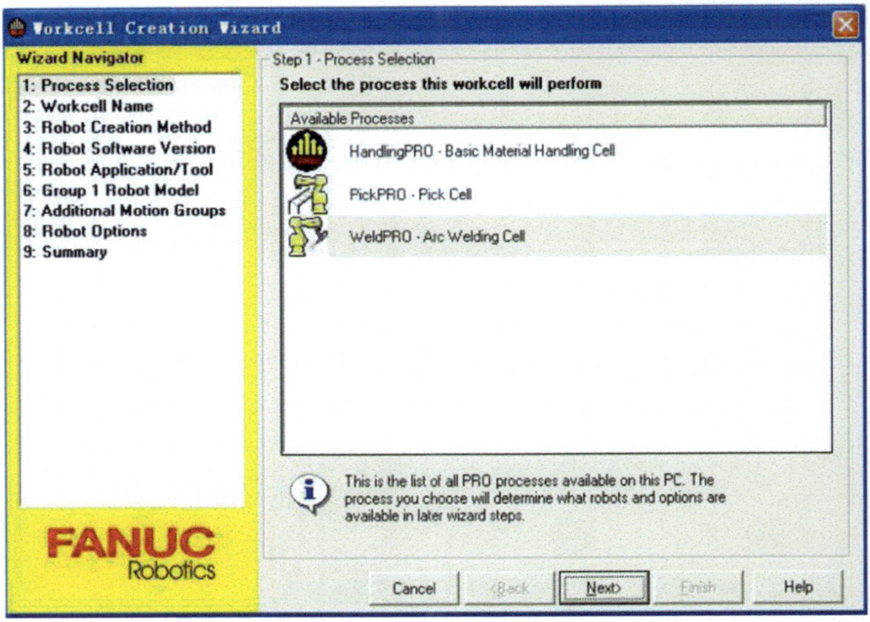

Fig. 7.110 Process selection window

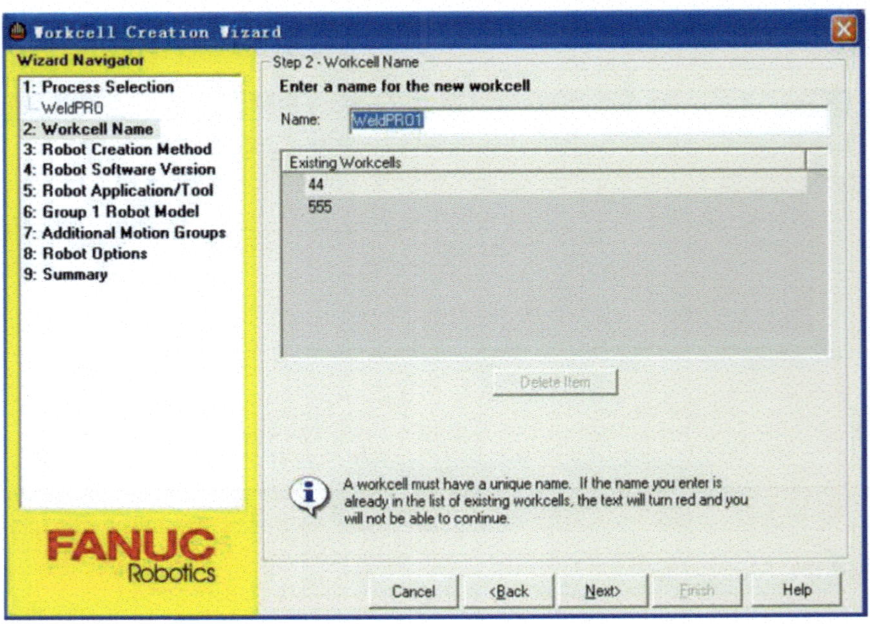

Fig. 7.111 Workstation naming window

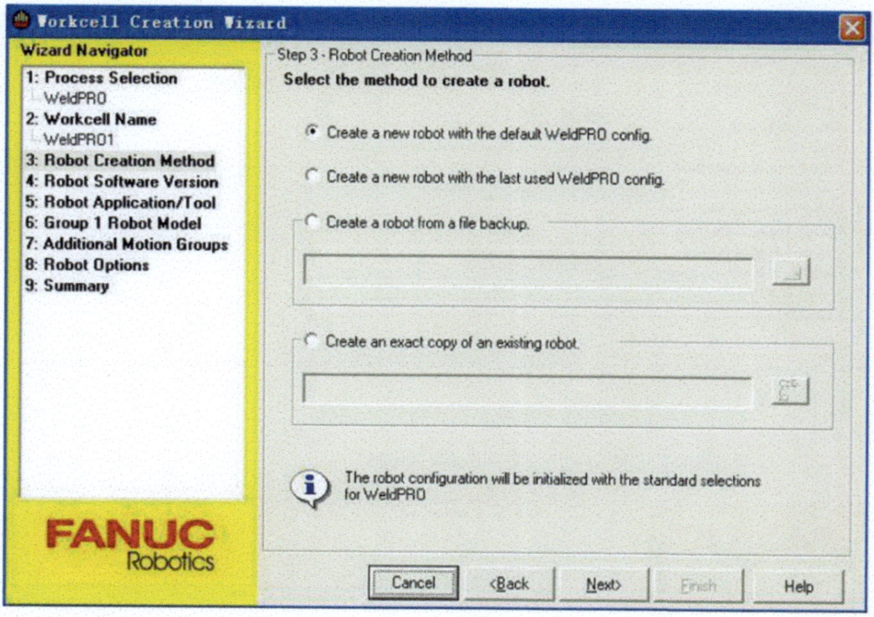

Fig. 7.112 Robot creation mode window

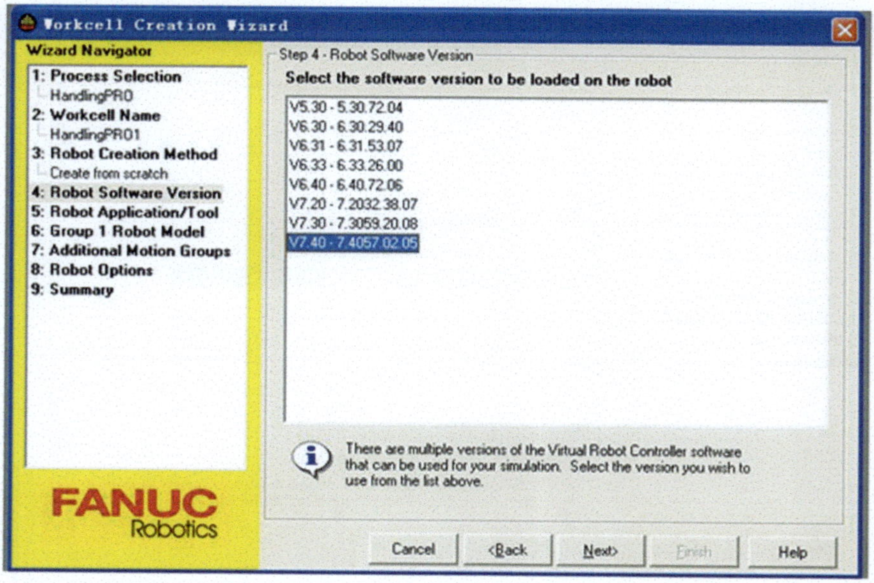

Fig. 7.113 Software version selection window

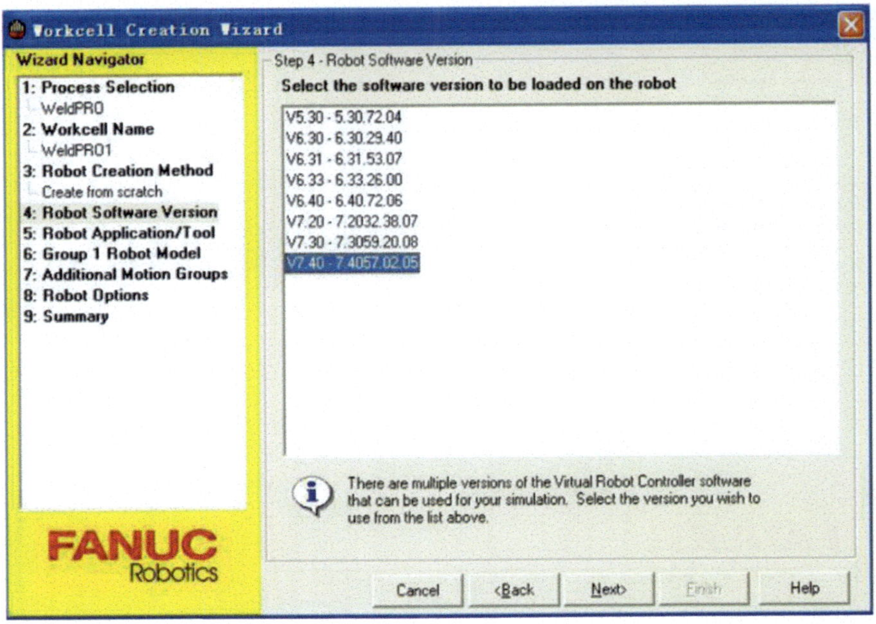

Fig. 7.114 Toolkit selection window

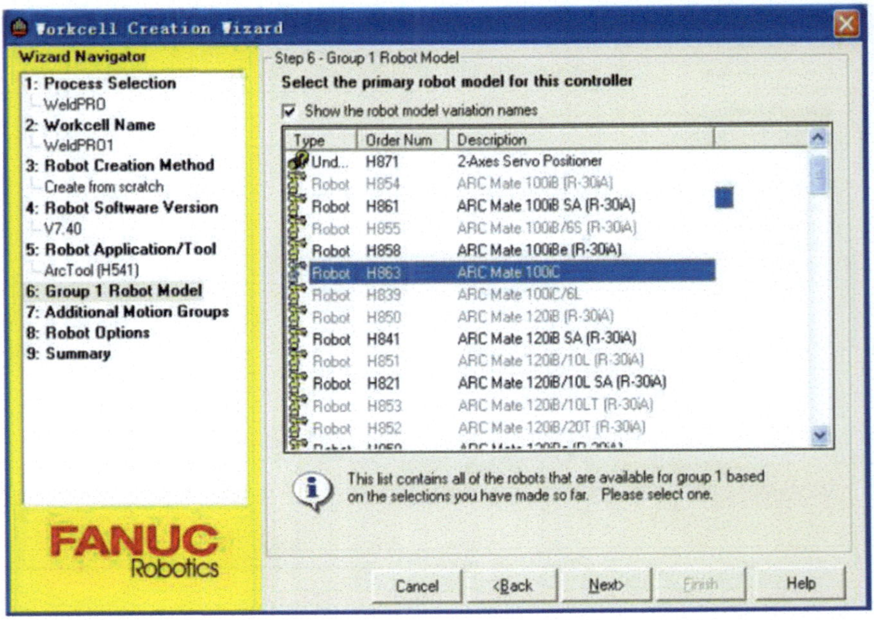

Fig. 7.115 Robot model selection window

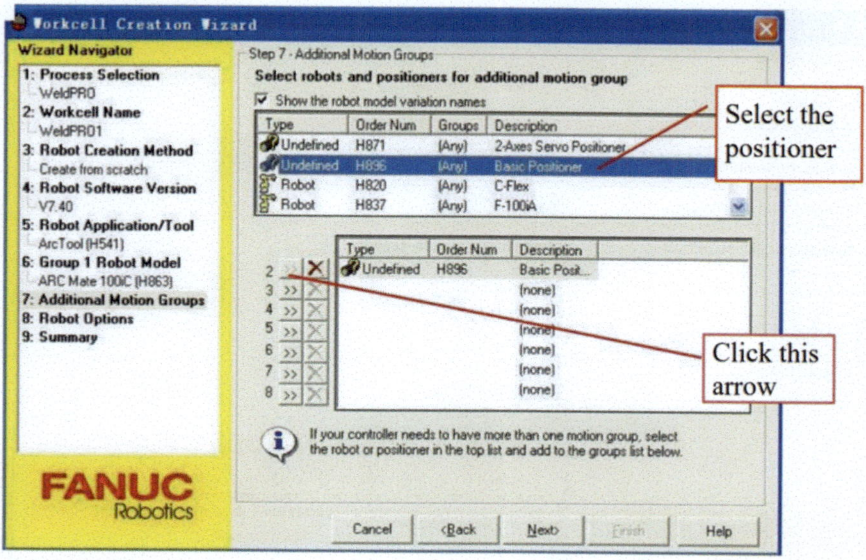

Fig. 7.116 Window for selecting robots and positioners of other motion groups

Select the robot model on the right side of the window, as shown in Fig. 7.115. The pane contains almost all the robot models. Select the robot model from the pane and click Next to go to the next selection window, as shown in Fig. 7.116.

The right side of the window shown in Fig. 7.116 is the pane for selecting the robots and positioners of other motion groups. When several robots need to be added, click "continue to add", and then click Next to enter the next selection window, as shown in Fig. 7.117.

The right side of the window shown in Fig. 7.117 is the selection pane for the optional robot software package. The window lists the commonly used workpiece software, such as 2D and 3D visual applications and additional axes. In addition, there is a "Languages" tab to set the language environment (The default language is English). After selecting the target software, click Next to open the next selection window, as shown in Fig. 7.118.

The right side of the window shown in Fig. 7.118 is the "check and confirm" pane. This pane lists all the options previously selected. If you are sure that all your previous selections are correct, click Finish to confirm. If you need to modify it, click Back to return to the previous step to modify it. Click Finish to set up the working environment and enter the simulation environment, as shown in Fig. 7.119.

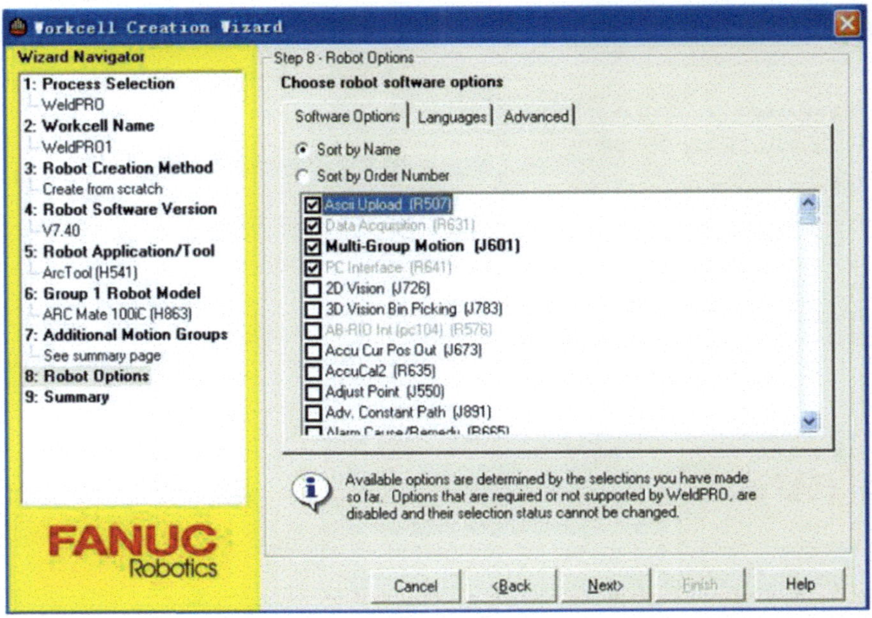

Fig. 7.117 Optional robot software package selection window

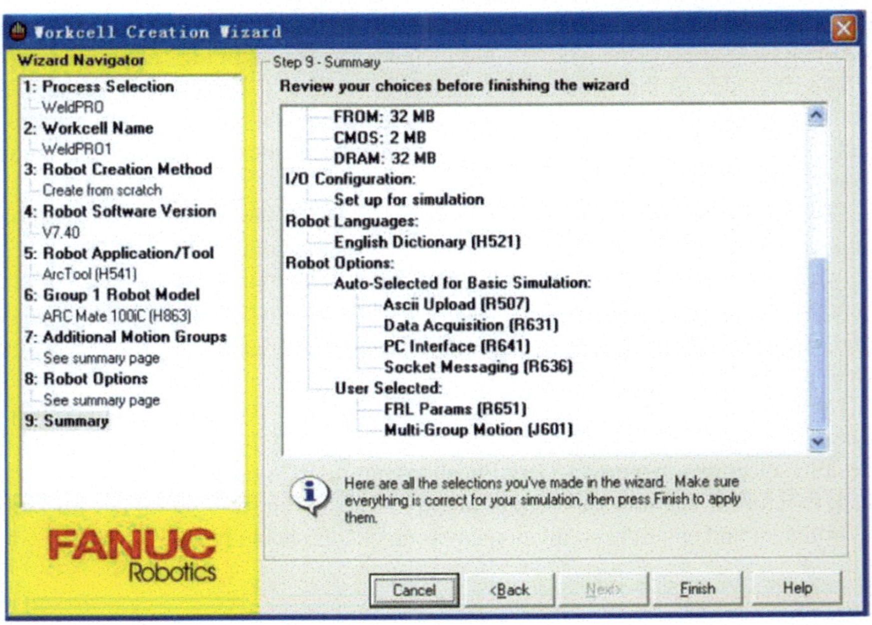

Fig. 7.118 Check and confirm window

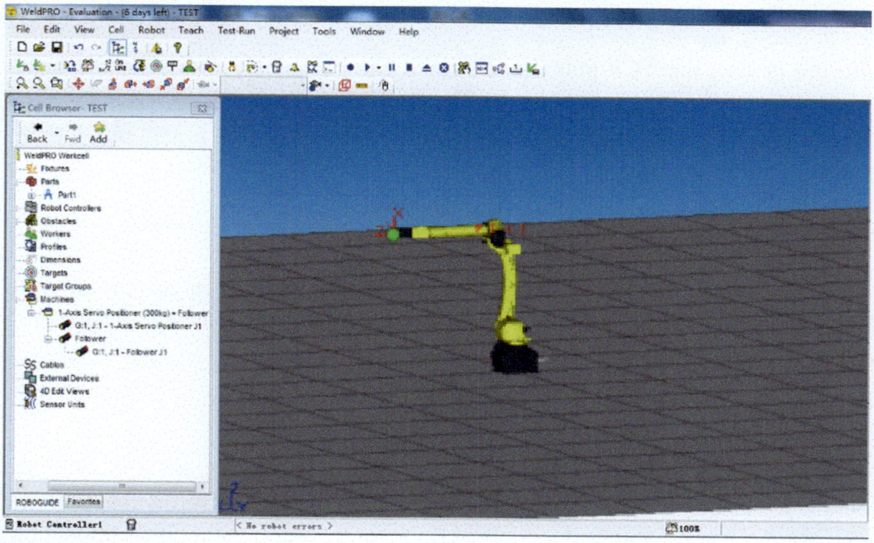

Fig. 7.119 Simulation environment

(3) Basic Operations

1) Operation on the simulation interface:

Move, rotate, zoom in and out the simulation interface with the mouse.
Move: Press the middle button and drag it.
Rotation: Press the right button and drag it.
Zoom in and out: Press the left and right keys simultaneously and move back and forth. Alternatively, directly roll the rollers;

2) Operations to change the position of a model:

Move: Place the mouse arrow on a green axis; when the arrow is in hand shape and axis number *X*, *Y* or *Z* appears, press the left button and drag, and the model will move in the direction of this axis. Alternatively, place the mouse on the coordinate, press the Ctrl key on the keyboard, and press and drag the left mouse button; in this way, the model will move in any direction.

Rotate: Press the Shift key on the keyboard, place the mouse on a coordinate axis, hold and drag the left button, and the model will rotate along this axis.

3) Operation of robot motion:

With the mouse, the TCP point of the robot can be quickly moved to the target surface, edge, point or center of circle. The method is as follows:

Move to the surface: Ctrl + Shift + left key;
Move to the edge: Ctrl + Alt + left key;
Move to the peak point: Ctrl + Alt + Shift + left key;
Move to the center: Alt + Ctrl + left key.

The TCP of the robot can also be dragged directly with the mouse to move the TCP point of the robot to the target position in the same way as changing the model position.

(4) Addition of equipment

1) Import of 3D models

Roboguide can be loaded with various entity objects, which can be divided into two parts, one is the self-contained models of Roboguide, and the other is the model file in igs or iges format that can be exported from other 3D software. The specific steps are as follows:

Click Cell-Add Fixture-CAD Library in the menu bar to display the dialog box shown in Fig. 7.120 for loading the self-contained library model files of ROBOGUIDE, including various welding torches, machining centers, and injection molding machines, etc. Alternatively, click the Cell-Add Fixture-Single CAD File on the menu bar to display the file browse dialog box for loading 3D models in igs format exported from other 3D software such as Solidworks, CATIA, and UG, etc.

2) Add the torch and set the TCP

After the creation of a new work cell is completed, the Cell Browse window will be automatically generated, as shown in Fig. 7.121. Sequentially click Robot Controllers-C: 1-Robot Controller 1-GP: 1-M-10iA/12 in the navigation bar until Tooling-UT: 1 (Eoat1) appears. Then, right click Tooling-UT: 1 (Eoat1), select Add Link-CAD Library from the window, as shown in Fig. 7.122, and enter the welding torch model database (Fig. 7.123).

Sequentially click the expansion button on the left of Library-EOATs-weld_torches in the Image Librarian window as shown in Fig. 7.123 and select the appropriate welding torch from the right pane.

After selecting the welding torch, click the OK button in the lower right corner. The welding torch will appear on the sixth axis of the robot, and the Link1, UT1 (Eoats) dialog box will pop up (can also be accessed by double-clicking Link1 in Cell Browser), as shown in Fig. 7.124.

Close the Link1, UT: 1 (Eoats) dialog box and double-click Cell Browser-UT: 1 (Eoats), and the dialog box in Fig. 7.125 pops up. Click UTOOL and check Edit UTOOL button to manually adjust the position data (*X/Y/Z*) and angle value (*W/P/R*)

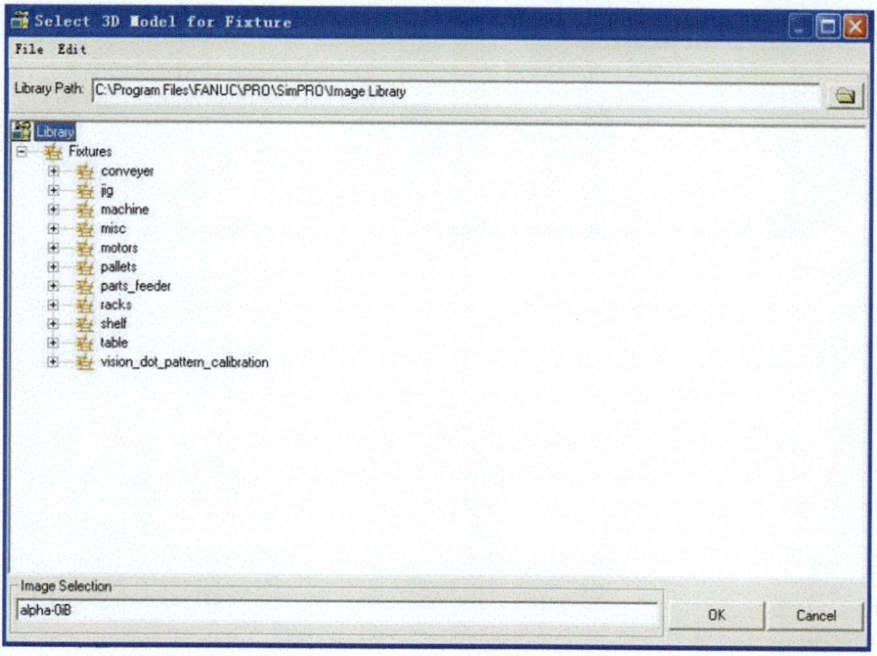

Fig. 7.120 Three-dimensional model import dialog box

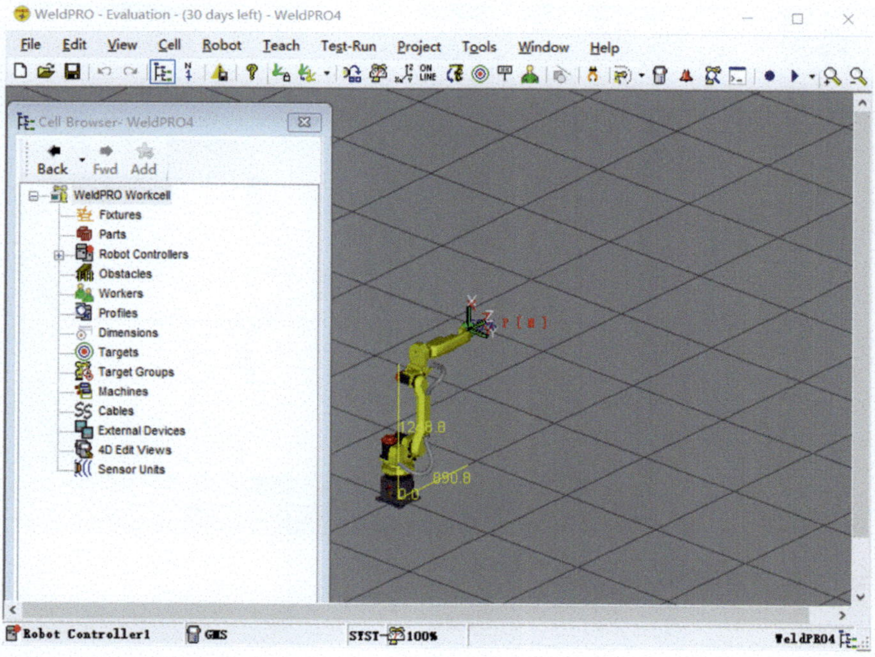

Fig. 7.121 Cell browser window (1)

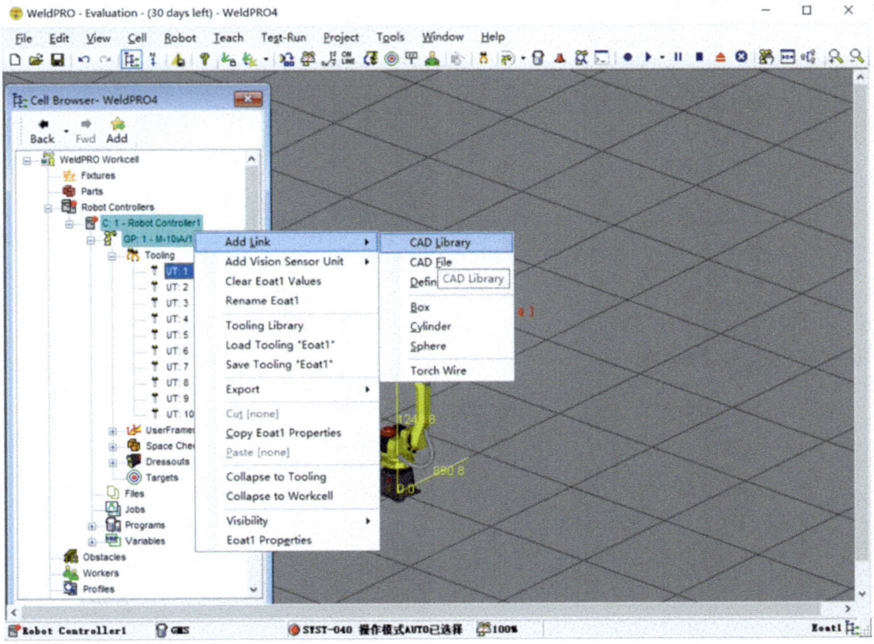

Fig. 7.122 Cell browser window (2)

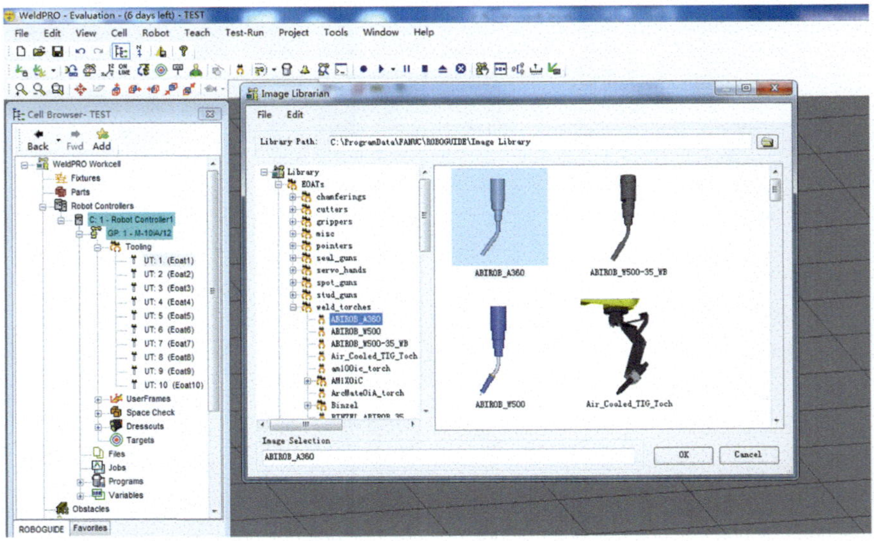

Fig. 7.123 Torch selection window

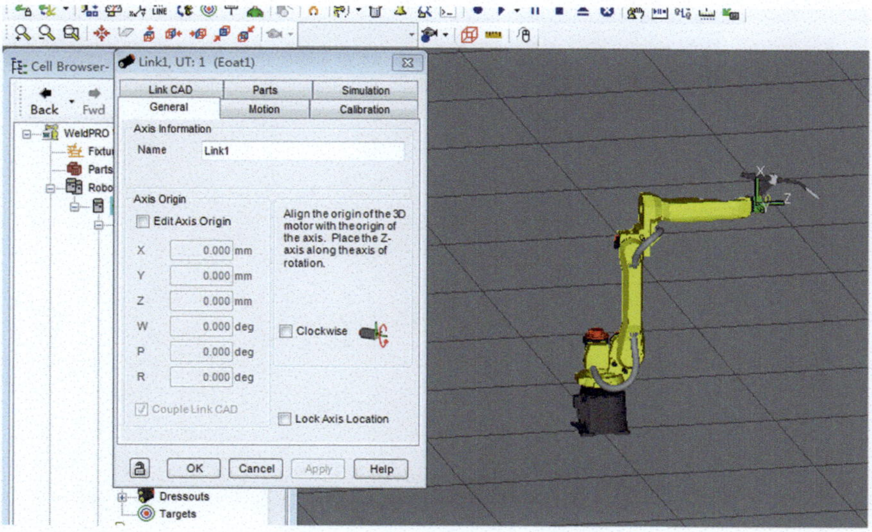

Fig. 7.124 Window after torch addition

of TCP, and then click the "Apply" button to complete the welding torch configuration. Alternatively, the position data (*X*/*Y*/*Z*) of TCP can also be input in the following method: Use the mouse to drag the green balloon to the welding wire tip and click "Use Current Triad Location" button. The system will automatically calculate the position data (*X*/*Y*/*Z*) of TCP and fill them into the input box.

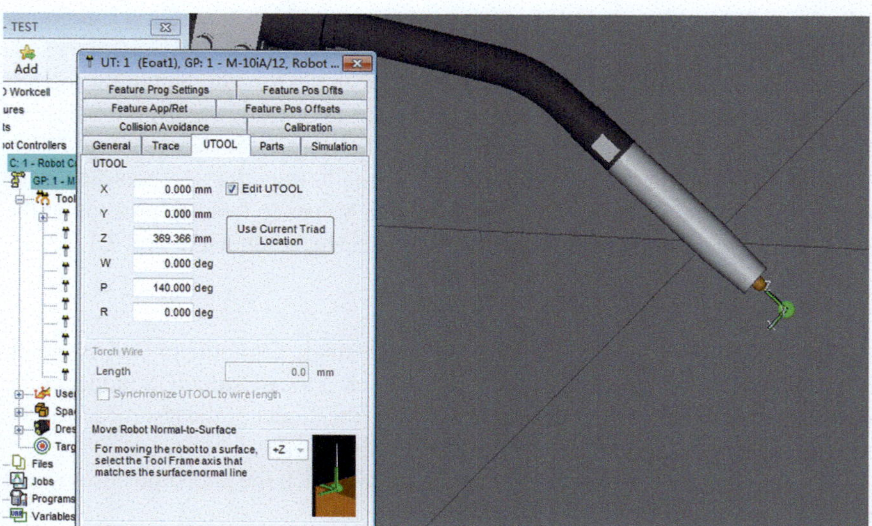

Fig. 7.125 TCP setting dialog box

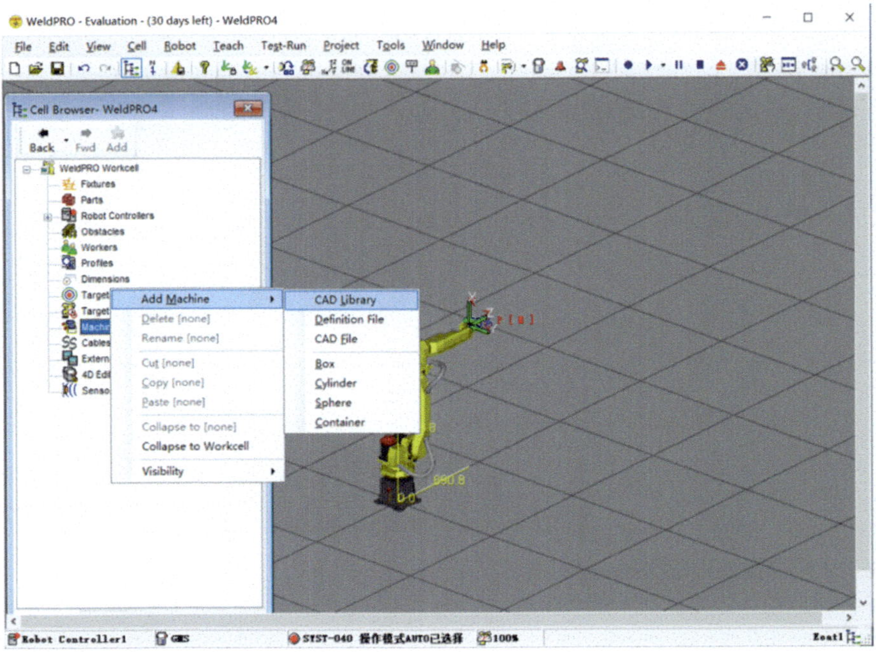

Fig. 7.126 Adding external shaft

3) Add external shaft

Taking a single-shaft positioner as an example, add the external shaft as follows:

Right click Machines in Cell Browser and select Add Machine-CAD Library, as shown in Fig. 7.126.

Enter the outer shaft selection interface, click the expansion button on the left of Library-Positioners, and the external shaft selection window pops up. Select an appropriate external shaft positioner and click "OK" button, as shown in Fig. 7.127.

Adjust the position of the positioner by manually dragging the coordinates or directly entering values. Click Lock All Location Values to lock the position of the positioner. The addition of the positioner is completed, as shown in Fig. 7.128.

4) Adding of test pieces

As shown in Fig. 7.129, click the Parts-Add Part-single CAD file in Cell Browser, and the IGS model of the test piece will pop up; select the existing IGS model of the test piece, and click the Open button. In this way, the model of the test piece will be added to the working group, as shown in Fig. 7.130.

Then add the test piece to the positioner, double-click the added positioner, and the dialog box shown in Fig. 7.131 will pop up.

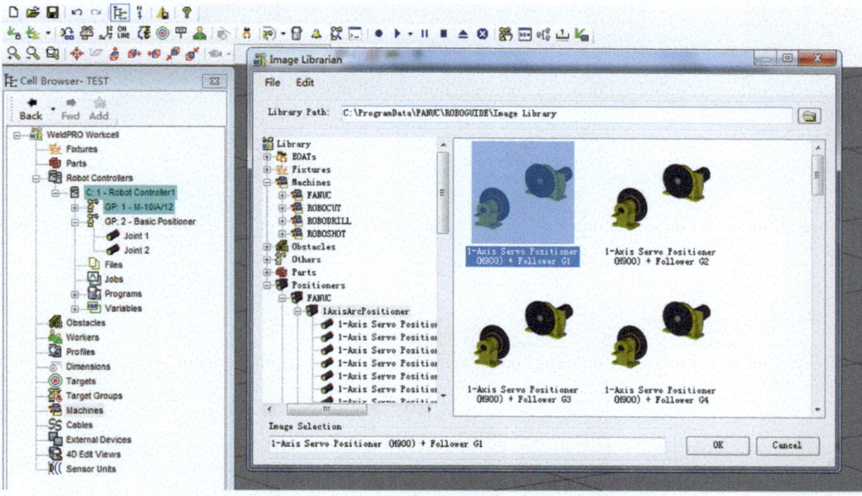

Fig. 7.127 External shaft selection interface

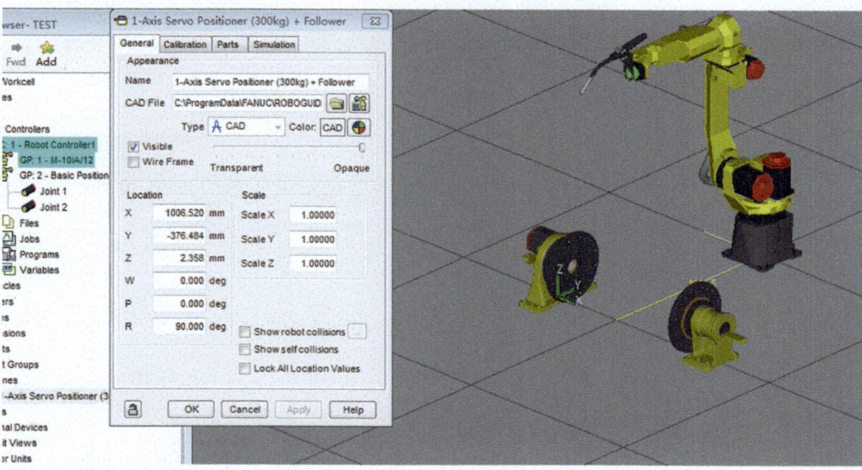

Fig. 7.128 Display window after adding an external axis

Click Parts to display the dialog box shown in Fig. 7.132. Check Part 1 in the dialog box and click Apply to link the workpiece with the positioner. Then click Edit Part Offset to adjust the position of the workpiece relative to the positioner. Then click Apply to install the workpiece on the positioner.

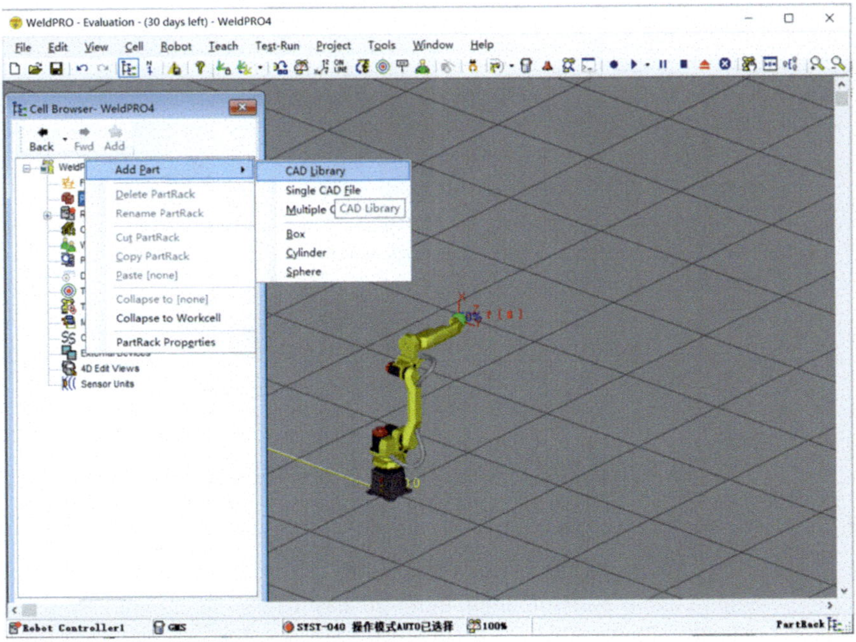

Fig. 7.129 Addition a test piece (1)

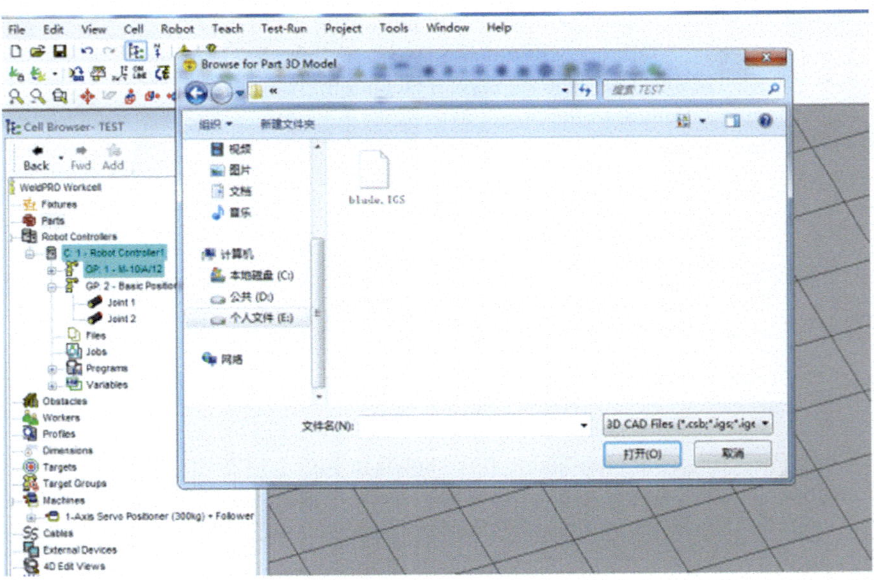

Fig. 7.130 Adding a test piece (2)

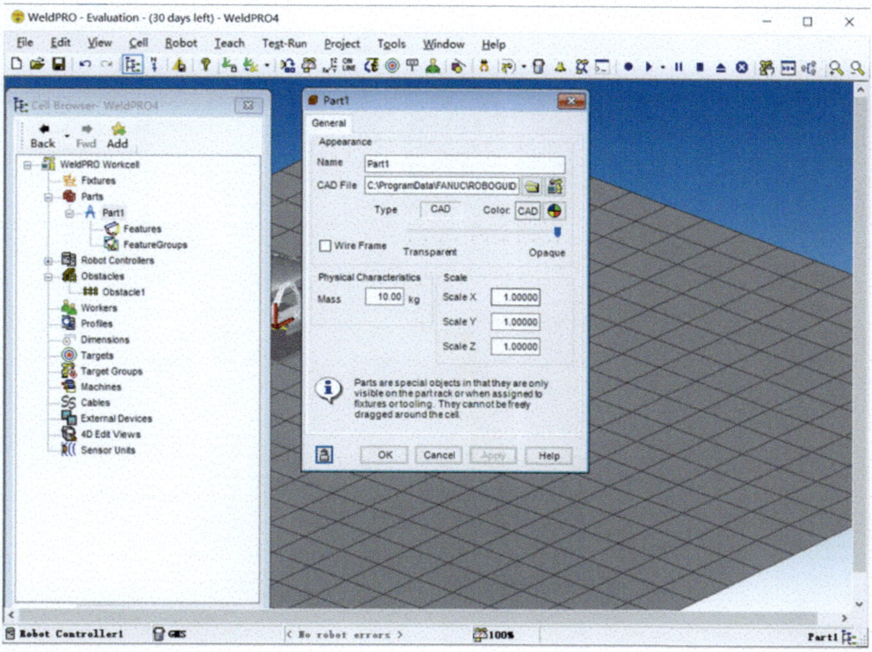

Fig. 7.131 Adding a test piece (3)

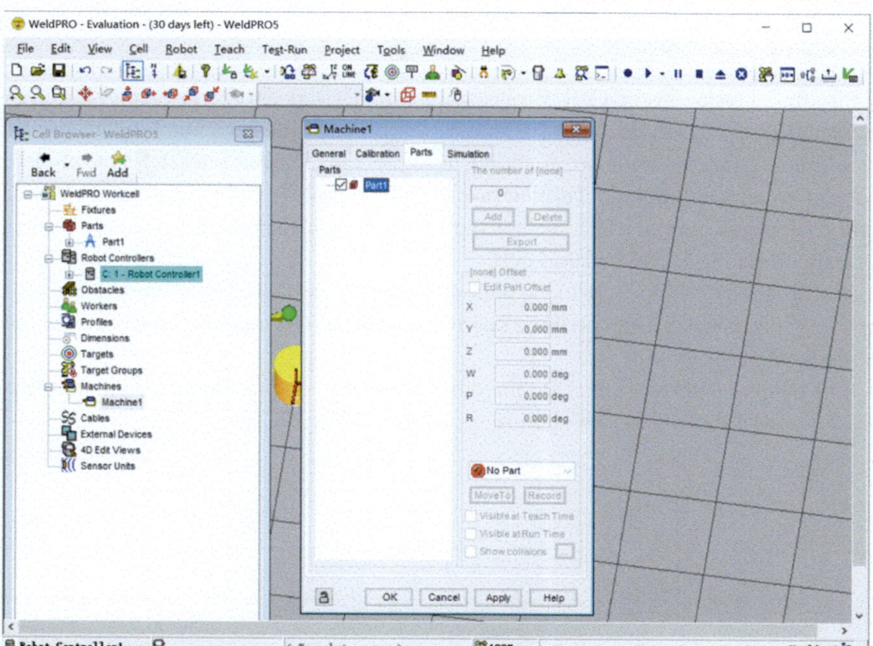

Fig. 7.132 Adjusting the position of the workpiece relative to the positioner

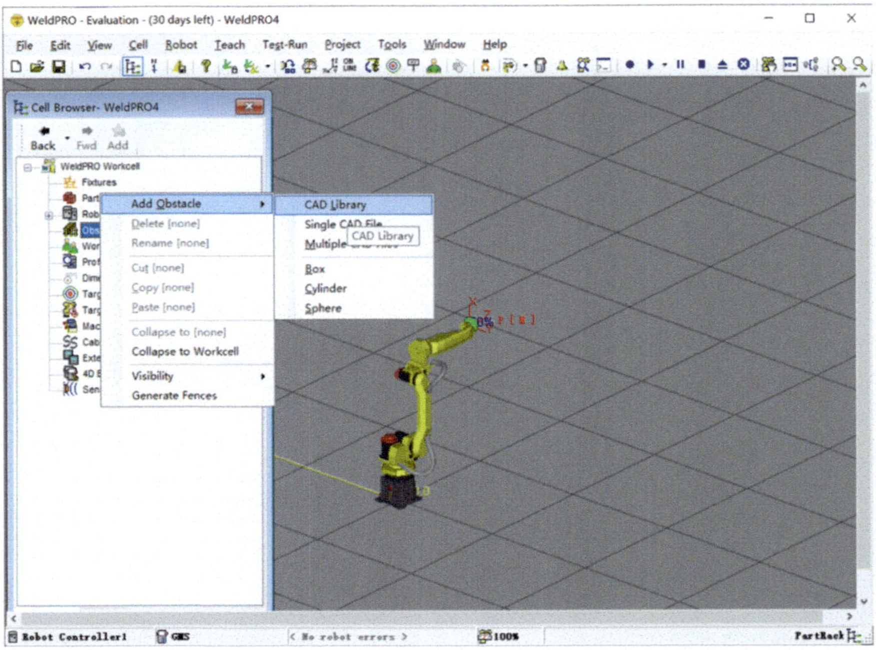

Fig. 7.133 Adding peripherals

5) Adding peripherals

Right-click Obstacles in Cell Browser and then Click Add Obstacle—CADLi-Library (Fig. 7.133) in the pop-up menu bar in turn, and the peripheral device model selection interface will pop up, as shown in Fig. 7.134. The peripheral device models that can be selected include welding power supply, protective fence, external button station, and gas cylinder, etc. Select the required model and click OK to add the corresponding model to the Workcell.

Modify the model position in the pop-up model setting dialog box. You can directly input position data or drag the three coordinate axes of the model with the mouse, as shown in Fig. 7.135.

Following the above steps, add the other peripheral devices required to complete the creation of the whole workstation, as shown in Fig. 7.136.

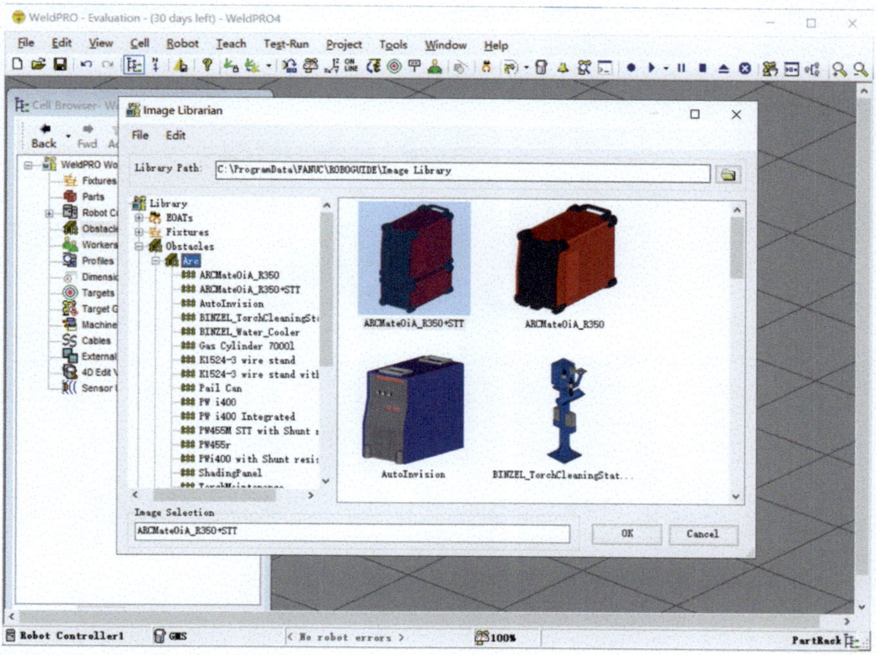

Fig. 7.134 Peripheral model selection interface

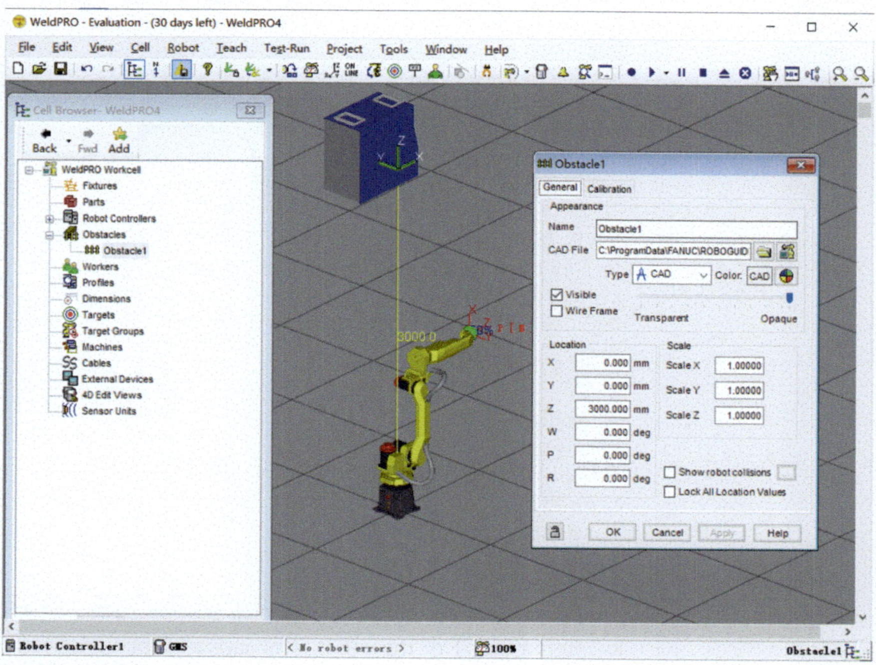

Fig. 7.135 Moving peripheral device location

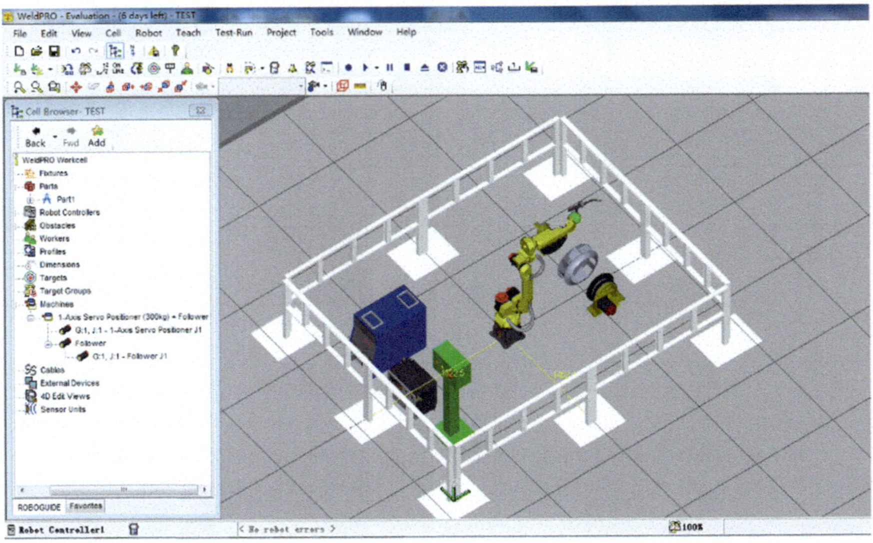

Fig. 7.136 Simulation workstation

Bipliography

1. Cai Zixing, "Robotics (4th edition)", Beijing: Tsinghua University Press, 2022
2. Xiong Youlun, Li Wenlong, Chen Wenbin, et al., "Robotics: Modeling, Control, and Vision (2nd Edition)", Wuhan: Huazhong University of Science and Technology Press, 2020
3. S K Saha, "Introduction to Robotics (2th Edition)", New Delhi: McGraw Hill Education (India) Private Limited, 2014
4. Zhang Tie, Xie Cunxi, "Robotics", Guangzhou: South China University of Technology Press, 2005
5. Zhan Suo, Wang Zhaoyu and Zhang Yue, "Thoroughly learning the comprehensive application of Siemens PLC, frequency converter, and touch screen", Beijing: China Electric Power Press, 2012
6. Li Fangyuan, "Illustrated Introduction to Siemens S7 1200 PLC-Practice", Beijing: Machinery Industry Press, 2010
7. Ding Tianhuai, Chen Xianglin. "Eddy current sensor array testing technology". Journal of Testing Technology, 2006, 20 (1): 1–5
8. Ling Baoming, Zhuge Xiangbin. Ling Yun, "Research on Temperature Stability of Eddy Current Sensors". Journal of Instrumentation, 1994 (4): 342–346
9. Tan Zugen, Chen Shouchuan. "Basic principle analysis and parameter selection of eddy current sensors. Journal of Instrumentation". 1980 (1): 116–125
10. Dong Chunlin, Li Jizhong, Luan Guohong, "The Development Status and Trends of Robot Friction Stir Welding:. Aviation Manufacturing Technology, 2014, 17: 76–79
11. Wang Yunpeng, "Welding Structure Production" , Beijing: Machinery Industry Press, 2004
12. Yuan Junmin, "MOTOMAN spot welding robot system and its application". Metal processing 2008,14: 35–38
13. Li Huawei. "Analysis of the characteristics of robot spot welding pliers", Proceedings of the 7th China Robot Welding Academic and Technical Exchange Conference, Changchun, April 2009.
14. Feng Jicai, Zhao Xihua, Wu Lin. "The application status and development of spot welding robot welding systems". Robots, 1991, 13 (2): 53–58
15. The Complete Equipment and Special Tools Branch of China Welding Society and the Robotics and Automation Professional Committee of China Mechanical Engineering Society Welding Society, "Welding Robot User Manual", Beijing: Machinery Industry Press, 2014
16. Chen Zhunian, Chen Maoai, "Welding Engineer Handbook", Beijing: Machinery Industry Press, 2019
17. J Norberto Pires, Altino Loureiro and Gunnar Bolmsjo, "Welding Robots - technology, system issues and applications", London: Springer, 2006
18. Lin Shangyang, Chen Shanben, Li Chengtong, "Welding robots and their applications", Beijing: China Standards Publishing House, 2000

M. Chen et al., *Technologies of Robotic Welding*, Advanced and Intelligent Manufacturing in China, https://doi.org/10.1007/978-981-97-3149-7

19. Hu Shengsun, “Welding Process Automation Technology and Its Applications (2nd Edition)”, Beijing: Machinery Industry Press, 2015
20. Saeed B. Niku. “Introduction to Robotics - Analysis, System, and Applications”, Upper Saddle River: Prentice Hall Press, 2001
21. Zhuo Yangwa, Bai Xiaocan, Chen Yongming, “Three Rule Curve Interpolation Algorithms for Robots”. Equipment Manufacturing Technology, 2009,11: 27–29
22. Reza N. Jazar, “Theory of Applied Robotics: Kinematics, Dynamics, and Control (second edition)”, London: Springer, 2010
23. Lin Yaoyao, Zhong Chongquan, “Servo driver speed control technology”. Electrical Transmission, 2014.44 (3): 21–26
24. Guo Tongying, An Dong, “Robotics and its intelligent control”, Beijing: People’s Posts and Telecommunications Publishing House, 2014
25. Huang Zhijian, “Robot Drive and Control and Application Examples”, Beijing: Chemical Industry Press, 2016
26. Chen Wanmi, “Robot control technology”, Beijing: Machinery Industry Press, 2017
27. Zhang Hao, Huang Yongde, Guo Yue et al., “Research status of friction stir welding technology and process suitable for robot welding”. Materials Review, 2018,32 (1): 128–133
28. Yan Jianan. “Application technology of servo motors”, Beijing: Science Press, 2017
29. Kou Baoquan, Cheng Shukang. AC servo motors and their control, Beijing: Machinery Industry Press, 2008
30. Zhan Qiang, “Robotics: Mechanisms, kinematics, dynamics, and motion planning”, Beijing: Tsinghua University Press, 2018
31. Bruno Siciliano, Lorenzo Sciavicco, Luigi Villani, Giuseppe Oriolo, “Robotics: Modeling, Planning, and Control”, London: Springer, 2011
32. Guo Tongying, Zhang Hui, “Robot sensors and their information fusion technology”, Beijing: Chemical Industry Press, 2015
33. Chen Shanben, Lin Tao, “Intelligent welding robot technology”, Beijing: Machinery Industry Press, 2006
34. American Welding Society (AWS) Welding Handbook Committee. 2001. Annette O’Brien, and Carlos Guzman, eds. Welding process, part 2. Vol. 3 of Welding handbook. 9th ed. Miami: American Welding Society.
35. American Welding Society (AWS) Welding Handbook Committee. 2001. Annette O’Brien ed. Welding process, part 1. Vol. 2 of Welding handbook. 9th ed. Miami: American Welding Society.
36. Chen Dongsheng, “Research on arc interaction and factors affecting weld formation in Tandem P-GMAW”, Master Thesis, Shandong University, School of Materials Science and Engineering, 2015
37. Cederberg P, Olsson M and Bolmsjö G, “Virtual triangulation sensor development, behavior simulation and CAR integration applied to robotic arc-welding”. Journal of Intelligent and Robotic Systems, 35(4): 365–379.
38. Chen Maoai, Zhangdong, Wu Chansong, “Current waveform effects on CMT welding of mild steel”. Journal of Materials Processing Technology, 2017, 243: 395–404
39. Chen Dongsheng, Chen Maoai, “Effects of phase difference on the behavior of arc and weld pool in tandem P-GMAW”. Journal of Materials Processing Technology”. 2015, 225: 45–55.
40. Wang Yuwen, Chen Maoai, Li Chaofan, “Influence of current waveform on C-GMAW of 6061 aluminum alloy with addition of ER4043 wire”. Welding in the World, 2022, 66:1093–1109
41. Zhang Dong, “Research on the Effect of Waveform Parameters on Welding Stability and Welding Formation in High Speed CMT Welding”. Master Thesis, Shandong University, School of Materials Science and Engineering, 2016.
42. Shi Meng, “Effect of different wire feeding methods on metal transfer, welding pool behavior and welding appearance of stainless steel TOP-TIG welding”. Master Thesis, Shandong University, School of Materials Science and Engineering, 2018.
43. Pickin C G, Young K, “Evaluation of cold metal transfer (CMT) process for welding aluminium alloy”. Science and Technology of Welding and Joining, 2006, 11(5): 583–585

44. Ueyama T, Ohnawa T, Tanaka M, et al., "Effects of torch configuration and welding current on weld bead formation in high speed tandem pulsed gas metal arc welding of steel sheets". Science and Technology of Welding & Joining, 2005, 10(6): 750–759
45. Ueyama T, Ohnawa T, Tanaka M, et al., "Occurrence of arc interaction in tandem pulsed gas metal arc welding". Science and Technology of Welding & Joining, 2007, 12(6): 523–529
46. Ueyama T, Uezono T, Era T, et al., "Solution to problems of arc interruption and arc length control in tandem pulsed gas metal arc welding". Science and Technology of Welding & Joining, 2009, 14(4): 305–314
47. Lassaline E, Zajaczkowski B, North T H,N, "arrow groove twin-wire GMAW of high-strength steel". Weld. J., 1989, 68(9): 53–58
48. Michie K, Blackman S, Ogunbiyi T E B, "Twin-wire GMAW: process characteristics and applications". Welding Journal, 1999, 78(5): 31–34
49. C. Rosellini & L. Jarvis, "The keyhole TIG welding process: a valid alternative for valuable metal joints, Welding International". 2009, 23(8): 616–621
50. Zu Ming Liu YueXiao Fang, "Stable keyhole welding process with K-TIG". Journal of Materials Processing Technology". 2016, 238: 65–72
51. Shuang Lin Cui, ZuMing Liu, YueXiao Fang, "Keyhole process in K-TIG welding on 4 mm thick 304 stainless steel ". Journal of Materials Processing Technology. 2017, 243: 217–228
52. Yueqiao Feng , Zhen Luo. "Keyhole gas tungsten arc welding of AISI 316L stainless steel". Materials and Design, 2015, 85: 24–31
53. T. Opderbecke, S. Guiheux. "TOP-TIG: robotic TIG welding with integrated wire feeder". Welding International, 2009, 23(7): 523–529
54. Zhouo Xiaochen, Li Huan, Song Chunguan, Zhang Yuchang, "Analysis of droplet transfer characteristics in TOPTIG welding". Transaction of The China Welding Insititue, 2017, 38(7): 46–52

SPRINGER NATURE

GPSR Compliance

The European Union's (EU) General Product Safety Regulation (GPSR) is a set of rules that requires consumer products to be safe and our obligations to ensure this.

If you have any concerns about our products, you can contact us on ProductSafety@springernature.com

In case Publisher is established outside the EU, the EU authorized representative is:

Springer Nature Customer Service Center GmbH
Europaplatz 3
69115 Heidelberg, Germany

The manufacturer's authorised representative in the EU is Springer Nature Customer Service Centre GmbH, Europaplatz 3, 69115 Heidelberg, Germany. If you have any concerns regarding our products, please contact ProductSafety@springernature.com

Printed and bound by CPI Group (UK) Ltd, Croydon, CR0 4YY
30/09/2026
02225298-0009